SCHAUM'S OUTLINE OF

THEORY AND PROBLEMS

OF

MODERN ABSTRACT ALGEBRA

•

BY

FRANK AYRES, JR., Ph.D.
Formerly Professor and Head,
Department of Mathematics
Dickinson College

•

SCHAUM'S OUTLINE SERIES
McGRAW-HILL, INC.
New York St. Louis San Francisco Auckland Bogotá Caracas Lisbon London Madrid Mexico City Milan Montreal New Delhi San Juan Singapore Sydney Tokyo Toronto

Formerly published as Modern Algebra.

ISBN 07-002655-6

25 26 27 28 29 30 SH SH 9 8 7 6 5 4

Cover design by Amy E. Becker.

This book is printed on recycled paper containing a minimum of 50% total recycled fiber with 10% postconsumer de-inked fiber. Soybean based inks are used on the cover and text.

Preface

This study of algebraic systems is designed to be used either as a supplement to current texts or as a text for a course in modern abstract algebra at the junior-senior level. As such, it is intended to provide a solid foundation for future study of a variety of systems rather than to be a study in depth of any one or more.

The basic ingredients of algebraic systems – sets of elements, relations, operations, mappings – are treated in the first two chapters. The pattern established here,

(i) a simple and concise presentation of each topic,
(ii) a wide variety of familiar examples,
(iii) proofs of most theorems included among the solved problems,
(iv) a carefully selected set of supplementary exercises,

is followed throughout the book.

Beginning in Chapter 3 with the Peano postulates for the natural numbers, the several number systems of elementary algebra are constructed in turn and their salient properties deduced. This not only introduces the reader to a detailed and rigorous development of these number systems but also provides him with much needed practice for the deduction of properties of the abstract systems which follow.

The first abstract algebraic system – the Group – is considered in Chapter 9. Cosets of a subgroup, invariant subgroups and their quotient groups are investigated, and the chapter ends with the Jordan-Hölder Theorem for finite groups.

Chapters 10-11 are concerned with Rings, Integral Domains and Fields. Polynomials over rings and fields are then considered in Chapter 12 along with a certain amount of elementary theory of equations. Throughout these chapters, considerable attention is given to finite rings.

Vector spaces are introduced in Chapter 13. The algebra of linear transformations on a vector space of finite dimension leads naturally to the algebra of matrices (Chapter 14). Matrices are then used to solve systems of linear equations and, thus, provide simpler solutions of a number of problems connected with vector spaces. Matrix polynomials are considered in Chapter 15 as an example of a non-commutative polynomial ring. The characteristic polynomial of a square matrix over a field is then defined. The characteristic roots and associated invariant vectors of real symmetric matrices are used to reduce the equations of conics and quadric surfaces to standard form. Linear algebras are formally defined in Chapter 16 and other examples briefly considered.

In the final chapter, Boolean algebras are introduced and the important application to simple electric circuits indicated.

The author wishes to take this opportunity to express his appreciation to the staff of the Schaum Publishing Company, especially to Jeffrey Albert and Alan Hopenwasser, for their unfailing cooperation.

FRANK AYRES, JR.

Carlisle, Penna.
June, 1965

CONTENTS

CONTENTS

Chapter 1

Sets

SETS

Any collection of objects as (*a*) the points of a given line segment, (*b*) the lines through a given point in ordinary space, (*c*) the natural numbers less than ten, (*d*) the five Jones boys and their dog, (*e*) the pages of this book, ... will be called a *set* or *class*. The individual points, lines, numbers, boys and dog, pages, ... will be called *elements* of the respective sets. Generally, sets will be denoted by capital letters and arbitrary elements of sets will be denoted by small letters.

Let A be a given set and let p and q denote certain objects. When p is an element of A, we shall indicate this fact by writing $p \in A$; when both p and q are elements of A, we shall write $p, q \in A$ instead of $p \in A$ and $q \in A$; when q is not an element of A, we shall write $q \notin A$.

Although in much of our study of sets we will not be concerned with the type of elements, sets of numbers will naturally appear in many of the examples and problems. For convenience, we shall now reserve

N to denote the set of all natural numbers
I to denote the set of all integers
Q to denote the set of all rational numbers
R to denote the set of all real numbers

Example 1: (*a*) $1 \in N$ and $205 \in N$ since 1 and 205 are natural numbers; $\frac{1}{2}, -5 \notin N$ since $\frac{1}{2}$ and -5 are not natural numbers.

(*b*) The symbol $\in$ indicates membership and may be translated as "in", "is in", "are in", "be in" according to context. Thus, "Let $r \in Q$" may be read as "Let r be in Q" and "For any $p, q \in I$" may be read as "For any p and q in I". We shall at times write $n \neq 0 \in I$ instead of $n \neq 0$, $n \in I$; also $p \neq 0, q \in I$ instead of $p, q \in I$ with $p \neq 0$.

The sets to be introduced here will always be *well-defined*, that is, it will always be possible to determine whether any given object does or does not belong to the particular set. The sets of the first paragraph were defined by means of precise statements in words. At times, a set will be given in tabular form by exhibiting its elements between a pair of braces; for example,

$A = \{a\}$ is the set consisting of the single element a.
$B = \{a, b\}$ is the set consisting of the two elements a and b.
$C = \{1, 2, 3, 4\}$ is the set of natural numbers less than 5.
$K = \{2, 4, 6, \ldots\}$ is the set of all even natural numbers.
$L = \{\ldots, -15, -10, -5, 0, 5, 10, 15, \ldots\}$ is the set of all integers having 5 as a factor.

The sets C, K, and L above may also be defined as follows:

$$C = \{x : x \in N,\ x < 5\}$$
$$K = \{x : x \in N,\ x \text{ is even}\}$$
$$L = \{x : x \in I,\ x \text{ is divisible by } 5\}$$

Here each set consists of *all* objects x satisfying the conditions following the colon.

See Problem 1.

EQUAL SETS

When two sets A and B consist of the same elements, they are called *equal* and we shall write $A = B$. To indicate that A and B are not equal, we shall write $A \neq B$.

Example 2: (i) When $A = \{\text{Mary, Helen, John}\}$ and $B = \{\text{Helen, John, Mary}\}$, then $A = B$. Note that a variation in the order in which the elements of a set are tabulated is immaterial.

(ii) When $A = \{2,3,4\}$ and $B = \{3,2,3,2,4\}$, then $A = B$ since each element of A is in B and each element of B is in A. Note that a set is not changed by repeating one or more of its elements.

(iii) When $A = \{1,2\}$ and $B = \{1,2,3,4\}$, then $A \neq B$ since 3 is an element of B but not of A.

SUBSETS OF A SET

Let S be a given set. Any set A, each of whose elements is also an element of S, is said to be *contained in* S and called a *subset* of S.

Example 3: The sets $A = \{2\}$, $B = \{1,2,3\}$, and $C = \{4,5\}$ are subsets of $S = \{1,2,3,4,5\}$. Also, $D = \{1,2,3,4,5\} = S$ is a subset of S.

The set $E = \{1,2,6\}$ is not a subset of S since $6 \in E$ but $6 \notin S$.

Let A be a subset of S. If $A \neq S$, we shall call A a *proper subset* of S and write $A \subset S$ (to be read "A is a proper subset of S" or "A is properly contained in S"). More often and in particular when the possibility $A = S$ is not excluded, we shall write $A \subseteq S$ (to be read "A is a subset of S" or "A is contained in S"). Of all the subsets of a given set S, only S itself is *improper*, that is, is not a proper subset of S.

Example 4: For the sets of Example 3 we may write $A \subseteq S$, $B \subseteq S$, $C \subseteq S$, $D \subseteq S$, $E \nsubseteq S$. The precise statements, of course, are $A \subset S$, $B \subset S$, $C \subset S$, $D = S$, $E \nsubseteq S$.

Note carefully that $\in$ connects an element and a set, while $\subset$ and $\subseteq$ connect two sets. Thus, $2 \in S$ and $\{2\} \subset S$ are correct statements, while $2 \subset S$ and $\{2\} \in S$ are incorrect.

Let A be a proper subset of S. Now S consists of the elements of A together with certain elements not in A. These latter elements, i.e. $(x : x \in S,\ x \notin A)$, constitute another proper subset of S called the *complement* of the subset A in S.

Example 5: For the set $S = \{1,2,3,4,5\}$ of Example 3, the complement of $A = \{2\}$ in S is $F = \{1,3,4,5\}$. Also, $B = \{1,2,3\}$ and $C = \{4,5\}$ are complementary subsets in S.

Our discussion of complementary subsets of a given set implies that these subsets be proper. The reason is simply that, thus far, we have been depending upon intuition regarding sets; that is, we have tacitly assumed that every set must have at least one element. In order to remove this restriction (also to provide a complement for the improper subset S in S), we introduce the *empty* or *null set* $\emptyset$ as the set having no elements. There follow readily

(i) $\emptyset$ is a subset of every set S.

(ii) $\emptyset$ is a proper subset of every set $S \neq \emptyset$.

Example 6: The subsets of $S = \{a,b,c\}$ are $\emptyset$, $\{a\}$, $\{b\}$, $\{c\}$, $\{a,b\}$, $\{a,c\}$, $\{b,c\}$, and $\{a,b,c\}$. The pairs of complementary subsets are:

$\{a,b,c\}$ and $\emptyset$ $\{a,b\}$ and $\{c\}$

$\{a,c\}$ and $\{b\}$ $\{b,c\}$ and $\{a\}$

There is an even number of subsets and, hence, an odd number of proper subsets of a set of 3 elements. Is this true for a set of 303 elements? of 303,000 elements?

UNIVERSAL SETS

If $U \neq \emptyset$ is a given set whose subsets are under consideration, the given set will often be referred to as a *universal set*.

Example 7: Consider the equation

$$(x+1)(2x-3)(3x+4)(x^2-2)(x^2+1) = 0$$

whose solution set, that is, the set whose elements are the roots of the equation, is $S = \{-1, 3/2, -4/3, \sqrt{2}, -\sqrt{2}, i, -i\}$ provided the universal set is the set of all complex numbers. However, if the universal set is R, the solution set is $A = \{-1, 3/2, -4/3, \sqrt{2}, -\sqrt{2}\}$. What is the solution set if the universal set is Q? is I? is N?

If, on the contrary, we are given two sets $A = \{1,2,3\}$ and $B = \{4,5,6,7\}$, and nothing more, we have little knowledge of the universal set U of which they are subsets. For example, U might be $\{1,2,3,\ldots,7\}$, $\{x: x \in N, x \leq 1000\}$, N, I, ... Nevertheless, when dealing with a number of sets $A, B, C, \ldots$, we shall always think of them as subsets of some universal set U not necessarily explicitly defined. With respect to this universal set, the complements of the subsets $A, B, C, \ldots$ will be denoted by $A', B', C', \ldots$ respectively.

INTERSECTION AND UNION OF SETS

Let A and B be given sets. The set of all elements which belong to both A and B is called the *intersection* of A and B. It will be denoted by $A \cap B$ (read either as "the intersection of A and B" or as "A cap B"). Thus,

$$A \cap B = \{x: x \in A \text{ and } x \in B\}$$

The set of all elements which belong to A alone or to B alone or to both A and B is called the *union* of A and B. It will be denoted by $A \cup B$ (read either as "the union of A and B" or as "A cup B"). Thus,

$$A \cup B = \{x: x \in A \text{ alone or } x \in B \text{ alone or } x \in A \cap B\}$$

More often, however, we shall write

$$A \cup B = \{x: x \in A \text{ or } x \in B\}$$

The two are equivalent since every element of $A \cap B$ is an element of A.

Example 8: Let $A = \{1,2,3,4\}$ and $B = \{2,3,5,8,10\}$; then $A \cup B = \{1,2,3,4,5,8,10\}$ and $A \cap B = \{2,3\}$.

See also Problems 2-4.

Two sets A and B will be called *disjoint* if they have no element in common, that is, if $A \cap B = \emptyset$. In Example 6, any two of the sets $\{a\}, \{b\}, \{c\}$ are disjoint; also the sets $\{a,b\}$ and $\{c\}$, the sets $\{a,c\}$ and $\{b\}$, and the sets $\{b,c\}$ and $\{a\}$ are disjoint.

VENN DIAGRAMS

The complement, intersection, and union of sets may be pictured by means of Venn diagrams. In the diagrams below the universal set U is represented by points (not indicated) in the interior of a rectangle, and any of its non-empty subsets by points in the interior of closed curves. (To avoid confusion, we shall agree that no element of U is represented by a point on the boundary of any of these curves.) In Fig. 1-1(a), the subsets A and B of U satisfy $A \subset B$; in Fig. 1-1(b), $A \cap B = \emptyset$; in Fig. 1-1(c), A and B have at least one element in common so that $A \cap B \neq \emptyset$.

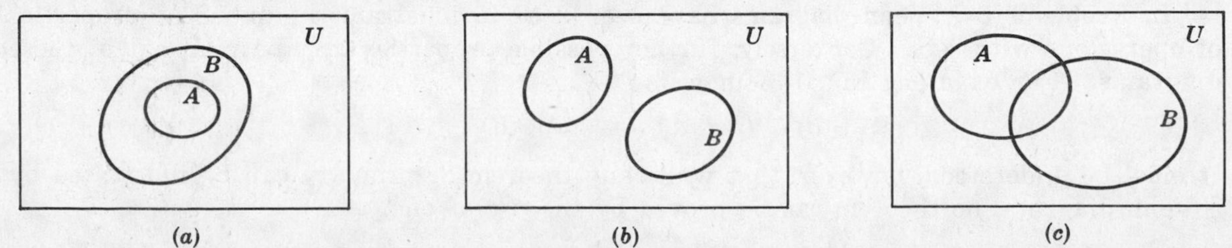

(a) (b) (c)

Fig. 1-1

Suppose now the interior of U, except for the interior of A, in the diagrams above be shaded. In each case, the shaded area will represent the complementary set A' of A in U.

The union $A \cup B$ and the intersection $A \cap B$ of the sets A and B of Fig. 1-1(c) are represented by the shaded area in Fig. 1-2(a) and (b) respectively. In Fig. 1-2(a), the unshaded area represents $(A \cup B)'$, the complement of $A \cup B$ in U; in Fig. 1-2(b), the unshaded area represents $(A \cap B)'$. From these diagrams, as also from the definitions of $\cup$ and $\cap$, it is clear that $A \cup B = B \cup A$ and $A \cap B = B \cap A$.

See Problems 5-7.

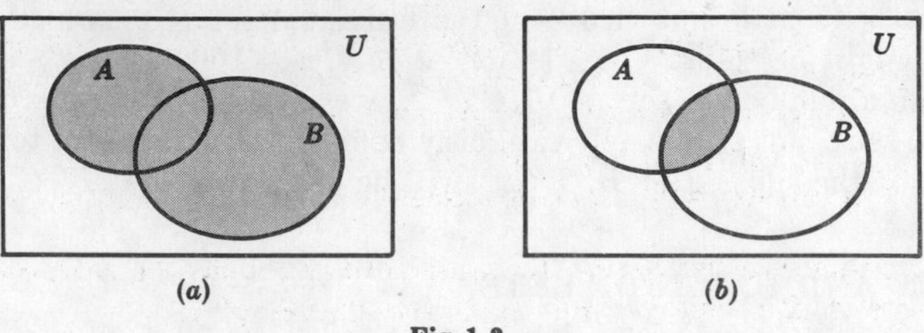

Fig. 1-2

OPERATIONS WITH SETS

In addition to complementation, union, and intersection, which we shall call operations with sets, we define:

The *difference* $A - B$, in that order, of two sets A and B is the set of all elements of A which do not belong to B, i.e.,

$$A - B = \{x:\ x \in A,\ x \notin B\}$$

In Fig. 1-3, $A - B$ is represented by the shaded area and $B - A$ by the cross-hatched area. There follow

$$A - B = A \cap B' = B' - A'$$

$$A - B = \emptyset \quad \text{if and only if } A \subseteq B$$

$$A - B = B - A \quad \text{if and only if } A = B$$

$$A - B = A \quad \text{if and only if } A \cap B = \emptyset$$

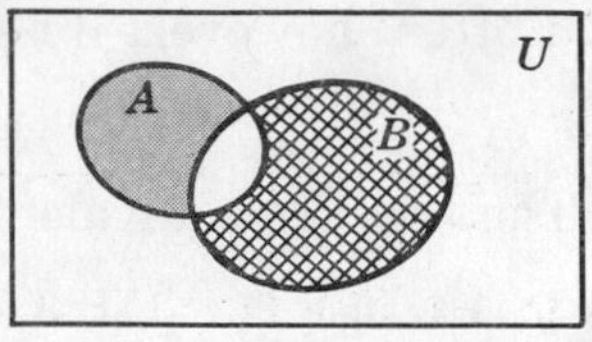

Fig. 1-3

Example 9: Prove: (a) $A - B = A \cap B' = B' - A'$; ($b$) $A - B = \emptyset$ if and only if $A \subseteq B$; (c) $A - B = A$ if and only if $A \cap B = \emptyset$.

(a) $A - B = \{x:\ x \in A,\ x \notin B\} = \{x:\ x \in A \text{ and } x \in B'\} = A \cap B'$
$= \{x:\ x \notin A',\ x \in B'\} = B' - A'$

(b) Suppose $A - B = \emptyset$. Then, by (a), $A \cap B' = \emptyset$, i.e., A and B' are disjoint. Now B and B' are disjoint; hence, since $B \cup B' = U$, we have $A \subseteq B$.

Conversely, suppose $A \subseteq B$. Then $A \cap B' = \emptyset$ and $A - B = \emptyset$.

(c) Suppose $A - B = A$. Then $A \cap B' = A$, i.e., $A \subseteq B'$. Hence, by (b),

$$A \cap (B')' = A \cap B = \emptyset$$

Conversely, suppose $A \cap B = \emptyset$. Then $A - B' = \emptyset$, $A \subseteq B'$, $A \cap B' = A$ and $A - B = A$.

In Problems 5-7, Venn diagrams have been used to illustrate a number of properties of operations with sets. Conversely, further possible properties may be read out of these diagrams. For example, Fig. 1-3 suggests

$$(A - B) \cup (B - A) = (A \cup B) - (A \cap B)$$

It must be understood, however, that while any theorem or property can be illustrated by a Venn diagram, no theorem can be proved by the use of one.

Example 10: Prove $(A-B) \cup (B-A) = (A \cup B) - (A \cap B)$.

The proof consists in showing that every element of $(A-B) \cup (B-A)$ is an element of $(A \cup B) - (A \cap B)$ and, conversely, every element of $(A \cup B) - (A \cap B)$ is an element of $(A-B) \cup (B-A)$. Each step follows from a previous definition and it will be left for the reader to substantiate these steps.

Let $x \in (A-B) \cup (B-A)$; then $x \in A-B$ or $x \in B-A$. If $x \in A-B$, then $x \in A$ but $x \notin B$; if $x \in B-A$, then $x \in B$ but $x \notin A$. In either case, $x \in A \cup B$ but $x \notin A \cap B$. Hence, $x \in (A \cup B) - (A \cap B)$ and

$$(A-B) \cup (B-A) \subseteq (A \cup B) - (A \cap B)$$

Conversely, let $x \in (A \cup B) - (A \cap B)$; then $x \in A \cup B$ but $x \notin A \cap B$. Now either $x \in A$ but $x \notin B$, i.e., $x \in A-B$, or $x \in B$ but $x \notin A$, i.e., $x \in B-A$. Hence, $x \in (A-B) \cup (B-A)$ and $(A \cup B) - (A \cap B) \subseteq (A-B) \cup (B-A)$.

Finally, $(A-B) \cup (B-A) \subseteq (A \cup B) - (A \cap B)$ and $(A \cup B) - (A \cap B) \subseteq (A-B) \cup (B-A)$ imply $(A-B) \cup (B-A) = (A \cup B) - (A \cap B)$

For future reference we list below the more important laws governing operations with sets. Here the sets A, B, C are subsets of U the universal set.

LAWS OF OPERATIONS WITH SETS			
(1.1)	$(A')' = A$		
(1.2)	$\emptyset' = U$	(1.2')	$U' = \emptyset$
(1.3)	$A - A = \emptyset$, $A - \emptyset = A$, $A - B = A \cap B'$		
(1.4)	$A \cup \emptyset = A$	(1.4')	$A \cap U = A$
(1.5)	$A \cup U = U$	(1.5')	$A \cap \emptyset = \emptyset$
(1.6)	$A \cup A = A$	(1.6')	$A \cap A = A$
(1.7)	$A \cup A' = U$	(1.7')	$A \cap A' = \emptyset$
Associative Laws			
(1.8)	$(A \cup B) \cup C = A \cup (B \cup C)$	(1.8')	$(A \cap B) \cap C = A \cap (B \cap C)$
Commutative Laws			
(1.9)	$A \cup B = B \cup A$	(1.9')	$A \cap B = B \cap A$
Distributive Laws			
(1.10)	$A \cup (B \cap C) = (A \cup B) \cap (A \cup C)$	(1.10')	$A \cap (B \cup C) = (A \cap B) \cup (A \cap C)$
De Morgan's Laws			
(1.11)	$(A \cup B)' = A' \cap B'$	(1.11')	$(A \cap B)' = A' \cup B'$
(1.12)	$A - (B \cup C) = (A-B) \cap (A-C)$	(1.12')	$A - (B \cap C) = (A-B) \cup (A-C)$

See Problems 8-16.

THE PRODUCT SET

Let $A = \{a, b\}$ and $B = \{b, c, d\}$. The set of distinct ordered pairs

$$C = \{(a,b), (a,c), (a,d), (b,b), (b,c), (b,d)\}$$

in which the first component of each pair is an element of A while the second is an element of B, is called the *product set* $C = A \times B$ (in that order) of the given sets. Thus, if A and B are arbitrary sets, we define

$$A \times B = \{(x,y) : x \in A, y \in B\}$$

Example 11: Identify the elements of $X = \{1,2,3\}$ as the coordinates of points on the x-axis (see Fig. 1-4), thought of as a number scale, and the elements of $Y = \{1,2,3,4\}$ as the coordinates of points on the y-axis, thought of as a number scale. Then the elements of $X \times Y$ are the rectangular coordinates of the 12 points shown. Similarly, when $X = Y = N$, the set $X \times Y$ are the coordinates of all points in the first quadrant having integral coordinates.

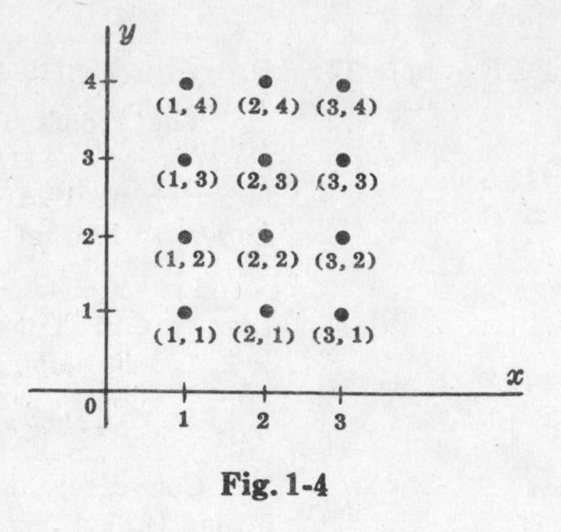

Fig. 1-4

MAPPINGS

Consider the set $H = \{h_1, h_2, h_3, \ldots, h_8\}$ of all houses on a certain block of Main Street and the set $C = \{c_1, c_2, c_3, \ldots, c_{39}\}$ of all children living in this block. We shall be concerned here with the natural association of each child of C with the house of H in which the child lives. Let us assume that this results in associating c_1 with h_2, c_2 with h_5, c_3 with h_2, c_4 with h_5, c_5 with h_8, $\ldots$, c_{39} with h_3. Such an association of or correspondence between the elements of C and H is called a *mapping of C into H*. The unique element of H associated with any element of C is called the *image* of that element (of C) in the mapping.

Now there are two possibilities for this mapping: (1) every element of H is an image, that is, in each house there lives at least one child; (2) at least one element of H is not an image, that is, in at least one house there live no children. In the case (1), we shall call the correspondence a *mapping of C onto H*. Thus, the use of 'onto' instead of 'into' calls attention to the fact that in the mapping every element of H is an image. In the case (2), we shall call the correspondence a *mapping of C into, but not onto, H*. Whenever we write "α is a mapping of A into B" the possibility that α may, in fact, be a mapping of A onto B is not excluded. Only when it is necessary to distinguish between cases will we write either "α is a mapping of A onto B" or "α is a mapping of A into, but not onto, B".

A particular mapping α of one set into another may be defined in various ways. For example, the mapping of C into H above may be defined by listing the ordered pairs

$$\alpha = \{(c_1, h_2), (c_2, h_5), (c_3, h_2), (c_4, h_5), (c_5, h_8), \ldots, (c_{39}, h_3)\}$$

It is now clear that α is simply a certain subset of the product set $C \times H$ of C and H. Hence, we define

> A mapping of a set A into a set B is a subset of $A \times B$ in which each element of A occurs once and only once as the first component in the elements of the subset.

In any mapping α of A into B, the set A is called the *domain* and the set B is called the *co-domain* of α. If the mapping is "onto", B is also called the *range* of α; otherwise, the range of α is the proper subset of B consisting of the images of all elements of A.

A mapping of a set A into a set B may also be displayed by the use of $\rightarrow$ to connect associated elements.

Example 12: Let $A = \{a, b, c\}$ and $B = \{1, 2\}$. Then

$$\alpha: \quad a \rightarrow 1, \; b \rightarrow 2, \; c \rightarrow 2$$

is a mapping of A onto B (every element of B is an image) while

$$\beta: \quad 1 \rightarrow a, \; 2 \rightarrow b$$

is a mapping of B into, but not onto, A (not every element of A is an image).

In the mapping α, A is the domain and B is both the co-domain and the range. In the mapping β, B is the domain, A is the co-domain, and $C = \{a, b\} \subset A$ is the range.

When the number of elements involved is small, Venn diagrams may be used to advantage. Fig. 1-5 below displays the mappings α and β of this example.

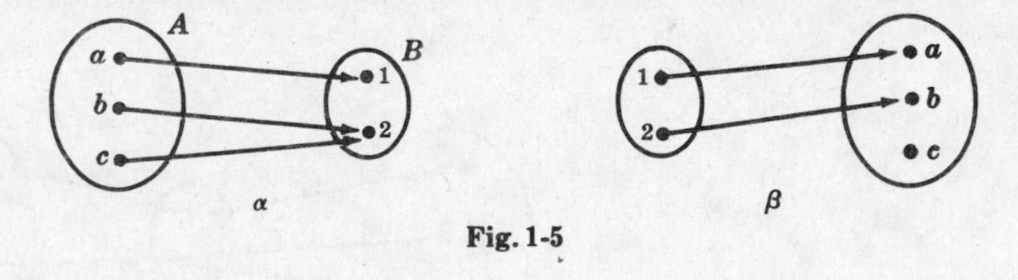

Fig. 1-5

A third way of denoting a mapping is discussed in

Example 13: Consider the mapping α of N into itself, that is, of N into N,

$$\alpha: \quad 1 \rightarrow 3, \ 2 \rightarrow 5, \ 3 \rightarrow 7, \ 4 \rightarrow 9, \ \ldots$$

or, more compactly, $$\alpha: \quad n \rightarrow 2n+1, \ n \in N$$

Such a mapping will frequently be defined by

$$\alpha: \quad 1\alpha = 3, \ 2\alpha = 5, \ 3\alpha = 7, \ 4\alpha = 9, \ \ldots$$

or, more compactly, by $$\alpha: \quad n\alpha = 2n+1, \ n \in N$$

Here N is the domain (also the co-domain) but not the range of the mapping. The range is the proper subset M of N given by

$$M = \{x: \ x = 2n+1, \ n \in N\}$$

or $$M = \{x: \ x \in N, \ x \text{ is odd}\}$$

Mappings of a set X into a set Y, especially when X and Y are sets of numbers, are better known to the reader as *functions*. For instance, defining $X = N$ and $Y = M$ in Example 13 and using f instead of α, the mapping (function) may be expressed in *functional notation* as

$$\text{(i)} \qquad y = f(x) \qquad 2x+1$$

We say here that y is defined as a *function of* x. It is customary nowadays to distinguish between "function" and "function of". Thus, in the example, we would define the function f by

$$f = \{(x, y): \ y = 2x+1, \ x \in X\}$$

or $$f = \{(x, \ 2x+1): \ x \in X\}$$

that is, as the particular subset of $X \times Y$, and consider (i) as the "rule" by which this subset is determined. Throughout much of this book we shall use the term mapping rather than function and, thus, find little use for the functional notation.

Let α be a mapping of A into B and β be a mapping of B into C. Now the effect of α is to map $a \in A$ into $a\alpha \in B$ and the effect of B is to map $a\alpha \in B$ into $(a\alpha)\beta \in C$. The net result of applying α followed by β is a mapping of A into C which we define by

$$\alpha\beta: \quad a(\alpha\beta) = (a\alpha)\beta, \ a \in A$$

We shall call $\alpha\beta$ the *product* of the mappings α and β in that order. Unfortunately, the notation has not been standardized so that $\alpha\beta$ is sometimes used to denote the effect of the mapping β followed by the mapping α. Note also that we have used the term product twice in this chapter with meanings quite different from the familiar product, say, of two integers. This is unavoidable unless we keep inventing new names.

Example 14: Let $A = \{a, b, c\}$, $B = \{d, e\}$, $C = \{f, g, h, i\}$ and

$$\alpha: \quad a\alpha = d, \; b\alpha = e, \; c\alpha = e$$

$$\beta: \quad d\beta = f, \; e\beta = h$$

Then $\quad \alpha\beta: \quad a(\alpha\beta) = (a\alpha)\beta = d\beta = f, \; b(\alpha\beta) = e\beta = h, \; c(\alpha\beta) = h$

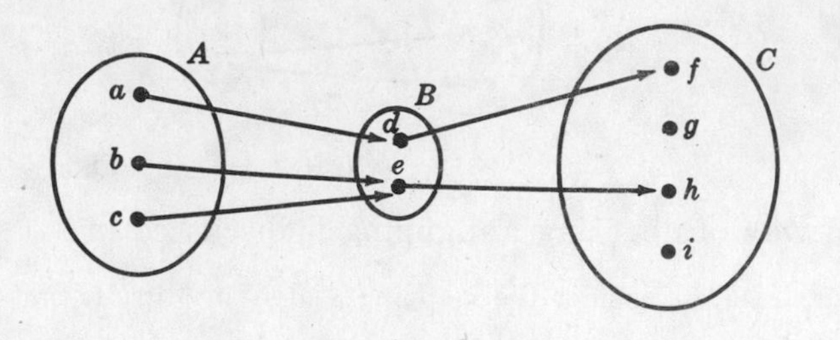

Fig. 1-6

ONE-TO-ONE MAPPINGS

A mapping $a \to a'$ of a set A into a set B is called a *one-to-one mapping of A into B* if the images of distinct elements of A are distinct elements of B; if, in addition, every element of B is an image, the mapping is called a *one-to-one mapping of A onto B*. In the latter case, it is clear that the mapping $a \to a'$ induces a mapping $a' \to a$ of B onto A. The two mappings are usually combined into $a \leftrightarrow a'$ and called a *one-to-one correspondence between A and B*.

Example 15: (*a*) The mapping α of Example 14 is not a one-to-one mapping of A into B (the distinct elements b and c of A have the same image).

(*b*) The mapping β of Example 14 is a one-to-one mapping of B into, but not onto, C ($g \in C$ is not an image).

(*c*) When $A = \{a, b, c, d\}$ and $B = \{p, q, r, s\}$,

(i) $\quad \alpha_1: \quad a \leftrightarrow p, \; b \leftrightarrow q, \; c \leftrightarrow r, \; d \leftrightarrow s$

and (ii) $\quad \alpha_2: \quad a \leftrightarrow r, \; b \leftrightarrow p, \; c \leftrightarrow q, \; d \leftrightarrow s$

are examples of one-to-one mappings of A onto B.

Two sets A and B are said to have the same number of elements if and only if a one-to-one mapping of A onto B exists. A set A is said to have n elements if there exists a one-to-one mapping of A onto the subset $S = \{1, 2, 3, \ldots, n\}$ of N. In this case, A is called a *finite* set.

The mapping $$\alpha: \quad n\alpha = 2n, \; n \in N$$
of N onto the proper subset $M = \{x : x \in N, \; x \text{ is even}\}$ of N is both one-to-one and onto. Now N is an *infinite set*; in fact, we may define an infinite set as one for which there exists a one-to-one correspondence between it and one of its proper subsets.

An infinite set is called *countable* or *denumerable* if there exists a one-to-one correspondence between it and the set N of all natural numbers.

ONE-TO-ONE MAPPING OF A SET ONTO ITSELF

Let $$\alpha: x \leftrightarrow x+1, \quad \beta: x \leftrightarrow 3x, \quad \gamma: x \leftrightarrow 2x-5, \quad \delta: x \leftrightarrow x-1$$
be one-to-one mappings of R onto itself. Since for any $x \in R$

$$x\alpha\beta = (x+1)\beta = 3(x+1)$$

while
$$x\beta\alpha = (3x)\alpha = 3x+1$$

we see that (i) $\alpha\beta \neq \beta\alpha$

However,
$$x\gamma\delta = (2x-5)\delta = 2x-6$$

and
$$x\alpha(\gamma\delta) = (x+1)\gamma\delta = 2(x+1)-6 = 2x-4$$

while
$$x\alpha\gamma = (x+1)\gamma = 2x-3$$

and
$$x(\alpha\gamma)\delta = (2x-3)-1 = 2x-4$$

Thus, (ii) $\alpha(\gamma\delta) = (\alpha\gamma)\delta$

Now
$$x\alpha\delta = (x+1)\delta = x$$

and
$$x\delta\alpha = (x-1)\alpha = x$$

that is, α followed by δ (also, δ followed by α) maps each $x \in R$ into itself. Denote by $\mathcal{J}$, the *identity* mapping,

$$\mathcal{J}: x \leftrightarrow x$$

Then (iii) $\alpha\delta = \delta\alpha = \mathcal{J}$

that is, δ undoes whatever α does (also, α undoes whatever δ does). In view of (iii), δ is called the *inverse mapping* of α and we write $\delta = \alpha^{-1}$; also, α is the inverse of δ and we write $\alpha = \delta^{-1}$.

See Problem 18.

In Problem 19, we prove

Theorem I. If α is a one-to-one mapping of a set S onto a set T, then α has a unique inverse, and conversely.

In Problem 20, we prove

Theorem II. If α is a one-to-one mapping of a set S onto a set T and β is a one-to-one mapping of T onto a set U, then $(\alpha\beta)^{-1} = \beta^{-1} \cdot \alpha^{-1}$.

Solved Problems

1. Exhibit in tabular form: (*a*) $A = \{a : a \in N,\ 2 < a < 6\}$, (*b*) $B = \{p : p \in N,\ p < 10,\ p \text{ is odd}\}$, (*c*) $C = \{x : x \in I,\ 2x^2 + x - 6 = 0\}$.

(*a*) Here A consists of all natural numbers $(a \in N)$ between 2 and 6; thus, $A = \{3, 4, 5\}$.

(*b*) B consists of the odd natural numbers less than 10; thus, $B = \{1, 3, 5, 7, 9\}$.

(*c*) The elements of C are the integral roots of $2x^2 + x - 6 = (2x-3)(x+2) = 0$; thus $C = \{-2\}$.

2. Let $A = \{a,b,c,d\}$, $B = \{a,c,g\}$, $C = \{c,g,m,n,p\}$. Find:
$A \cup B = \{a,b,c,d,g\}$, $A \cup C = \{a,b,c,d,g,m,n,p\}$, $B \cup C = \{a,c,g,m,n,p\}$;
$A \cap B = \{a,c\}$, $A \cap C = \{c\}$, $B \cap C = \{c,g\}$; $A \cap (B \cup C) = \{a,c\}$;
$(A \cap B) \cup C = \{a,c,g,m,n,p\}$, $(A \cup B) \cap C = \{c,g\}$, $(A \cap B) \cup (A \cap C) = A \cap (B \cup C)$.

3. Consider the subsets $K = \{2,4,6,8\}$, $L = \{1,2,3,4\}$, $M = \{3,4,5,6,8\}$ of $U = \{1,2,3,\ldots,10\}$. (*a*) Exhibit K', L', M' in tabular form. (*b*) Show that $(K \cup L)' = K' \cap L'$.

(*a*) $K' = \{1,3,5,7,9,10\}$, $L' = \{5,6,7,8,9,10\}$, $M' = \{1,2,7,9,10\}$.

(*b*) $K \cup L = \{1,2,3,4,6,8\}$ so that $(K \cup L)' = \{5,7,9,10\}$. Then $K' \cap L' = \{5,7,9,10\} = (K \cup L)'$.

4. For the sets of Problem 2, show: (*a*) $(A \cup B) \cup C = A \cup (B \cup C)$, (*b*) $(A \cap B) \cap C = A \cap (B \cap C)$.

(*a*) Since $A \cup B = \{a,b,c,d,g\}$ and $C = \{c,g,m,n,p\}$, we have $(A \cup B) \cup C = \{a,b,c,d,g,m,n,p\}$. Since $A = \{a,b,c,d\}$ and $B \cup C = \{a,c,g,m,n,p\}$, we have $A \cup (B \cup C) = \{a,b,c,d,g,m,n,p\} = (A \cup B) \cup C$.

(*b*) Since $A \cap B = \{a,c\}$, we have $(A \cap B) \cap C = \{c\}$. Since $B \cap C = \{c,g\}$, we have $A \cap (B \cap C) = \{c\} = (A \cap B) \cap C$.

5. In Fig. 1-1(*c*), let $C = A \cap B$, $D = A \cap B'$, $E = B \cap A'$ and $F = (A \cup B)'$. Verify: (*a*) $(A \cup B)' = A' \cap B'$, (*b*) $(A \cap B)' = A' \cup B'$.

(*a*) $A' \cap B' = (E \cup F) \cap (D \cup F) = F = (A \cup B)'$

(*b*) $A' \cup B' = (E \cup F) \cup (D \cup F) = (E \cup F) \cup D = C' = (A \cap B)'$

6. Use the Venn diagram of Fig. 1-7 to verify:

(*a*) $E = (A \cap B) \cap C'$

(*b*) $A \cup B \cup C = (A \cup B) \cup C = A \cup (B \cup C)$

(*c*) $A \cup B \cap C$ is ambiguous

(*d*) $A' \cap C' = G \cup L$

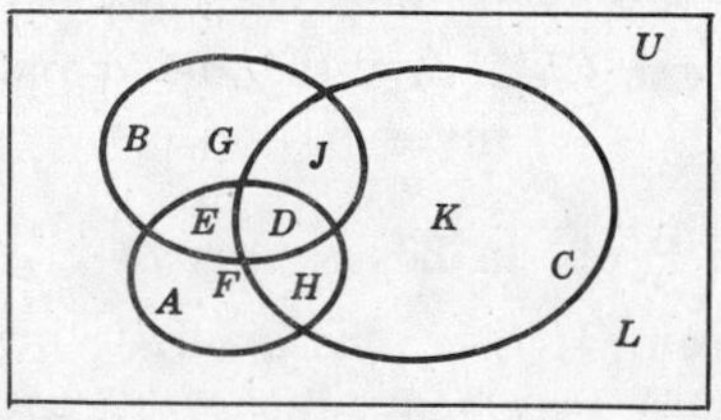

(*a*) $A \cap B = D \cup E$ and $C' = E \cup F \cup G \cup L$; then
$$(A \cap B) \cap C' = E$$

(*b*) $A \cup B \cup C = E \cup F \cup G \cup D \cup H \cup J \cup K$. Now
$$A \cup B = E \cup F \cup G \cup D \cup H \cup J$$
and
$$C = D \cup H \cup J \cup K$$
so that
$$(A \cup B) \cup C = E \cup F \cup G \cup D \cup H \cup J \cup K = A \cup B \cup C$$

Also, $B \cup C = E \cup G \cup D \cup H \cup J \cup K$ and $A = E \cup F \cup D \cup H$

so that $A \cup (B \cup C) = E \cup F \cup G \cup D \cup H \cup J \cup K = A \cup B \cup C$

Fig. 1-7

(*c*) $A \cup B \cap C$ could be interpreted either as $(A \cup B) \cap C$ or as $A \cup (B \cap C)$. Now $(A \cup B) \cap C = D \cup H \cup J$, while $A \cup (B \cap C) = A \cup (D \cup J) = A \cup J$. Thus, $A \cup B \cap C$ is ambiguous.

(*d*) $A' = G \cup J \cup K \cup L$ and $C' = E \cup F \cup G \cup L$; hence, $A' \cap C' = G \cup L$.

7. Let A and B be subsets of U. Use Venn diagrams to illustrate: $A \cap B' = A$ if and only if $A \cap B = \emptyset$.

Suppose $A \cap B = \emptyset$ and refer to Fig. 1-1(*b*). Now $A \subset B'$; hence $A \cap B' = A$.
Suppose $A \cap B \neq \emptyset$ and refer to Fig. 1-1(*c*). Now $A \not\subset B'$; hence $A \cap B' \neq A$.
Thus, $A \cap B' = A$ if and only if $A \cap B = \emptyset$.

8. Prove: $(A \cup B) \cup C = A \cup (B \cup C)$.

Let $x \in (A \cup B) \cup C$. Then $x \in A \cup B$ or $x \in C$, so that $x \in A$ or $x \in B$ or $x \in C$. When $x \in A$, then $x \in A \cup (B \cup C)$; when $x \in B$ or $x \in C$, then $x \in B \cup C$ and hence $x \in A \cup (B \cup C)$. Thus $(A \cup B) \cup C \subseteq A \cup (B \cup C)$.

Let $x \in A \cup (B \cup C)$. Then $x \in A$ or $x \in B \cup C$, so that $x \in A$ or $x \in B$ or $x \in C$. When $x \in A$ or $x \in B$, then $x \in A \cup B$ and hence $x \in (A \cup B) \cup C$; when $x \in C$, then $x \in (A \cup B) \cup C$. Thus $A \cup (B \cup C) \subseteq (A \cup B) \cup C$.

Now $(A \cup B) \cup C \subseteq A \cup (B \cup C)$ and $A \cup (B \cup C) \subseteq (A \cup B) \cup C$ imply $(A \cup B) \cup C = A \cup (B \cup C)$ as required. Thus, $A \cup B \cup C$ is unambiguous.

9. Prove: $(A \cap B) \cap C = A \cap (B \cap C)$.

Let $x \in (A \cap B) \cap C$. Then $x \in A \cap B$ and $x \in C$, so that $x \in A$ and $x \in B$ and $x \in C$. Since $x \in B$ and $x \in C$, then $x \in B \cap C$; since $x \in A$ and $x \in B \cap C$, then $x \in A \cap (B \cap C)$. Thus $(A \cap B) \cap C \subseteq A \cap (B \cap C)$.

Let $x \in A \cap (B \cap C)$. Then $x \in A$ and $x \in B \cap C$, so that $x \in A$ and $x \in B$ and $x \in C$. Since $x \in A$ and $x \in B$, then $x \in A \cap B$; since $x \in A \cap B$ and $x \in C$, then $x \in (A \cap B) \cap C$. Thus $A \cap (B \cap C) \subseteq (A \cap B) \cap C$ and $(A \cap B) \cap C = A \cap (B \cap C)$ as required. Thus, $A \cap B \cap C$ is unambiguous.

10. Prove: $A \cap (B \cup C) = (A \cap B) \cup (A \cap C)$.

Let $x \in A \cap (B \cup C)$. Then $x \in A$ and $x \in B \cup C$ ($x \in B$ or $x \in C$), so that $x \in A$ and $x \in B$ or $x \in A$ and $x \in C$. When $x \in A$ and $x \in B$, then $x \in A \cap B$ and so $x \in (A \cap B) \cup (A \cap C)$; similarly, when $x \in A$ and $x \in C$, then $x \in A \cap C$ and so $x \in (A \cap B) \cup (A \cap C)$. Thus $A \cap (B \cup C) \subseteq (A \cap B) \cup (A \cap C)$.

Let $x \in (A \cap B) \cup (A \cap C)$, so that $x \in A \cap B$ or $x \in A \cap C$. When $x \in A \cap B$, then $x \in A$ and $x \in B$ so that $x \in A$ and $x \in B \cup C$; similarly, when $x \in A \cap C$, then $x \in A$ and $x \in C$ so that $x \in A$ and $x \in B \cup C$. Thus $x \in A \cap (B \cup C)$ and $(A \cap B) \cup (A \cap C) \subseteq A \cap (B \cup C)$. Finally $A \cap (B \cup C) = (A \cap B) \cup (A \cap C)$ as required.

11. Prove: $(A \cup B)' = A' \cap B'$.

Let $x \in (A \cup B)'$. Now $x \notin A \cup B$, so that $x \notin A$ and $x \notin B$. Then $x \in A'$ and $x \in B'$, that is, $x \in A' \cap B'$; hence $(A \cup B)' \subseteq A' \cap B'$.

Let $x \in A' \cap B'$. Now $x \in A'$ and $x \in B'$, so that $x \notin A$ and $x \notin B$. Then $x \not\subset A \cup B$, so that $x \in (A \cup B)'$; hence $A' \cap B' \subseteq (A \cup B)'$. Thus $(A \cup B)' = A' \cap B'$ as required.

12. Prove: $(A \cap B) \cup C = (A \cup C) \cap (B \cup C)$.

$C \cup (A \cap B) = (C \cup A) \cap (C \cup B)$ by (1.10), Page 5.

Then $(A \cap B) \cup C = (A \cup C) \cap (B \cup C)$ by (1.9), Page 5.

13. Prove: $A - (B \cup C) = (A - B) \cap (A - C)$.

Let $x \in A - (B \cup C)$. Now $x \in A$ and $x \notin B \cup C$, that is, $x \in A$ but $x \notin B$ and $x \notin C$. Then $x \in A - B$ and $x \in A - C$, so that $x \in (A - B) \cap (A - C)$ and $A - (B \cup C) \subseteq (A - B) \cap (A - C)$.

Let $x \in (A - B) \cap (A - C)$. Now $x \in A - B$ and $x \in A - C$, that is, $x \in A$ but $x \notin B$ and $x \notin C$. Then $x \in A$ but $x \notin B \cup C$, so that $x \in A - (B \cup C)$ and $(A - B) \cap (A - C) \subseteq A - (B \cup C)$. Thus $A - (B \cup C) = (A - B) \cap (A - C)$ as required.

14. Prove: $(A \cup B) \cap B' = A$ if and only if $A \cap B = \emptyset$.

Using (1.10′) and (1.7′), Page 5, we find

$$(A \cup B) \cap B' = (A \cap B') \cup (B \cap B') = A \cap B'$$

We are then to prove: $A \cap B' = A$ if and only if $A \cap B = \emptyset$.

(a) Suppose $A \cap B = \emptyset$. Then $A \subseteq B'$ and $A \cap B' = A$.

(b) Suppose $A \cap B' = A$. Then $A \subseteq B'$ and $A \cap B = \emptyset$.

Thus, $(A \cup B) \cap B' = A$ if (by (a)) and only if (by (b)) $A \cap B = \emptyset$.

15. Prove: $X \subseteq Y$ if and only if $Y' \subseteq X'$.

(i) Suppose $X \subseteq Y$. Let $y' \in Y'$. Then $y' \notin X$ since $y' \notin Y$; hence, $y' \in X'$ and $Y' \subseteq X'$.

(ii) Conversely, suppose $Y' \subseteq X'$. Now, by (i), $(X')' \subseteq (Y')'$; hence, $X \subseteq Y$ as required.

16. Prove the identity $(A - B) \cup (B - A) = (A \cup B) - (A \cap B)$ of Example 10 using the identity $A - B = A \cap B'$ of Example 9.

We have

$$\begin{aligned}
(A - B) \cup (B - A) &= (A \cap B') \cup (B \cap A') \\
&= [(A \cap B') \cup B] \cap [(A \cap B') \cup A'] && \text{by (1.10), Page 5} \\
&= [(A \cup B) \cap (B' \cup B)] \cap [(A \cup A') \cap (B' \cup A')] && \text{by (1.10)} \\
&= [(A \cup B) \cap U] \cap [U \cap (B' \cup A')] && \text{by (1.7)} \\
&= (A \cup B) \cap (B' \cup A') && \text{by (1.4')} \\
&= (A \cup B) \cap (A' \cup B') && \text{by (1.9)} \\
&= (A \cup B) \cap (A \cap B)' && \text{by (1.11')} \\
&= (A \cup B) - (A \cap B)
\end{aligned}$$

17. Show that any two line segments have the same number of points.

Let the line segments be AB and $A'B'$ of Fig. 1-8. We are to show that it is always possible to establish a one-to-one correspondence between the points of the two segments. Denote the intersection of AB' and BA' by P. On AB take any point C and denote the intersection of CP and $A'B'$ by C'. The mapping

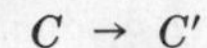

$$C \to C'$$

is the required correspondence, since each point of AB has a unique image on $A'B'$ and each point of $A'B'$ is the image of a unique point on AB.

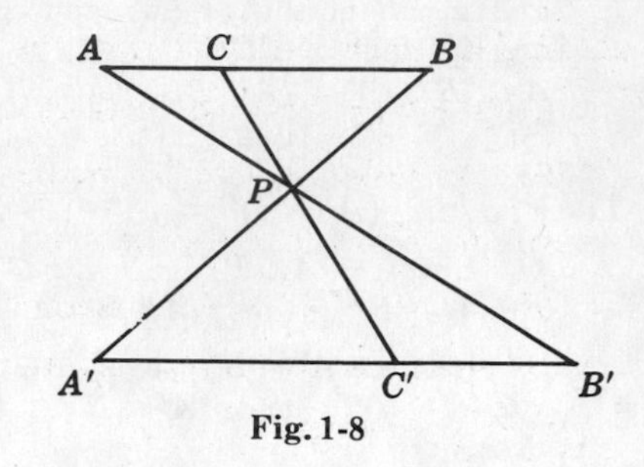

Fig. 1-8

18. Prove: (a) $x \to x + 2$ is a mapping of N into, but not onto, N. (b) $x \to 3x - 2$ is a one-to-one mapping of Q onto Q, (c) $x \to x^3 - 3x^2 - x$ is a mapping of R onto R but is not one-to-one.

(a) Clearly $x + 2 \in N$ when $x \in N$. The mapping is not onto since 2 is not an image.

(b) Clearly $3x - 2 \in Q$ when $x \in Q$. Also, each $r \in Q$ is the image of $x = (r + 2)/3 \in Q$.

(c) Clearly $x^3 - 3x^2 - x \in R$ when $x \in R$. Also, when $r \in R$, $x^3 - 3x^2 - x = r$ always has a real root x whose image is r. When $r = -3$, $x^3 - 3x^2 - x = r$ has 3 real roots $x = -1, 1, 3$. Since each has $r = -3$ as image, the mapping is not one-to-one.

19. Prove: If α is a one-to-one mapping of a set S onto a set T, then α has a unique inverse and conversely.

Suppose α is a one-to-one mapping of S onto T; then for any $s \in S$, we have

$$s \to s\alpha = t \in T$$

Since t is unique, it follows that α induces a one-to-one mapping

$$\beta: \quad t\beta \to s$$

Now $s(\alpha\beta) = (s\alpha)\beta = t\beta = s$; hence, $\alpha\beta = \mathcal{J}$ and β is an inverse of α. Suppose this inverse is not unique; in particular, suppose β and γ are inverses of α. Since

$$\alpha\beta = \beta\alpha = \mathcal{J} \qquad \text{and} \qquad \alpha\gamma = \gamma\alpha = \mathcal{J}$$

it follows that

$$\beta\alpha\gamma = \beta(\alpha\gamma) = \beta \cdot \mathcal{J} = \beta$$

and
$$\beta\alpha\gamma = (\beta\alpha)\gamma = \mathcal{J} \cdot \gamma = \gamma$$

Thus $\beta = \gamma$; the inverse of α is unique.

Conversely, let the mapping α of S into T have a unique inverse α^{-1}. Suppose for $s_1, s_2 \in S$, with $s_1 \neq s_2$, we have $s_1\alpha = s_2\alpha$. Then $(s_1\alpha)\alpha^{-1} = (s_2\alpha)\alpha^{-1}$, so that $s_1(\alpha \cdot \alpha^{-1}) = s_2(\alpha \cdot \alpha^{-1})$ and $s_1 = s_2$, a contradiction. Thus α is a one-to-one mapping. Now, for any $t \in T$, we have $(t\alpha^{-1})\alpha = t(\alpha^{-1} \cdot \alpha) = t \cdot \mathcal{J} = t$; hence, t is the image of $s = t\alpha^{-1} \in S$ and the mapping is onto.

20. Prove: If α is a one-to-one mapping of a set S onto a set T and β is a one-to-one mapping of T onto a set U, then $(\alpha\beta)^{-1} = \beta^{-1} \cdot \alpha^{-1}$.

Since $(\alpha\beta)(\beta^{-1} \cdot \alpha^{-1}) = \alpha(\beta \cdot \beta^{-1})\alpha^{-1} = \alpha \cdot \alpha^{-1} = \mathcal{J}$, $\beta^{-1} \cdot \alpha^{-1}$ is an inverse of $\alpha\beta$. By Problem 19 such an inverse is unique; hence, $(\alpha\beta)^{-1} = \beta^{-1} \cdot \alpha^{-1}$.

Supplementary Problems

21. Exhibit each of the following in tabular form:

(a) the set of negative integers greater than -6,

(b) the set of integers between -3 and 4,

(c) the set of integers whose squares are less than 20,

(d) the set of all positive factors of 18,

(e) the set of all common positive factors of 16 and 24,

(f) $\{p : p \in N, p^2 < 10\}$

(g) $\{b : b \in N, 3 \leq b \leq 8\}$

(h) $\{x : x \in I, 3x^2 + 7x + 2 = 0\}$

(i) $\{x : x \in Q, 2x^2 + 5x + 3 = 0\}$

Partial answer: (a) $\{-5, -4, -3, -2, -1\}$, (d) $\{1, 2, 3, 6, 9, 18\}$, (f) $\{1, 2, 3\}$, (h) $\{-2\}$

22. Verify: (a) $\{x : x \in N, x < 1\} = \emptyset$, (b) $\{x : x \in I, 6x^2 + 5x - 4 = 0\} = \emptyset$.

23. Exhibit the 15 proper subsets of $S = \{a, b, c, d\}$.

24. Show that the number of proper subsets of $S = \{a_1, a_2, \ldots, a_n\}$ is $2^n - 1$.

25. Using the sets of Problem 2, verify:
(a) $(A \cup B) \cup C = A \cup (B \cup C)$, (b) $(A \cap B) \cap C = A \cap (B \cap C)$, (c) $(A \cup B) \cap C \neq A \cup (B \cap C)$.

26. Using the sets of Problem 3, verify: (a) $(K')' = K$, (b) $(K \cap L)' = K' \cup L'$, (c) $(K \cup L \cup M)' = K' \cap L' \cap M'$, (d) $K \cap (L \cup M) = (K \cap L) \cup (K \cap M)$.

27. Let "$n \mid m$" mean "n is a factor of m". Given $A = \{x : x \in N, 3 \mid x\}$ and $B = \{x : x \in N, 5 \mid x\}$, list 4 elements of each of the sets A', B', $A \cup B$, $A \cap B$, $A \cup B'$, $A \cap B'$, $A' \cup B'$ where A' and B' are the respective complements of A and B in N.

28. Prove those laws of (1.8)-(1.12′), Page 5, which were not treated in Problems 8-13.

29. Let A and B be subsets of a universal set U. Prove:

(a) $A \cup B = A \cap B$ if and only if $A = B$,

(b) $A \cap B = A$ if and only if $A \subseteq B$,

(c) $(A \cap B') \cup (A' \cap B) = A \cup B$ if and only if $A \cap B = \emptyset$.

30. Given $n(U) = 692$, $n(A) = 300$, $n(B) = 230$, $n(C) = 370$, $n(A \cap B) = 150$, $n(A \cap C) = 180$, $n(B \cap C) = 90$, $n(A \cap B' \cap C') = 10$ where $n(S)$ is the number of distinct elements in the set S, find:

(a) $n(A \cap B \cap C) = 40$ (c) $n(A' \cap B' \cap C') = 172$

(b) $n(A' \cap B \cap C') = 30$ (d) $n((A \cap B) \cup (A \cap C) \cup (B \cap C)) = 340$

31. Given the mappings $\alpha: n \rightarrow n^2 + 1$ and $\beta: n \rightarrow 3n + 2$ of N into N, find: $\alpha\alpha = n^4 + 2n^2 + 2$, $\beta\beta$, $\alpha\beta = 3n^2 + 5$, and $\beta\alpha$.

32. Which of the following mappings of I into I:

(a) $x \rightarrow x + 2$, (b) $x \rightarrow 3x$, (c) $x \rightarrow x^2$, (d) $x \rightarrow 4 - x$, (e) $x \rightarrow x^3$, (f) $x \rightarrow x^2 - x$

are (i) mappings of I onto I, (ii) one-to-one mappings of I onto I? *Ans.* (i), (ii); (a), (d)

33. Same as Problem 32 with I replaced by Q. *Ans.* (i), (ii); (a), (b), (d)

34. Same as Problem 32 with I replaced by R. *Ans.* (i), (ii); (a), (b), (d), (e)

35. (a) If E is the set of all even positive integers, show that $x \rightarrow x + 1$, $x \in E$ is not a mapping of E onto the set F of all odd positive integers.

(b) If E^* is the set consisting of zero and all even positive integers (i.e., the non-negative integers), show that $x \rightarrow x + 1$, $x \in E^*$ is a mapping of E^* onto F.

36. Given the one-to-one mappings

$$\begin{array}{ll} \mathcal{J}: & 1\mathcal{J} = 1,\ 2\mathcal{J} = 2,\ 3\mathcal{J} = 3,\ 4\mathcal{J} = 4 \\ \alpha: & 1\alpha = 2,\ 2\alpha = 3,\ 3\alpha = 4,\ 4\alpha = 1 \\ \beta: & 1\beta = 4,\ 2\beta = 1,\ 3\beta = 2,\ 4\beta = 3 \\ \gamma: & 1\gamma = 3,\ 2\gamma = 4,\ 3\gamma = 1,\ 4\gamma = 2 \\ \delta: & 1\delta = 1,\ 2\delta = 4,\ 3\delta = 3,\ 4\delta = 2 \end{array}$$

of $S = \{1, 2, 3, 4\}$ onto itself, verify:

(a) $\alpha\beta = \beta\alpha = \mathcal{J}$, hence $\beta = \alpha^{-1}$; (b) $\alpha\gamma = \gamma\alpha = \beta$; (c) $\alpha\delta \neq \delta\alpha$; (d) $\alpha^2 = \alpha\alpha = \gamma$; (e) $\gamma^2 = \mathcal{J}$, hence $\gamma^{-1} = \gamma$; (f) $\alpha^4 = \mathcal{J}$, hence $\alpha^3 = \alpha^{-1}$; (g) $(\alpha^2)^{-1} = (\alpha^{-1})^2$.

Chapter 2

Relations and Operations

RELATIONS

Consider the set $P = \{a, b, c, \ldots, t\}$ of all persons living on a certain block of Main Street. We shall be concerned in this section with statements such as "a is the brother of p", "c is the father of g", ..., called relations on (or in) the set P. Similarly, "is parallel to", "is perpendicular to", "makes an angle of 45° with", ..., are relations on the set L of all lines in a plane.

Suppose in the set P above that the only fathers are c, d, g and that

$$c \text{ is the father of } a, g, m, p, q$$
$$d \text{ is the father of } f$$
$$g \text{ is the father of } h, n$$

Then, with $\mathcal{R}$ meaning "is the father of", we may write

$$c\,\mathcal{R}\,a,\ c\,\mathcal{R}\,g,\ c\,\mathcal{R}\,m,\ c\,\mathcal{R}\,p,\ c\,\mathcal{R}\,q,\ d\,\mathcal{R}\,f,\ g\,\mathcal{R}\,h,\ g\,\mathcal{R}\,n$$

Now $c\,\mathcal{R}\,a$ may be thought of as determining an ordered pair, either (a, c) or (c, a), of the product set $P \times P$. Although both will be found in use, we shall *always* associate

$$c\,\mathcal{R}\,a \text{ with the ordered pair } (a, c)$$

With this understanding, $\mathcal{R}$ determines on P the set of ordered pairs

$$(a, c),\ (g, c),\ (m, c),\ (p, c),\ (q, c),\ (f, d),\ (h, g),\ (n, g)$$

As in the case of the term function in Chapter 1, we define this subset of $P \times P$ to be the relation $\mathcal{R}$. Thus,

> A *relation* $\mathcal{R}$ on a set S (more precisely, a *binary relation* on S since it will be a relation between pairs of elements of S) is a subset of $S \times S$.

Example 1: (*a*) Let $S = \{2, 3, 5, 6\}$ and let $\mathcal{R}$ mean "divides".

Since $2\,\mathcal{R}\,2,\ 2\,\mathcal{R}\,6,\ 3\,\mathcal{R}\,3,\ 3\,\mathcal{R}\,6,\ 5\,\mathcal{R}\,5,\ 6\,\mathcal{R}\,6$, we have

$$\mathcal{R} = \{(2, 2), (6, 2), (3, 3), (6, 3), (5, 5), (6, 6)\}$$

(*b*) Let $S = \{1, 2, 3, \ldots, 20\}$ and $\mathcal{R}$ mean "is three times".

Then $3\,\mathcal{R}\,1,\ 6\,\mathcal{R}\,2,\ 9\,\mathcal{R}\,3,\ 12\,\mathcal{R}\,4,\ 15\,\mathcal{R}\,5,\ 18\,\mathcal{R}\,6$ and

$$\mathcal{R} = \{(1, 3), (2, 6), (3, 9), (4, 12), (5, 15), (6, 18)\}$$

(*c*) Consider $\mathcal{R} = \{(x, y) : 2x - y = 6,\ x \in R\}$. Geometrically, each $(x, y) \in \mathcal{R}$ is a point on the graph of the equation $2x - y = 6$. Thus, while the choice

$$c\,\mathcal{R}\,a \quad \text{means} \quad (a, c) \in \mathcal{R} \quad \text{rather than} \quad (c, a) \in \mathcal{R}$$

may have appeared strange at the time, it is now seen to be in keeping with the idea that any equation $y = f(x)$ is merely a special type of binary relation.

PROPERTIES OF BINARY RELATIONS

A relation $\mathcal{R}$ on a set S is called *reflexive* if $a\,\mathcal{R}\,a$ for every $a \in S$.

Example 2: (*a*) Let T be the set of all triangles in a plane and $\mathcal{R}$ mean "is congruent to". Now any triangle $t \in T$ is congruent to itself; thus, $t \mathcal{R} t$ for every $t \in T$, and $\mathcal{R}$ is reflexive.

(*b*) For the set T let $\mathcal{R}$ mean "has twice the area of". Clearly, $t \not\mathcal{R} t$ and $\mathcal{R}$ is not reflexive.

A relation $\mathcal{R}$ on a set S is called *symmetric* if whenever $a \mathcal{R} b$ then $b \mathcal{R} a$.

Example 3: (*a*) Let P be the set of all persons living on a block of Main Street and $\mathcal{R}$ mean "has the same surname as". When $x \in P$ has the same surname as $y \in P$, then y has the same surname as x; thus, $x \mathcal{R} y$ implies $y \mathcal{R} x$ and $\mathcal{R}$ is symmetric.

(*b*) For the same set P, let $\mathcal{R}$ mean "is the brother of" and suppose $x \mathcal{R} y$. Now y may be the brother or sister of x; thus, $x \mathcal{R} y$ does not necessarily imply $y \mathcal{R} x$ and $\mathcal{R}$ is not symmetric.

A relation $\mathcal{R}$ on a set S is called *transitive* if whenever $a \mathcal{R} b$ and $b \mathcal{R} c$ then $a \mathcal{R} c$.

Example 4: (*a*) Let S be the set of all lines in a plane and $\mathcal{R}$ mean "is parallel to". Clearly, if line a is parallel to line b and if b is parallel to line c, then a is parallel to c and $\mathcal{R}$ is transitive.

(*b*) For the same set S, let $\mathcal{R}$ mean "is perpendicular to". Now if line a is perpendicular to line b and if b is perpendicular to line c, then a is parallel to c. Thus, $\mathcal{R}$ is not transitive.

EQUIVALENCE RELATIONS

A relation $\mathcal{R}$ on a set S is called an *equivalence relation* on S when $\mathcal{R}$ is (i) reflexive, (ii) symmetric, and (iii) transitive.

Example 5: The relation "=" on the set R is undoubtedly the most familiar equivalence relation.

Example 6: Is the relation "has the same surname as" on the set P of Example 3 an equivalence relation?

Here we must check the validity of each of the following statements involving arbitrary $x,y,z \in P$:

(i) x has the same surname as x.

(ii) If x has the same surname as y, then y has the same surname as x.

(iii) If x has the same surname as y and if y has the same surname as z, then x has the same surname as z.

Since each of these is valid, "has the same surname as" is (i) reflexive, (ii) symmetric, (iii) transitive and, hence, is an equivalence relation on P.

Example 7: It follows from Example 3(*b*) that "is the brother of" is not symmetric and, hence, is not an equivalence relation on P.

See Problems 1-3.

EQUIVALENCE SETS

Let S be a set and $\mathcal{R}$ be an equivalence relation on S. If $a \in S$, the elements $y \in S$ satisfying $y \mathcal{R} a$ constitute a subset, $[a]$, of S, called an *equivalence set* or *equivalence class*. Thus, formally,

$$[a] \quad = \quad \{y:\ y \in S,\ y \mathcal{R} a\}$$

(Note the use of brackets here to denote equivalence classes.)

Example 8: Consider the set T of all triangles in a plane and the equivalence relation (see Problem 1) "is congruent to". When $a,b \in T$ we shall mean by $[a]$ the set or class of all triangles of T congruent to the triangle a, and by $[b]$ the set or class of all triangles of T congruent to the triangle b. We note, in passing, that triangle a is included in $[a]$ and that if triangle c is included in both $[a]$ and $[b]$ then $[a]$ and $[b]$ are merely two other ways of indicating the class $[c]$.

A set $\{A, B, C, \ldots\}$ of non-empty subsets of a set S will be called a *partition* of S provided (i) $A \cup B \cup C \cup \cdots = S$ and (ii) the intersection of every pair of distinct subsets is the empty set. The principal result of this section is

Theorem I. An equivalence relation $\mathcal{R}$ on a set S effects a partition of S and, conversely, a partition of S defines an equivalence relation on S.

Example 9: Two integers will be said to have the same parity if both are even or both are odd. The relation "has the same parity as" on I is an equivalence relation. (Prove this.) The relation establishes two subsets of I:

$$A = \{x: x \in I, x \text{ is even}\} \quad \text{and} \quad B = \{x: x \in I, x \text{ is odd}\}$$

Now every element of I will be found either in A or in B but never in both. Hence, $A \cup B = I$ and $A \cap B = \emptyset$, and the relation effects a partition of I.

Example 10: Consider the subsets $A = \{3, 6, 9, \ldots, 24\}$, $B = \{1, 4, 7, \ldots, 25\}$ and $C = \{2, 5, 8, \ldots, 23\}$ of $S = \{1, 2, 3, \ldots, 25\}$. Clearly, $A \cup B \cup C = S$ and $A \cap B = A \cap C = B \cap C = \emptyset$, so that $\{A, B, C\}$ is a partition of S. The equivalence relation which yields this partition is "has the same remainder when divided by 3 as".

In proving Theorem I, (see Problem 6), use will be made of the following properties of equivalence sets:

(1) $a \in [a]$

(2) If $b \in [a]$, then $[b] = [a]$.

(3) If $[a] \cap [b] \neq \emptyset$, then $[a] = [b]$.

The first of these follows immediately from the reflexive property $a \mathcal{R} a$ of an equivalence relation. For proofs of the others, see Problems 4-5.

ORDERING IN SETS

Consider the subset $A = \{2, 1, 3, 12, 4\}$ of N. In writing this set we have purposely failed to follow a natural inclination to give it as $A = \{1, 2, 3, 4, 12\}$ so as to point out that the latter version results from the use of the binary relation $(\leq)$ defined on N. This ordering of the elements of A (also, of N) is said to be *total*, since for every $a, b \in A$ $(m, n \in N)$ either $a < b$, $a = b$, or $a > b$ $(m < n,\ m = n,\ m > n)$. On the other hand, the binary relation $(\,|\,)$, (see Problem 27, Chapter 1) effects only a *partial ordering* on A, i.e., $2 \mid 4$ but $2 \nmid 3$. These orderings of A can best be illustrated by means of diagrams. Fig. 2-1 shows the ordering of A effected by $(\leq)$. We begin at the lowest point of the diagram and follow the arrows to obtain

$$1 \leq 2 \leq 3 \leq 4 \leq 12$$

Fig. 2-1

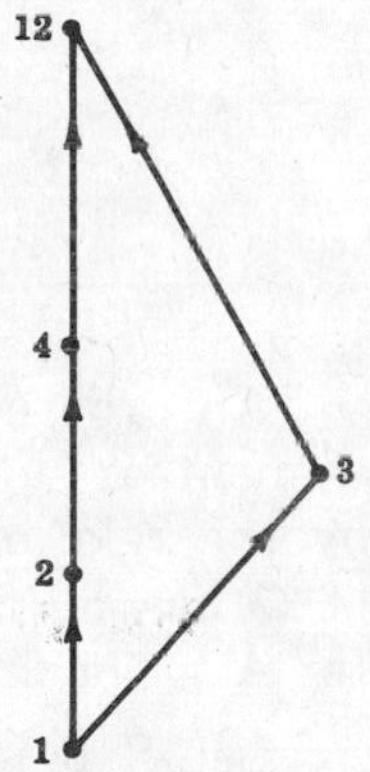

Fig. 2-2

It is to be expected that the diagram for a totally ordered set is always a straight line. Fig. 2-2 shows the partial ordering of A effected by the relation $(\,|\,)$.

See also Problem 7.

A set S will be said to be *partially ordered* (the possibility of a total ordering is not excluded) by a binary relation $\mathcal{R}$ if for arbitrary $a, b, c \in S$,

(i) $\mathcal{R}$ is reflexive, i.e., $a \mathcal{R} a$

(ii′) $\mathcal{R}$ is anti-symmetric, i.e., $a \mathcal{R} b$ and $b \mathcal{R} a$ if and only if $a = b$

(iii) $\mathcal{R}$ is transitive, i.e., $a \mathcal{R} b$ and $b \mathcal{R} c$ implies $a \mathcal{R} c$.

It will be left for the reader to check that these properties are satisfied by each of the relations $(\leq)$ and $(|)$ on A and also to verify the properties contain a redundancy in that (ii′) implies (i). The redundancy has been introduced to make perfectly clear the essential difference between the relations of this and the previous section.

Let S be a partially ordered set with respect to $\mathcal{R}$. Then:

(1) every subset of S is also partially ordered with respect to $\mathcal{R}$ while some subsets may be totally ordered. For example, in Fig. 2-2 the subset $\{1, 2, 3\}$ is partially ordered while the subset $\{1, 2, 4\}$ is totally ordered by the relation $(|)$.

(2) the element $a \in S$ is called a *first element* of S if $a \mathcal{R} x$ for every $x \in S$.

(3) the element $g \in S$ is called a *last element* of S if $x \mathcal{R} g$ for every $x \in S$.

(The first (last) element of an ordered set, assuming there is one, is unique.)

(4) the element $a \in S$ is called a *minimal element* of S if $x \mathcal{R} a$ implies $x = a$ for every $x \in S$.

(5) the element $g \in S$ is called a *maximal element* of S if $g \mathcal{R} x$ implies $g = x$ for every $x \in S$.

Example 11: (*a*) In the orderings of A of Fig. 2-1 and 2-2 the first element is 1 and the last element is 12. Also, 1 is a minimal element and 12 is a maximal element.

(*b*) In Fig. 2-3, $S = \{a, b, c, d\}$ has a first element a but no last element. Here, a is a minimal element while c and d are maximal elements.

(*c*) In Fig. 2-4, $S = \{a, b, c, d, e\}$ has a last element e but no first element. Here, a and b are minimal elements while e is a maximal element.

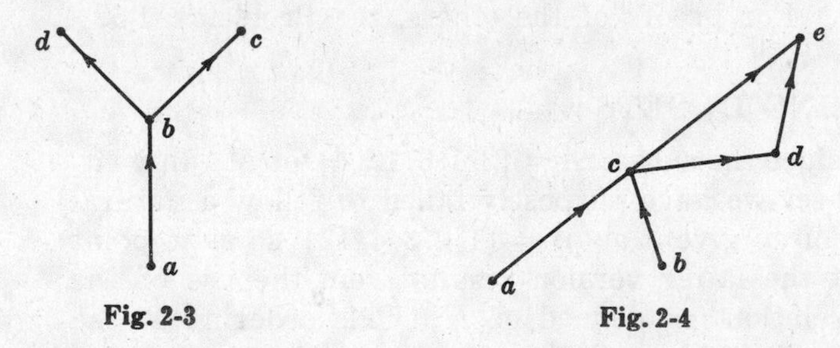

Fig. 2-3 Fig. 2-4

An ordered set S having the property that each of its non-empty subsets has a first element, is said to be *well-ordered*. For example, consider the sets N and Q each ordered by the relation $(\leq)$. Clearly, N is well-ordered but, since the subset $\{x : x \in Q, x > 2\}$ of Q has no first element, Q is not well-ordered. Is I well-ordered by the relation $(\leq)$? Is $A = \{1, 2, 3, 4, 12\}$ well-ordered by the relation $(|)$?

Let S be well-ordered by the relation $\mathcal{R}$. Then for arbitrary $a, b \in S$, the subset $\{a, b\}$ of S has a first element and so either $a \mathcal{R} b$ or $b \mathcal{R} a$. We have proved

Theorem II. Every well-ordered set is totally ordered.

OPERATIONS

Let $Q^+ = \{x : x \in Q, x > 0\}$. For every $a, b \in Q^+$, we have

$$a + b,\ b + a,\ a \cdot b,\ b \cdot a,\ a \div b,\ b \div a \ \in \ Q^+$$

Addition, multiplication and division are examples of *binary operations* on Q^+. (Note that such operations are simply mappings of $Q^+ \times Q^+$ into Q^+.) For example, addition associates with each pair $a, b \in Q^+$ an element $a + b \in Q^+$. Now $a + b = b + a$ but, in general, $a \div b \neq b \div a$; hence, to insure a unique image it is necessary to think of these operations as defined on ordered pairs of elements. Thus

A *binary operation* "$\circ$" on a non-empty set S is a mapping which associates with each ordered pair (a, b) of elements of S a uniquely defined element $a \circ b$ of S. In brief, a binary operation on a set S is a mapping of $S \times S$ into S.

Example 12: (*a*) Addition is a binary operation on the set of even natural numbers (the sum of two even natural numbers is an even natural number) but is not a binary operation on the set of odd natural numbers (the sum of two odd natural numbers is an even natural number).

(*b*) Neither addition nor multiplication are binary operations on $S = \{0, 1, 2, 3, 4\}$ since, for example, $2 + 3 = 5 \notin S$ and $2 \cdot 3 = 6 \notin S$.

(*c*) The adjoining Table 2-1, defining a certain binary operation $\circ$ on the set $A = \{a, b, c, d, e\}$ is to be read as follows: For every ordered pair (x, y) of $A \times A$, we find $x \circ y$ as the entry common to the row labeled x and the column labeled y. For example, the encircled element is $d \circ e$ (not $e \circ d$).

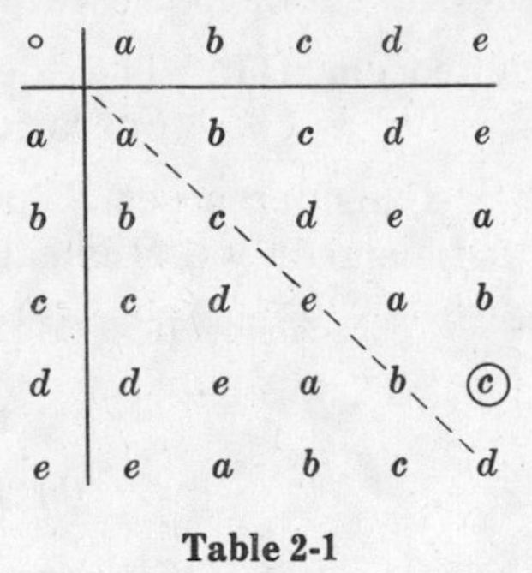

$\circ$	a	b	c	d	e
a	a	b	c	d	e
b	b	c	d	e	a
c	c	d	e	a	b
d	d	e	a	b	(c)
e	e	a	b	c	d

Table 2-1

The fact that $\circ$ is a binary operation on a set S is frequently indicated by the equivalent statement: The set S is *closed* with respect to the operation $\circ$. Example 12(*a*) may then be expressed: The set of even natural numbers is closed with respect to addition; the set of odd natural numbers is not closed with respect to addition.

TYPES OF BINARY OPERATIONS

A binary operation $\circ$ on a set S is called *commutative* whenever $x \circ y = y \circ x$ for all $x, y \in S$.

Example 13: (*a*) Addition and multiplication are commutative binary operations while division is not a commutative binary operation on Q^+.

(*b*) The operation $\circ$ on A of Table 2-1 is commutative. This may be checked readily by noting that (i) each row (b, c, d, e, a in the second row, for example) and the same numbered column (b, c, d, e, a in the second column) read exactly the same or that (ii) the elements of S are symmetrically placed with respect to the principal diagonal (dotted line) extending from the upper left to the lower right of the table.

A binary operation $\circ$ on a set S is called *associative* whenever $(x \circ y) \circ z = x \circ (y \circ z)$ for all $x, y, z \in S$.

Example 14: (*a*) Addition and multiplication are associative binary operations on Q^+.

(*b*) The operation $\circ$ on A of Table 2-1 is associative. We find, for instance, $(b \circ c) \circ d = d \circ d = b$ and $b \circ (c \circ d) = b \circ a = b$; $(d \circ e) \circ d = c \circ d = a$ and $d \circ (e \circ d) = d \circ c = a$; .... Completing the proof here becomes exceedingly tedious but it is suggested that the reader check a few other random choices.

(*c*) Let $\circ$ be a binary operation on R defined by

$$a \circ b = a + 2b \quad \text{for all} \quad a, b \in R$$

Since $\quad (a \circ b) \circ c = (a + 2b) \circ c = a + 2b + 2c$

while $\quad a \circ (b \circ c) = a \circ (b + 2c) = a + 2(b + 2c) = a + 2b + 4c$

the operation is not associative.

A set S is said to have an *identity* (*unit* or *neutral*) element with respect to a binary operation $\circ$ on S if there exists an element $u \in S$ with the property $u \circ x = x \circ u = x$ for every $x \in S$.

Example 15: (*a*) An identity element of Q with respect to addition is 0 since $0 + x = x + 0 = x$ for every $x \in Q$; an identity element of Q with respect to multiplication is 1 since $1 \cdot x = x \cdot 1 = x$ for every $x \in Q$.

(*b*) N has no identity element with respect to addition, but 1 is an identity element with respect to multiplication.

(c) An identity element of the set A of Example 12(c) with respect to $\circ$ is a. Note that there is only one.

In Problem 8, we prove:

Theorem III. The identity element, if one exists, of a set S with respect to a binary operation on S is unique.

Consider a set S having the identity element u with respect to a binary operation $\circ$. An element $y \in S$ is called an *inverse* of $x \in S$ provided $x \circ y = y \circ x = u$.

Example 16: (a) The inverse with respect to addition, or *additive inverse* of $x \in I$ is $-x$ since $x + (-x) = 0$, the additive identity element of I. In general, $x \in I$ does not have a multiplicative inverse.

(b) In Example 12(c), the inverses of a, b, c, d, e are respectively a, e, d, c, b.

It is not difficult to prove

Theorem IV. Let $\circ$ be a binary operation on a set S. The inverse with respect to $\circ$ of $x \in S$, if one exists, is unique.

Finally, let S be a set on which two binary operations $\square$ and $\circ$ are defined. The operation $\square$ is said to be *left distributive* with respect to $\circ$ if

$$a \square (b \circ c) = (a \square b) \circ (a \square c) \quad \text{for all } a,b,c \in S \tag{a}$$

and is said to be *right distributive* with respect to $\circ$ if

$$(b \circ c) \square a = (b \square a) \circ (c \square a) \quad \text{for all } a,b,c \in S \tag{b}$$

When both (a) and (b) hold, we say simply that $\square$ is *distributive* with respect to $\circ$. Note that the right members of (a) and (b) are equal whenever $\square$ is commutative.

Example 17: (a) For the set of all integers, multiplication ($\square = \cdot$) is distributive with respect to addition ($\circ = +$) since $x \cdot (y + z) = x \cdot y + x \cdot z$ for all $x,y,z \in I$.

(b) For the set of all integers, let $\circ$ be ordinary addition and $\square$ be defined by

$$x \square y = x^2 \cdot y = x^2 y \quad \text{for all } x,y \in I$$

Since $\quad a \square (b + c) = a^2 b + a^2 c = (a \square b) + (a \square c)$

$\square$ is left distributive with respect to $+$. Since

$$(b + c) \square a = ab^2 + 2abc + ac^2 \neq (b \square a) + (c \square a) = b^2 a + c^2 a$$

$\square$ is not right distributive with respect to $+$.

WELL-DEFINED OPERATIONS

Let $S = \{a, b, c, \ldots\}$ be a set on which a binary operation $\circ$ is defined and let the relation $\mathcal{R}$ partition S into a set $E = \{[a], [b], [c], \ldots\}$ of equivalence classes. Let a binary operation $\oplus$ on E be defined by

$$[a] \oplus [b] = [a \circ b] \quad \text{for every } [a], [b] \in E$$

Now it is not immediately clear that, for arbitrary $p,q \in [a]$ and $r,s \in [b]$, we have

$$[p \circ r] = [q \circ s] = [a \circ b] \tag{c}$$

We shall say that $\oplus$ is *well-defined* on E, that is,

$$[p] \oplus [r] = [q] \oplus [s] = [a] \oplus [b]$$

if and only if (c) holds.

Example 18: The relation "has the same remainder when divided by 9 as" partitions N into nine equivalence classes $[1], [2], [3], \ldots, [9]$. If $\circ$ is interpreted as addition on N, it is easy to show that $\oplus$ as defined above is well-defined. For example, when $x,y \in N$, $9x + 2 \in [2]$ and $9y + 5 \in [5]$; then $[2] \oplus [5] = [(9x + 2) + (9y + 5)] = [9(x + y) + 7] = [7] = [2 + 5]$, etc.

ISOMORPHISMS

Throughout this section we shall be using two sets:

$$A = \{1, 2, 3, 4\} \quad \text{and} \quad B = \{p, q, r, s\}$$

Now that ordering relations have been introduced, there will be a tendency to note here the familiar ordering used in displaying the elements of each set. We point this out in order to warn the reader against giving to any set properties which are not explicitly stated. In (1) below we consider A and B as arbitrary sets of four elements each and nothing more; for instance, we might have used $\{*, +, \$, \%\}$ as A or B; in (2) we introduce ordering relations on A and B but not the ones mentioned above; in (3) we define binary operations on the unordered sets A and B; in (4) we define binary operations on the ordered sets of (2).

(1) The mapping $\alpha: \; 1 \leftrightarrow p,\; 2 \leftrightarrow q,\; 3 \leftrightarrow r,\; 4 \leftrightarrow s$

is one of twenty-four establishing a 1-1 correspondence between A and B.

(2) Let A be ordered by the relation $\mathcal{R} = (|)$ and B be ordered by the relation $\mathcal{R}'$ as indicated in the diagram of Fig. 2-5. Since the diagram for A is as shown in Fig. 2-6, it is clear that the mapping

$$\beta: \; 1 \leftrightarrow r,\; 2 \leftrightarrow s,\; 3 \leftrightarrow q,\; 4 \leftrightarrow p$$

is a 1-1 correspondence between A and B which preserves the order relations, that is, for $u, v \in A$ and $x, y \in B$ with $u \leftrightarrow x$ and $v \leftrightarrow y$ then

$$u \mathcal{R}' v \quad \text{implies} \quad x \mathcal{R} y$$

and conversely.

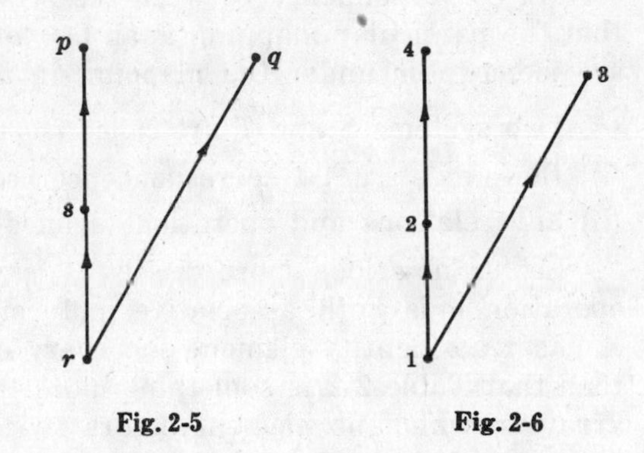

Fig. 2-5 Fig. 2-6

(3) On the unordered sets A and B, define the respective binary operations $\circ$ and $\Box$ with operation tables

A

$\circ$	1	2	3	4
1	1	2	3	4
2	2	4	1	3
3	3	1	4	2
4	4	3	2	1

Table 2-2

and

B

$\Box$	p	q	r	s
p	q	r	s	p
q	r	s	p	q
r	s	p	q	r
s	p	q	r	s

Table 2-3

It may be readily verified that the mapping

$$\gamma: \; 1 \leftrightarrow s,\; 2 \leftrightarrow p,\; 3 \leftrightarrow r,\; 4 \leftrightarrow q$$

is a 1-1 correspondence between A and B which preserves the operations, that is, whenever

$$w \in A \leftrightarrow x \in B \quad \text{and} \quad v \in A \leftrightarrow y \in B$$

(to be read "w in A corresponds to x in B and v in A corresponds to y in B"), then

$$w \circ v \leftrightarrow x \Box y$$

(4) On the ordered sets A and B of (2), define the respective binary operations $\circ$ and $\Box$ with operation tables

A

$\circ$	1	2	3	4
1	1	2	3	4
2	2	4	1	3
3	3	1	4	2
4	4	3	2	1

Table 2-4

and

B

$\square$	p	q	r	s
p	r	s	p	q
q	s	p	q	r
r	p	q	r	s
s	q	r	s	p

Table 2-5

It may be readily verified that the mapping

$$\beta: \quad 1 \leftrightarrow r,\ 2 \leftrightarrow s,\ 3 \leftrightarrow q,\ 4 \leftrightarrow p$$

is a 1-1 correspondence between A and B which preserves both the order relations *and* the operations.

By an *algebraic system* S we shall mean a set S together with any relations and operations defined on S. In each of the cases (1)-(4) above, we are concerned then with a certain correspondence between the sets of the two systems. In each case we shall say that the particular mapping is an *isomorphism* of A onto B or that the systems A and B are isomorphic under the mapping in accordance with:

Two systems S and T are called *isomorphic* provided

(i) there exists a 1-1 correspondence between the sets S and T, and

(ii) any relations and operations defined on the sets are preserved in the correspondence.

Let us now look more closely at, say, the two systems A and B of (3). The binary operation $\circ$ is both associative and commutative; also, with respect to this operation, A has 1 as identity element and every element of A has an inverse. One might suspect then that Table 2-2 is something more than the vacuous exercise of constructing a square array in which no element occurs twice in the same row or column. Considering the elements of A as digits rather than abstract symbols, it is easy to verify that the binary operation $\circ$ may be defined as: for every $x, y \in A$, $x \circ y$ is the remainder when $x \cdot y$ is divided by 5. (For example, $2 \cdot 4 = 8 = 1 \cdot 5 + 3$ and $2 \circ 4 = 3$.) Moreover, the system B is merely a disguised or coded version of A, the particular code being the 1-1 correspondence γ.

We shall make use of isomorphisms between algebraic systems in two ways:

(*a*) having discovered certain properties of one system (for example, those of A listed above) we may without further ado restate them as properties of any other system isomorphic with it.

(*b*) whenever more convenient, we may replace one system by any other isomorphic with it. Examples of this will be met with in Chapters 4 and 6.

PERMUTATIONS

Let $S = \{1, 2, 3, \ldots, n\}$ and consider the set S_n of the $n!$ permutations of these n symbols. (No significance is to be given the fact that they are natural numbers.) The definition of the product of mappings in Chapter 1 leads naturally to the definition of a "permutation operation" $\circ$ on the elements of S_n. First, however, we shall introduce more useful notations for permutations.

Let $i_1, i_2, i_3, \ldots, i_n$ be some arrangement of the elements of S. We now introduce a two-line notation for the permutation

$$\alpha = \begin{pmatrix} 1 & 2 & 3 & \ldots & n \\ i_1 & i_2 & i_3 & \ldots & i_n \end{pmatrix}$$

which is simply a variation of the notation for the mapping

$$\alpha: \quad 1\alpha = i_1, \ 2\alpha = i_2, \ 3\alpha = i_3, \ \ldots, \ n\alpha = i_n$$

Similarly, if $j_1, j_2, j_3, \ldots, j_n$ is another arrangement of the elements of S, we write

$$\beta = \begin{pmatrix} 1 & 2 & 3 & \ldots & n \\ j_1 & j_2 & j_3 & \ldots & j_n \end{pmatrix}$$

By the product $\alpha \circ \beta$ we shall mean that α and β are to be performed in that order. Now a rearrangement of any arrangement of the elements of S is simply another arrangement of these elements. Thus, for every $\alpha, \beta \in S_n$, $\alpha \circ \beta \in S_n$ and $\circ$ is a binary operation on S_n.

Example 19: Let $\alpha = \begin{pmatrix} 1 & 2 & 3 & 4 & 5 \\ 2 & 3 & 4 & 5 & 1 \end{pmatrix}$, $\beta = \begin{pmatrix} 1 & 2 & 3 & 4 & 5 \\ 1 & 3 & 2 & 5 & 4 \end{pmatrix}$, and $\gamma = \begin{pmatrix} 1 & 2 & 3 & 4 & 5 \\ 1 & 2 & 4 & 5 & 3 \end{pmatrix}$ be three of the 5! permutations in the set S_5 of all permutations on $S = \{1, 2, 3, 4, 5\}$.

Since the order of the columns of any permutation is immaterial, we may rewrite β as $\beta = \begin{pmatrix} 2 & 3 & 4 & 5 & 1 \\ 3 & 2 & 5 & 4 & 1 \end{pmatrix}$ in which the upper line of β is the lower line of α. Then

$$\alpha \circ \beta = \begin{pmatrix} 1 & 2 & 3 & 4 & 5 \\ 2 & 3 & 4 & 5 & 1 \end{pmatrix} \circ \begin{pmatrix} 2 & 3 & 4 & 5 & 1 \\ 3 & 2 & 5 & 4 & 1 \end{pmatrix} = \begin{pmatrix} 1 & 2 & 3 & 4 & 5 \\ 3 & 2 & 5 & 4 & 1 \end{pmatrix}$$

Similarly, rewriting α as $\begin{pmatrix} 1 & 3 & 2 & 5 & 4 \\ 2 & 4 & 3 & 1 & 5 \end{pmatrix}$, we find $\beta \circ \alpha = \begin{pmatrix} 1 & 2 & 3 & 4 & 5 \\ 2 & 4 & 3 & 1 & 5 \end{pmatrix}$. Thus, $\circ$ is not commutative.

Writing γ as $\begin{pmatrix} 3 & 2 & 5 & 4 & 1 \\ 4 & 2 & 3 & 5 & 1 \end{pmatrix}$, we find

$$(\alpha \circ \beta) \circ \gamma = \begin{pmatrix} 1 & 2 & 3 & 4 & 5 \\ 3 & 2 & 5 & 4 & 1 \end{pmatrix} \circ \begin{pmatrix} 3 & 2 & 5 & 4 & 1 \\ 4 & 2 & 3 & 5 & 1 \end{pmatrix} = \begin{pmatrix} 1 & 2 & 3 & 4 & 5 \\ 4 & 2 & 3 & 5 & 1 \end{pmatrix}$$

It is left for the reader to obtain $\beta \circ \gamma = \begin{pmatrix} 2 & 3 & 4 & 5 & 1 \\ 4 & 2 & 3 & 5 & 1 \end{pmatrix}$ and show that $(\alpha \circ \beta) \circ \gamma = \alpha \circ (\beta \circ \gamma)$. Thus, $\circ$ is associative in this example. It is now easy to show that $\circ$ is associative on S_5 and also on S_n.

The identity permutation is $\mathcal{J} = \begin{pmatrix} 1 & 2 & 3 & 4 & 5 \\ 1 & 2 & 3 & 4 & 5 \end{pmatrix}$ since clearly

$$\mathcal{J} \circ \alpha = \alpha \circ \mathcal{J} = \alpha, \quad \ldots$$

Finally, interchanging the two lines of α, we have

$$\begin{pmatrix} 2 & 3 & 4 & 5 & 1 \\ 1 & 2 & 3 & 4 & 5 \end{pmatrix} = \begin{pmatrix} 1 & 2 & 3 & 4 & 5 \\ 5 & 1 & 2 & 3 & 4 \end{pmatrix} = \alpha^{-1}$$

since $\alpha \circ \alpha^{-1} = \alpha^{-1} \circ \alpha = \mathcal{J}$. Moreover, it is evident that every element of S_5 has an inverse.

Another notation for permutations will now be introduced. The permutation

$$\alpha = \begin{pmatrix} 1 & 2 & 3 & 4 & 5 \\ 2 & 3 & 4 & 5 & 1 \end{pmatrix}$$

of Example 19 can be written in *cyclic* notation as (12345) where the cycle (12345) is interpreted to mean: 1 is replaced by 2, 2 is replaced by 3, 3 by 4, 4 by 5, and 5 by 1. The permutation

$$\gamma = \begin{pmatrix} 1 & 2 & 3 & 4 & 5 \\ 1 & 2 & 4 & 5 & 3 \end{pmatrix}$$

can be written as (345) where the cycle (345) is interpreted to mean: 1 and 2, the missing symbols, are unchanged while 3 is replaced by 4, 4 by 5, and 5 by 3. The permutation β can be written as (23)(45). The interpretation is clear: 1 is unchanged; 2 is replaced by 3 and 3 by 2; 4 is replaced by 5 and 5 by 4. We shall call (23)(45) a product of cycles. Note that these cycles are disjoint, i.e., have no symbols in common. Thus, in cyclic notation we shall expect a permutation on n symbols to consist of a single cycle or the product of two or more mutually disjoint cycles. Now (23) and (45) are themselves permutations of $S = \{1,2,3,4,5\}$ and, hence, $\beta = (23) \circ (45)$ but we shall continue to use juxtaposition to indicate the product of disjoint cycles. The reader will check that $\alpha \circ \beta = (135)$ and $\beta \circ \alpha = (124)$. In this notation the identity permutation $\mathcal{J}$ will be denoted by (1).

See Problem 11.

TRANSPOSITIONS

A permutation such as (12), (25), ... which involves interchanges of only two of the n symbols of $S = \{1,2,3,\ldots,n\}$ is called a *transposition*. Any permutation can be expressed, but not uniquely, as a product of transpositions.

Example 20: Express each of the following permutations

(*a*) (23), (*b*) (135), (*c*) (2345), (*d*) (12345)

on $S = \{1,2,3,4,5\}$ as products of transpositions.

(*a*) $(23) = (12) \circ (23) \circ (13) = (12) \circ (13) \circ (12)$

(*b*) $(135) = (13) \circ (15) = (15) \circ (35) = (15) \circ (13) \circ (15) \circ (13)$

(*c*) $(2345) = (23) \circ (24) \circ (25) = (25) \circ (34) \circ (35)$

(*d*) $(12345) = (12) \circ (13) \circ (14) \circ (15)$

The example above illustrates

Theorem V. Let a permutation α on n symbols be expressed as the product of r transpositions and also as a product of s transpositions. Then r and s are either both even or both odd.

For a proof, see Problem 12.

A permutation will be called *even* (*odd*) if it can be expressed as a product of an even (odd) number of transpositions. In Problem 13, we prove

Theorem VI. Of the $n!$ permutations of n symbols, half are even and half are odd.

Example 20 also illustrates

Theorem VII. A cycle of m symbols can be written as a product of $m-1$ transpositions.

ALGEBRAIC SYSTEMS

Much of the remainder of this book will be devoted to the study of various algebraic systems. Such systems may be studied in either of two ways:

(*a*) we begin with a set of elements (for example, the natural numbers or a set isomorphic to it), define the binary operations addition and multiplication, and derive the familiar laws governing operations with these numbers.

(*b*) we begin with a set S of elements (not identified); define a binary operation $\circ$; lay down certain postulates, for example, (i) $\circ$ is associative, (ii) there exists in S an identity element with respect to $\circ$, (iii) there exists in S an inverse with respect to $\circ$ of each element of S; and establish a number of theorems which follow.

We shall use both procedures here. In the next chapter we shall follow (*a*) in the study of the natural numbers.

Solved Problems

1. Show that "is congruent to" on the set T of all triangles in a plane is an equivalence relation.

(i) "a is congruent to a for all $a \in T$" is valid.

(ii) "If a is congruent to b, then b is congruent to a" is valid.

(iii) "If a is congruent to b and b is congruent to c, then a is congruent to c" is valid.

Thus "is congruent to" is an equivalence relation on T.

2. Show that "$<$" on I is not an equivalence relation.

(i) "$a < a$ for all $a \in I$" is not valid.

(ii) "If $a < b$, then $b < a$" is not valid.

(iii) "If $a < b$ and $b < c$, then $a < c$" is valid.

Thus "$<$" on I is not an equivalence relation. (Note that (i) or (ii) is sufficient.)

3. Let $\mathcal{R}$ be an equivalence relation and assume $c\,\mathcal{R}\,a$ and $c\,\mathcal{R}\,b$. Prove $a\,\mathcal{R}\,b$.

Since $c\,\mathcal{R}\,a$, then $a\,\mathcal{R}\,c$ (by the symmetric property). Since $a\,\mathcal{R}\,c$ and $c\,\mathcal{R}\,b$, then $a\,\mathcal{R}\,b$ (by the transitive property).

4. Prove: If $b \in [a]$, then $[b] = [a]$.

Denote by $\mathcal{R}$ the equivalence relation defining $[a]$. By definition, $b \in [a]$ implies $b\,\mathcal{R}\,a$ and $x \in [b]$ implies $x\,\mathcal{R}\,b$. Then $x\,\mathcal{R}\,a$ for every $x \in [b]$ (by the transitive property) and $[b] \subseteq [a]$. A repetition of the argument using $a\,\mathcal{R}\,b$ (which follows by the symmetric property of $\mathcal{R}$) and $y\,\mathcal{R}\,a$ (whenever $y \in [a]$) yields $[a] \subseteq [b]$. Thus $[b] = [a]$, as required.

5. Prove: If $[a] \cap [b] \neq \emptyset$, then $[a] = [b]$.

Suppose $[a] \cap [b] = \{r, s, \ldots\}$. Then $[r] = [a]$ and $[r] = [b]$ (by Problem 4), and $[a] = [b]$ (by the transitive property of $=$).

6. Prove: An equivalence relation $\mathcal{R}$ on a set S effects a partition of S and, conversely, a partition of S defines an equivalence relation on S.

Let $\mathcal{R}$ be an equivalence relation on S and define for each $p \in S$,

$$T_p = [p] = \{x : x \in S,\ x\,\mathcal{R}\,p\}$$

Since $p \in [p]$, it is clear that S is the union of all the distinct subsets $T_a, T_b, T_c, \ldots$ induced by $\mathcal{R}$. Now for any pair of these subsets, as T_b and T_c, we have $T_b \cap T_c = \emptyset$ since, otherwise, $T_b = T_c$ by Problem 5. Thus, $\{T_a, T_b, T_c, \ldots\}$ is the partition of S effected by $\mathcal{R}$.

Conversely, let $\{T_a, T_b, T_c, \ldots\}$ be any partition of S. On S define the binary relation $\mathcal{R}$ by $p\,\mathcal{R}\,q$ if and only if there is a T_i in the partition such that $p, q \in T_i$. It is clear that $\mathcal{R}$ is both reflexive and symmetric. Suppose $p\,\mathcal{R}\,q$ and $q\,\mathcal{R}\,r$; then by the definition of $\mathcal{R}$ there exist subsets T_j and T_k (not necessarily distinct) for which $p, q \in T_j$ and $q, r \in T_k$. Now $T_j \cap T_k \neq \emptyset$ and so $T_j = T_k$. Since $p, r \in T_j$, then $p\,\mathcal{R}\,r$ and $\mathcal{R}$ is transitive. This completes the proof that $\mathcal{R}$ is an equivalence relation.

7. Diagram the partial ordering of (*a*) the set of subsets of $S = \{a, b, c\}$ effected by the binary relation ($\subseteq$), (*b*) the set $B = \{2, 4, 5, 8, 15, 45, 60\}$ effected by the binary relation ($|$).

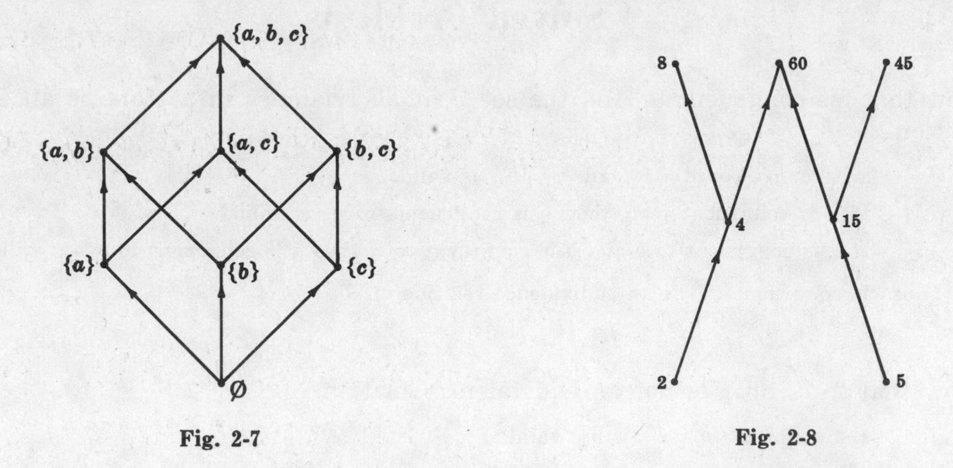

Fig. 2-7

Fig. 2-8

These figures need no elaboration once it is understood a minimum number of line segments is to be used. In Fig. 2-7, for example, $\emptyset$ is not joined directly to $\{a, b, c\}$ since $\emptyset \subseteq \{a, b, c\}$ is indicated by the path $\emptyset \subseteq \{a\} \subseteq \{a, b\} \subseteq \{a, b, c\}$. Likewise, in Fig. 2-8, segments joining 2 to 8 and 5 to 45 are unnecessary.

8. Prove: The identity element, if one exists, with respect to a binary operation $\circ$ on a set S is unique.

Assume the contrary, that is, assume q_1 and q_2 to be identity elements of S. Since q_1 is an identity element we have $q_1 \circ q_2 = q_2$, while since q_2 is an identity element we have $q_1 \circ q_2 = q_1$. Thus $q_1 = q_1 \circ q_2 = q_2$; the identity element is unique.

9. Show that multiplication is a binary operation on $S = \{1, -1, i, -i\}$ where $i = \sqrt{-1}$.

This can be done most easily by forming the adjoining table and noting that each entry is a unique element of S.

The order in which the elements of S are placed as labels on the rows and columns is immaterial; there is some gain, however, in using the same order for both.

The reader may show easily that multiplication on S is both associative and commutative, that 1 is the identity element and that the inverses of $1, -1, i, -i$ are respectively $1, -1, -i, i$.

$\circ$	1	-1	i	$-i$
1	1	-1	i	$-i$
-1	-1	1	$-i$	i
i	i	$-i$	-1	1
$-i$	$-i$	i	1	-1

Table 2-6

10. Determine the properties of the binary operations $\circ$ and $\square$ defined on $S = \{a, b, c, d\}$ by the given tables:

$\circ$	a	b	c	d
a	a	b	c	d
b	b	c	d	a
c	c	d	a	b
d	d	a	b	c

Table 2-7

$\square$	a	b	c	d
a	d	a	c	b
b	a	c	b	d
c	b	d	a	c
d	c	b	d	a

Table 2-8

The binary operation $\circ$ defined by Table 2-7 is commutative (check that the entries are symmetrically placed with respect to the principal diagonal) and associative (again, a chore). There is an identity element a (the column labels are also the elements of the first column and the row labels are also the elements of the first row). The inverses of a, b, c, d are respectively a, d, c, b $(a \circ a = b \circ d = c \circ c = d \circ b = a)$.

The binary operation $\square$ defined by Table 2-8 is neither commutative $(a \square c = c,\ c \square a = b)$ nor associative $(a \square (b \square c) = a \square b = a,\ (a \square b) \square c = a \square c = c)$. There is no identity element and, hence, no element of S has an inverse.

Since $a \square (d \circ c) = a \neq d = (a \square d) \circ (a \square c)$ and $(d \circ c) \square a \neq (d \square a) \circ (c \square a)$, $\square$ is neither left nor right distributive with respect to $\circ$; since $d \circ (c \square b) = c \neq a = (d \circ c) \square (d \circ b)$ and $\circ$ is commutative, $\circ$ is neither left nor right distributive with respect to $\square$.

11. (*a*) Write the permutations (23) and (13)(245) on 5 symbols in two line notation.
(*b*) Express the products $(23) \circ (13)(245)$ and $(13)(245) \circ (23)$ in cyclic notation.
(*c*) Express in cyclic notation the inverses of (23) and of (13)(245).

(*a*)
$$(23) = \begin{pmatrix} 1 & 2 & 3 & 4 & 5 \\ 1 & 3 & 2 & 4 & 5 \end{pmatrix} \quad \text{and} \quad (13)(245) = \begin{pmatrix} 1 & 2 & 3 & 4 & 5 \\ 3 & 4 & 1 & 5 & 2 \end{pmatrix}$$

(*b*)
$$(23) \circ (13)(245) = \begin{pmatrix} 1 & 2 & 3 & 4 & 5 \\ 1 & 3 & 2 & 4 & 5 \end{pmatrix} \circ \begin{pmatrix} 1 & 3 & 2 & 4 & 5 \\ 3 & 1 & 4 & 5 & 2 \end{pmatrix} = \begin{pmatrix} 1 & 2 & 3 & 4 & 5 \\ 3 & 1 & 4 & 5 & 2 \end{pmatrix} = (13452)$$

and
$$(13)(245) \circ (23) = \begin{pmatrix} 1 & 2 & 3 & 4 & 5 \\ 3 & 4 & 1 & 5 & 2 \end{pmatrix} \circ \begin{pmatrix} 3 & 4 & 1 & 5 & 2 \\ 2 & 4 & 1 & 5 & 3 \end{pmatrix} = \begin{pmatrix} 1 & 2 & 3 & 4 & 5 \\ 2 & 4 & 1 & 5 & 3 \end{pmatrix} = (12453)$$

(*c*) The inverse of (23) is
$$\begin{pmatrix} 1 & 3 & 2 & 4 & 5 \\ 1 & 2 & 3 & 4 & 5 \end{pmatrix} = \begin{pmatrix} 1 & 2 & 3 & 4 & 5 \\ 1 & 3 & 2 & 4 & 5 \end{pmatrix} = (23).$$

The inverse of (13)(245) is
$$\begin{pmatrix} 3 & 4 & 1 & 5 & 2 \\ 1 & 2 & 3 & 4 & 5 \end{pmatrix} = \begin{pmatrix} 1 & 2 & 3 & 4 & 5 \\ 3 & 5 & 1 & 2 & 4 \end{pmatrix} = (13)(254).$$

12. Prove: Let a permutation α on n symbols be expressed as the product of r transpositions and also as the product of $s > r$ transpositions. Then r and s are either both even or both odd.

Using the distinct symbols $x_1, x_2, x_3, \ldots, x_n$, we form the product

$$\begin{aligned} A = {} & (x_1 - x_2)(x_1 - x_3)(x_1 - x_4) \cdots (x_1 - x_n) \\ & (x_2 - x_3)(x_2 - x_4) \cdots (x_2 - x_n) \\ & \cdots\cdots\cdots\cdots\cdots\cdots \\ & (x_{n-1} - x_n) \end{aligned}$$

A transposition (u, v), where $u < v$, on A has the following effect: (1) any factor which involves neither x_u nor x_v is unchanged, (2) the single factor $x_u - x_v$ changes sign, (3) the remaining factors, each of which contains either x_u or x_v but not both, can be grouped into pairs, $(x_u - x_w)(x_v - x_w)$ where $u < v < w$, $(x_u - x_w)(x_w - x_v)$ where $u < w < v$, and $(x_w - x_u)(x_w - x_v)$ where $w < u < v$, which are all unchanged. Thus, the effect of the transposition on A is to change its sign.

Now the effect of α on A is to produce $(-1)^r A$ or $(-1)^s A$ according as α is written as the product of r or s transpositions. Since $(-1)^r A = (-1)^s A$, we have $A = (-1)^{s-r} A$ so that $s - r$ is even. Thus, r and s are either both even or both odd.

13. Prove: Of the $n!$ permutations on n symbols, half are even and half are odd.

Denote the even permutations by $p_1, p_2, p_3, \ldots, p_u$ and the odd permutations by $q_1, q_2, q_3, \ldots, q_v$. Let t be any transposition. Now $t \circ p_1, t \circ p_2, t \circ p_3, \ldots, t \circ p_u$ are permutations on n symbols. They are distinct since $p_1, p_2, p_3, \ldots, p_u$ are distinct and they are odd; thus, $u \leq v$. Also, $t \circ q_1, t \circ q_2, t \circ q_3, \ldots, t \circ q_v$ are distinct and even; thus, $v \leq u$. Hence $u = v = \frac{1}{2}n!$

Supplementary Problems

14. Which of the following are equivalence relations?

(*a*) "Is similar to" for the set T of all triangles in a plane.

(*b*) "Has the same radius as" for the set of all circles in a plane.

(*c*) "Is the square of" for the set N.

(*d*) "Has the same number of vertices as" for the set of all polygons in a plane.

(*e*) "$\subseteq$" for the set of sets $S = \{A, B, C, \ldots\}$.

(*f*) "$\leq$" for the set R.

Ans. (*a*), (*b*), (*d*).

15. (*a*) Show that "is a factor of" on N is reflexive and transitive but is not symmetric.

(*b*) Show that "costs within one dollar of" for men's shoes is reflexive and symmetric but not transitive.

(*c*) Give an example of a relation which is symmetric and transitive but not reflexive.

(*d*) Conclude from (*a*), (*b*), (*c*) that no two of the properties reflexive, symmetric, transitive of a binary relation implies the other.

16. Diagram the partial ordering of

(*a*) $A = \{1, 2, 3, 6\}$

(*b*) $B = \{1, 2, 3, 5, 30, 60\}$ and (*c*) $C = \{1, 3, 5, 15, 30, 45\}$

effected on each by the relation (|).

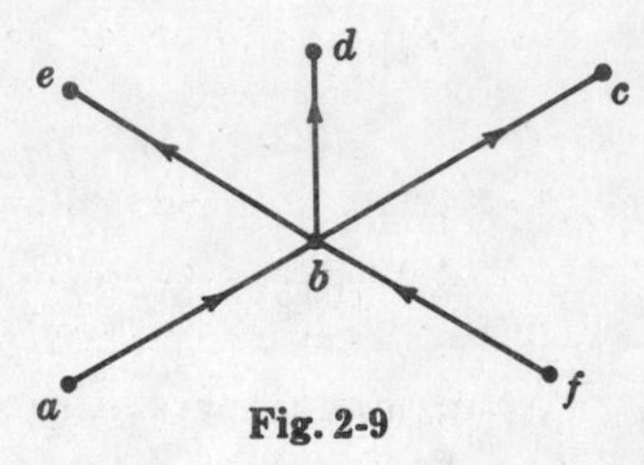

Fig. 2-9

17. Let $S = \{a, b, c, d, e, f\}$ be ordered by the relation $\mathcal{R}$ as shown in Fig. 2-9. (*a*) List all pairs $x, y \in S$ for which $x \not\mathcal{R} y$. (*b*) List all subsets of three elements each of which are totally ordered.

18. Verify:

(*a*) the ordered set of subsets of S in Problem 7(*a*) has $\emptyset$ as first element (also, as minimal element) and S as last element (also, as maximal element).

(*b*) the ordered set B of Problem 7(*b*) has neither a first nor last element. What are its minimal and maximal elements?

(*c*) the subset $C = \{2, 4, 5, 15, 60\}$ of B of Problem 7(*b*) has a last element but no first element. What are its minimal and maximal elements?

19. Show:

(*a*) multiplication is a binary operation on $S = \{1, -1\}$ but not on $T = \{1, 2\}$,

(*b*) addition is a binary relation on $S = \{x : x \in I, x < 0\}$ but multiplication is not.

20. Let $S = \{A, B, C, D\}$ where $A = \emptyset$, $B = \{a, b\}$, $C = \{a, c\}$, $D = \{a, b, c\}$. Construct tables to show that $\cup$ is a binary relation on S but $\cap$ is not.

21. For the binary operations $\circ$ and $\square$ defined on $S = \{a, b, c, d, e\}$ by Tables 2-9 and 2-10, assume associativity and investigate for all other properties.

$\circ$	a	b	c	d	e
a	a	d	a	d	e
b	d	b	b	d	e
c	a	b	c	d	e
d	d	d	d	d	e
e	e	e	e	e	e

Table 2-9

$\square$	a	b	c	d	e
a	a	c	c	a	a
b	c	c	c	b	b
c	c	c	c	c	c
d	a	b	c	d	d
e	a	b	c	d	e

Table 2-10

22. Let $S = \{A, B, C, D\}$ where $A = \emptyset$, $B = \{a\}$, $C = \{a, b\}$, $D = \{a, b, c\}$.

(*a*) Construct tables to show that $\cup$ and $\cap$ are binary operations on S.

(*b*) Assume associativity for each operation and investigate all other properties.

23. For the binary operation on $S = \{a, b, c, d, e, f, g, h\}$ defined by Table 2-11, assume associativity and investigate all other properties.

$\circ$	a	b	c	d	e	f	g	h
a	a	b	c	d	e	f	g	h
b	b	c	d	a	h	g	e	f
c	c	d	a	b	f	e	h	g
d	d	a	b	c	g	h	f	e
e	e	g	f	h	a	c	b	d
f	f	h	e	g	c	a	d	b
g	g	f	h	e	d	b	a	c
h	h	e	g	f	b	d	c	a

Table 2-11

24. Show that $\circ$ defined in Problem 23 is a binary operation on the subsets $S_0 = \{a\}$, $S_1 = \{a, c\}$, $S_2 = \{a, e\}$, $S_3 = \{a, f\}$, $S_4 = \{a, g\}$, $S_5 = \{a, h\}$, $S_6 = \{a, b, c, d\}$, $S_7 = \{a, c, e, f\}$, $S_8 = \{a, c, g, h\}$ but not on the subsets $T_1 = \{a, b\}$ and $T_2 = \{a, f, g\}$ of S.

25. Prove Theorem IV. *Hint*: Assume y and z to be inverses of x and consider $z \circ (x \circ y) = (z \circ x) \circ y$.

26. (*a*) Show that the set N of all natural numbers under addition and the set $M = \{2x : x \in N\}$ under addition are isomorphic. *Hint*: Use $n \in N \leftrightarrow 2n \in M$.

 (*b*) Is the set N under addition isomorphic to the set $P = \{2x - 1 : x \in N\}$ under addition?

 (*c*) Is the set M of (*a*) isomorphic to the set P of (*b*)?

27. Let A and B be sets with respective operations $\circ$ and $\square$. Suppose A and B are isomorphic and show:

 (*a*) if the associative (commutative) law holds in A it also holds in B.

 (*b*) if A has an identity element u, then its correspondent u' is the identity element in B.

 (*c*) if each element in A has an inverse with respect to $\circ$, the same is true of the elements of B with respect to $\square$.

 Hint: In (*a*), let $a \in A \leftrightarrow a' \in B$, $b \leftrightarrow b'$, $c \leftrightarrow c'$. Then

$$a \circ (b \circ c) \leftrightarrow a' \square (b' \square c'), \qquad (a \circ b) \circ c \leftrightarrow (a' \square b') \square c'$$

 and

$$a \circ (b \circ c) = (a \circ b) \circ c \quad \text{implies} \quad a' \square (b' \square c') = (a' \square b') \square c'$$

28. Express each of the following permutations on 8 symbols as a product of disjoint cycles and as a product of transpositions (minimum number).

 (*a*) $\begin{pmatrix} 1 & 2 & 3 & 4 & 5 & 6 & 7 & 8 \\ 2 & 3 & 4 & 1 & 5 & 6 & 7 & 8 \end{pmatrix}$ (*b*) $\begin{pmatrix} 1 & 2 & 3 & 4 & 5 & 6 & 7 & 8 \\ 3 & 4 & 5 & 6 & 7 & 8 & 1 & 2 \end{pmatrix}$ (*c*) $\begin{pmatrix} 1 & 2 & 3 & 4 & 5 & 6 & 7 & 8 \\ 3 & 4 & 1 & 6 & 8 & 2 & 7 & 5 \end{pmatrix}$

 (*d*) $(2468) \circ (348)$ (*e*) $(15)(2468) \circ (37)(15468)$ (*f*) $(135) \circ (3456) \circ (4678)$

 Partial answer. (*a*) $(1234) = (12)(13)(14)$

 (*c*) $(13)(246)(58) = (13)(24)(26)(58)$

 (*d*) $(28)(346) = (28)(34)(36)$

 (*f*) $(1637845) = (16)(13)(17)(18)(14)(15)$

 Note: For convenience, $\circ$ has been suppressed in indicating the products of transpositions.

29. Show that the cycles (1357) and (2468) of Problem 28(*b*) are commutative. State the theorem covering this.

30. Write in cyclic notation the 6 permutations on $S = \{1, 2, 3\}$, denote them in some order by $p_1, p_2, p_3, \ldots, p_6$, and form a table of products $p_i \circ p_j$.

31. Form the operation (product) table for the set $S = \{(1), (1234), (1432), (13)(24)\}$ of permutations on four symbols. Using Table 2-3, show that S is isomorphic to B.

Chapter 3

The Natural Numbers

THE PEANO POSTULATES

Thus far we have assumed those properties of the number systems necessary to provide examples and exercises in the earlier chapters. In this chapter we propose to develop the system of natural numbers assuming only a few of its simpler properties. These simple properties, known as the Peano Postulates (Axioms) after the Italian mathematician who in 1899 inaugurated the program, may be stated as follows:

Let there exist a non-empty set N such that

Postulate I : $1 \in N$

Postulate II : For each $n \in N$ there exists a unique $n^* \in N$, called the *successor* of n.

Postulate III : For each $n \in N$ we have $n^* \neq 1$.

Postulate IV : If $m,n \in N$ and $m^* = n^*$, then $m = n$.

Postulate V : Any subset K of N having the properties

(a) $1 \in K$

(b) $k^* \in K$ whenever $k \in K$

is equal to N.

First, we shall check to see that these are in fact well-known properties of the natural numbers. Postulates I and II need no elaboration; III states that there is a first natural number 1; IV states that distinct natural numbers m and n have distinct successors $m+1$ and $n+1$; V states essentially that any natural number can be reached by beginning with 1 and counting consecutive successors.

It will be noted that, in the definitions of addition and multiplication on N which follow, nothing beyond these postulates is used.

ADDITION ON N

Addition on N is defined by

(i) $n+1 = n^*$, for every $n \in N$

(ii) $n+m^* = (n+m)^*$ whenever $n+m$ is defined.

It can be shown that addition is then subject to the following laws:

For all $m,n,p \in N$,

A_1. Closure Law $\quad n+m \in N$

A_2. Commutative Law $\quad n+m = m+n$

A_3. Associative Law $\quad m+(n+p) = (m+n)+p$

A_4. Cancellation Law $\quad$ If $m+p = n+p$, then $m = n$.

MULTIPLICATION ON N

Multiplication on N is defined by

(iii) $n \cdot 1 = n$

(iv) $n \cdot m^* = n \cdot m + n$ whenever $n \cdot m$ is defined.

It can be shown that multiplication is then subject to the following laws:

For all $m, n, p \in N$,

$\mathbf{M}_1$. Closure Law $n \cdot m \in N$

$\mathbf{M}_2$. Commutative Law $m \cdot n = n \cdot m$

$\mathbf{M}_3$. Associative Law $m \cdot (n \cdot p) = (m \cdot n) \cdot p$

$\mathbf{M}_4$. Cancellation Law If $m \cdot p = n \cdot p$, then $m = n$.

Addition and multiplication are subject to the Distributive Laws:

For all $m, n, p \in N$,

$\mathbf{D}_1$. $m \cdot (n + p) = m \cdot n + m \cdot p$

$\mathbf{D}_2$. $(n + p) \cdot m = n \cdot m + p \cdot m$

MATHEMATICAL INDUCTION

Consider the proposition

$$P(m): \quad m^* \neq m, \text{ for every } m \in N$$

We shall now show how this proposition may be established using only the Postulates I-V. Define

$$K = \{k : k \in N, P(k) \text{ is true}\}$$

Now $1 \in N$ (Postulate I)

and $1^* \neq 1$ (Postulate III)

Thus $P(1)$ is true and $1 \in K$

Next, let k be any element of K; then

(*a*) $P(k): \quad k^* \neq k$

is true. Now if $(k^*)^* = k^*$ it follows by Postulate IV that $k^* = k$, a contradiction of (*a*). Hence

$$P(k^*): \quad (k^*)^* \neq k^*$$

is true and so $k^* \in K$. Now K has the two properties stated in Postulate V; thus, $K = N$ and the proposition is valid for every $m \in N$.

In establishing the validity of the above proposition, we have at the same time established the following

Principle of Mathematical Induction

A proposition $P(m)$ is true for all $m \in N$ provided:

$$P(1) \text{ is true}$$

and, for each $k \in N$, $P(k)$ is true implies $P(k^*)$ is true

The several laws $\mathbf{A}_1$-$\mathbf{A}_4$, $\mathbf{M}_1$-$\mathbf{M}_4$, $\mathbf{D}_1$-$\mathbf{D}_2$ can be established by mathematical induction. $\mathbf{A}_1$ is established in Example 1, $\mathbf{A}_3$ in Problem 1, $\mathbf{A}_2$ in Problems 2 and 3, and $\mathbf{D}_2$ in Problem 5.

Example 1: Prove the Closure Law: $n+m \in N$ for all $m,n \in N$.

We are to prove that $n + m$ is defined (is a natural number) by **(i)** and **(ii)** for all $m,n \in N$. Suppose n to be some fixed natural number and consider the proposition

$$P(m): \quad n+m \in N, \text{ for every } m \in N$$

Now $\quad P(1): \quad n+1 \in N$

is true since $n+1 = n^*$ (by **(i)**) and $n^* \in N$ (by Postulate II). Suppose next that for some $k \in N$

(a) $\quad P(k): \quad n+k \in N$ is true

It then follows that $\quad P(k^*): \quad n+k^* \in N$

is true since $n+k^* = (n+k)^*$ (by **(ii)**) and $(n+k)^* \in N$ whenever $n+k \in N$ (by Postulate II). Thus, by induction, $P(m)$ is true for all $m \in N$ and, since n was *any* natural number, the Closure Law for Addition is established.

In view of the Closure Laws $\mathbf{A}_1$ and $\mathbf{M}_1$, addition and multiplication are (see Chapter 2) binary operations on N. The laws $\mathbf{A}_3$ and $\mathbf{M}_3$ suggest as definitions for the sum and product of three elements $a_1, a_2, a_3 \in N$,

(v) $$a_1 + a_2 + a_3 = (a_1 + a_2) + a_3 = a_1 + (a_2 + a_3)$$

and

(vi) $$a_1 \cdot a_2 \cdot a_3 = (a_1 \cdot a_2) \cdot a_3 = a_1 \cdot (a_2 \cdot a_3)$$

Note that for a sum or product of three natural numbers, parentheses may be inserted at will. The sum of four natural numbers is considered in Problem 4. The general case is left as an exercise.

In Problem 6, we prove

Theorem I. Every element $n \neq 1$ of N is the successor of some other element of N.

THE ORDER RELATIONS

For each $m,n \in N$, we define "<" by

(vii) $m < n$ if and only if there exists some $p \in N$ such that $m+p = n$.

In Problem 8 it is shown that the relation $<$ is transitive but neither reflexive nor symmetric. By Theorem I,

$$1 < n \quad \text{for all} \quad n \neq 1$$

and, by **(i)** and **(vii)**, $$n < n^* \quad \text{for all} \quad n \in N$$

For each $m,n \in N$, we define ">" by

(viii) $$m > n \quad \text{if and only if} \quad n < m$$

There follows

The Trichotomy Law: For any $m,n \in N$ one and only one of the following

$$(a)\ m = n, \quad (b)\ m < n, \quad (c)\ m > n$$

is true.

For a proof, see Problem 10.

Further consequences of the order relations are Theorems II and II′:

Theorem II. If $m,n \in N$ and $m < n$, then for each $p \in N$,

$$(a) \quad m+p < n+p \qquad (b) \quad m \cdot p < n \cdot p$$

and, conversely, (a) or (b) with $m, n, p < N$ implies $m < n$.

Theorem II′. If $m,n \in N$ and $m > n$, then for each $p \in N$,

$$(a) \quad m+p > n+p$$
$$(b) \quad m \cdot p > n \cdot p$$

and, conversely, (a) or (b) with $m,n,p > N$ implies $m > n$.

Since Theorem II′ is merely Theorem II with m and n interchanged, it is clear that the proof of any part of Theorem II (see Problem 11) establishes the corresponding part of Theorem II′.

The relations "less than or equal to" ($\leqq$) and "equal to or greater than" ($\geqq$) are defined as follows:

For $m,n \in N$, $\quad m \leqq n$ if either $m < n$ or $m = n$

$\quad m \geqq n$ if either $m = n$ or $m > n$

holds.

Let A be any subset of N (i.e., $A \subseteq N$). An element p of A is called the *least element* of A provided $p \leqq a$ for every $a \in A$. Notice that in the language of sets, p is the first element of A with respect to the ordering $\leqq$. In Problem 12, we prove

Theorem III. The set N is well-ordered.

MULTIPLES AND POWERS

Let $a \in S$, on which binary operations $+$ and $\cdot$ have been defined, and define

$$1a = a \qquad a^1 = a$$

and

$$(k+1)a = ka + a \qquad a^{k+1} = a^k \cdot a$$

whenever ka and a^k, for $k \in N$, are defined.

Example 2: Since $1a = a$ and $a^1 = a$, we have

$2a = (1+1)a = 1a + 1a = a + a$ and $a^2 = a^{1+1} = a^1 \cdot a = a \cdot a$

$3a = (2+1)a = 2a + 1a = a + a + a$ and $a^3 = a^{2+1} = a^2 \cdot a = a \cdot a \cdot a$

etc.

It must be understood here that in Example 2 the $+$ in $1+1$ and the $+$ in $a+a$ are to be presumed quite different, since the first denotes addition on N and the second on S. (It might be helpful to denote the operations on S by $\oplus$ and $\odot$). In particular, $ka = a+a+\cdots+a$ is a multiple of a, and can be written as $k \cdot a$ only if $k \in S$.

Using the induction principle, the following properties may be established for all $a,b \in S$ and all $m,n \in N$:

(ix) $ma + na = (m+n)a$ **(ix)′** $a^m \cdot a^n = a^{m+n}$

(x) $m(na) = (m \cdot n)a$ **(x)′** $(a^n)^m = a^{m \cdot n}$

and, when $+$ and $\cdot$ are commutative on S,

(xi) $na + nb = n(a+b)$ **(xi)′** $a^n \cdot b^n = (ab)^n$

ISOMORPHIC SETS

It should be evident by now that the set $\{1, 1^*, (1^*)^*, \ldots\}$, together with the operations and relations defined on it developed here differs from the familiar set $\{1,2,3,\ldots\}$ with operations and relations in vogue at the present time only in the symbols used. Had a Roman written this chapter he would, of course, have reached the same conclusion with his system $\{\text{I}, \text{II}, \text{III}, \ldots\}$. We say simply that the three are isomorphic.

Solved Problems

1. Prove the Associative Law $\mathbf{A}_3$: $m+(n+p) = (m+n)+p$ for all $m, n, p \in N$.

Let m and n be fixed natural numbers and consider the proposition

$$P(p): \quad m+(n+p) = (m+n)+p \quad \text{for all } p \in N$$

We first check the validity of $P(1)$: $m+(n+1) = (m+n)+1$. By (i) and (ii), Page 30,

$$m+(n+1) = m+n^* = (m+n)^* = (m+n)+1$$

and $P(1)$ is true.

Next, suppose that for some $k \in N$,

$$P(k): \quad m+(n+k) = (m+n)+k$$

is true. We need to show that this assures

$$P(k^*): \quad m+(n+k^*) = (m+n)+k^*$$

is true. By (ii), $\quad m+(n+k^*) = m+(n+k)^* = [m+(n+k)]^*$

and $\quad (m+n)+k^* = [(m+n)+k]^*$

Then, whenever $P(k)$ is true,

$$m+(n+k)^* = [m+(n+k)]^* = [(m+n)+k]^* = (m+n)+k^*$$

and $P(k^*)$ is true. Thus, $P(p)$ is true for all $p \in N$ and, since m and n were any natural numbers, $\mathbf{A}_3$ follows.

2. Prove $P(n)$: $n+1 = 1+n$ for all $n \in N$.

Clearly $P(1): 1+1 = 1+1$ is true. Next, suppose that for some $k \in N$,

$$P(k): \quad k+1 = 1+k$$

is true. We are to show that this assures

$$P(k^*): \quad k^*+1 = 1+k^*$$

is true. Using in turn the definition of k^*, $\mathbf{A}_3$, the assumption that $P(k)$ is true, and the definition of k^*, we have

$$1+k^* = 1+(k+1) = (1+k)+1 = (k+1)+1 = k^*+1$$

Thus $P(k^*)$ is true and $P(n)$ is established.

3. Prove the Commutative Law $\mathbf{A}_2$: $m+n = n+m$ for all $m, n \in N$.

Let n be a fixed but arbitrary natural number and consider

$$P(m): \quad m+n = n+m \quad \text{for all } m \in N$$

By Problem 2, $P(1)$ is true. Suppose that for some $k \in N$,

$$P(k): \quad k+n = n+k$$

is true. Now

$$k^*+n = (k+1)+n = k+(1+n) = k+(n+1) = k+n^* = (k+n)^* = (n+k)^* = n+k^*$$

Thus $\quad P(k^*): \quad k^*+n = n+k^*$

is true and $\mathbf{A}_2$ follows.

The reader will check carefully that in obtaining the sequence of equalities above, only definitions, postulates, such laws of addition as have been proved and, of course, the critical assumption $P(k)$ is true for some arbitrary $k \in N$ have been used. Both here and in later proofs, it will be understood that when the evidence supporting a step in a proof is not cited, the reader is expected to supply it.

4. (*a*) Let $a_1, a_2, a_3, a_4 \in N$ and define $a_1 + a_2 + a_3 + a_4 = (a_1 + a_2 + a_3) + a_4$. Show that in $a_1 + a_2 + a_3 + a_4$ we may insert parentheses at will.

Using **(v)**, we have $a_1 + a_2 + a_3 + a_4 = (a_1 + a_2 + a_3) + a_4 = (a_1 + a_2) + a_3 + a_4 = (a_1 + a_2) + (a_3 + a_4) = a_1 + a_2 + (a_3 + a_4) = a_1 + (a_2 + a_3 + a_4)$, etc.

(*b*) For $b, a_1, a_2, a_3 \in N$, prove $b \cdot (a_1 + a_2 + a_3) = b \cdot a_1 + b \cdot a_2 + b \cdot a_3$.

$$b \cdot (a_1 + a_2 + a_3) = b \cdot [(a_1 + a_2) + a_3] = b \cdot (a_1 + a_2) + b \cdot a_3 = b \cdot a_1 + b \cdot a_2 + b \cdot a_3$$

5. Prove the Distributive Law $\mathbf{D}_2$: $(n+p) \cdot m = n \cdot m + p \cdot m$ for all $m, n, p \in N$.

Let n and p be fixed and consider

$$P(m): \quad (n+p) \cdot m = n \cdot m + p \cdot m \quad \text{for all } m \in N$$

Using $\mathbf{A}_1$ and **(iii)**, we find

$$P(1): \quad (n+p) \cdot 1 = n + p = n \cdot 1 + p \cdot 1$$

is true. Suppose that for some $k \in N$,

$$P(k): \quad (n+p) \cdot k = n \cdot k + p \cdot k$$

is true. Then

$$\begin{aligned}(n+p) \cdot k^* &= (n+p) \cdot k + (n+p) = n \cdot k + p \cdot k + n + p \\ &= n \cdot k + (p \cdot k + n) + p = n \cdot k + (n + p \cdot k) + p \\ &= (n \cdot k + n) + (p \cdot k + p) = n \cdot k^* + p \cdot k^*\end{aligned}$$

Thus,

$$P(k^*): \quad (n+p) \cdot k^* = n \cdot k^* + p \cdot k^*$$

is true and $\mathbf{D}_2$ is established.

6. Prove: Every element $n \neq 1$ of N is the successor of some other element of N.

First, we note that Postulate III excludes 1 as a successor. Denote by K the set consisting of the element 1 and all elements of N which are successors, i.e.,

$$K = \{k: k \in N, k = 1 \text{ or } k = m^* \text{ for some } m \in N\}$$

Now every $k \in K$ has a unique successor $k^* \in N$ (Postulate II) and, since k^* is a successor, we have $k^* \in K$. Then $K = N$ (Postulate V). Hence, for any $n \in N$ we have either $n = 1$ or $n = m^*$ for some $m \in N$.

7. Prove: $m + n \neq m$ for all $m, n \in N$.

Let n be fixed and consider $P(m)$: $m + n \neq m$ for all $m \in N$. By Postulate III, $P(1)$: $1 + n \neq 1$ is true. Suppose that for some $k \in N$,

(*a*) $$P(k): \quad k + n \neq k$$

is true. Now $(k+n)^* \neq k^*$ since, by Postulate IV, $(k+n)^* = k^*$ implies $k + n = k$, a contradiction of (*a*). Thus

$$P(k^*): \quad k^* + n \neq k^*$$

is true and the theorem is established.

8. Show that $<$ is transitive but neither reflexive nor symmetric.

Let $m, n, p \in N$ and suppose that $m < n$ and $n < p$. By **(vii)** there exist $r, s \in N$ such that $m + r = n$ and $n + s = p$. Then

$$n + s = (m+r) + s = m + (r+s) = p$$

Thus, $m < p$ and $<$ is transitive.

Let $n \in N$. Now $n < n$ is false since, if it were true, there would exist some $k \in N$ such that $n + k = n$, contrary to the result in Problem 7. Thus, $<$ is not reflexive.

Finally, let $m, n \in N$ and suppose $m < n$ and $n < m$. Since $<$ is transitive, it follows that $m < m$, contrary to the result in the paragraph immediately above. Thus, $<$ is not symmetric.

9. Prove: $1 \leqq n$, for every $n \in N$.

When $n = 1$ the equality holds; otherwise, by Problem 6, $n = m^* = m + 1$, for some $m \in N$, and the inequality holds.

10. Prove the Trichotomy Law: For any $m, n \in N$, one and only one of

$$(a)\ m = n \qquad (b)\ m < n \qquad (c)\ m > n$$

is true.

Let m be any element of N and construct the subsets

$$N_1 = \{m\}, \qquad N_2 = \{x : x \in N, x < m\}, \qquad N_3 = \{x : x \in N, x > m\}$$

We are to show that $\{N_1, N_2, N_3\}$ is a partition of N relative to $\{=, <, >\}$.

(1) Suppose $m = 1$; then $N_1 = \{1\}$, $N_2 = \emptyset$ (Problem 9) and $N_3 = \{x : x \in N, x > 1\}$. Clearly $N_1 \cup N_2 \cup N_3 = N$. Thus, to complete the proof for this case, there remains only to check $N_1 \cap N_2 = N_1 \cap N_3 = N_2 \cap N_3 = \emptyset$.

(2) Suppose $m \neq 1$. Since $1 \in N_2$, it follows that $1 \in N_1 \cup N_2 \cup N_3$. Now select any $n \neq 1 \in N_1 \cup N_2 \cup N_3$. There are three cases to be considered:

(i) $n \in N_1$. Here, $n = m$ and so $n^* \in N_3$.

(ii) $n \in N_2$ so that $n + p = m$ for some $p \in N$. If $p = 1$, then $n^* = m \in N_1$; if $p \neq 1$ so that $p = 1 + q$ for some $q \in N$, then $n^* + q = m$ and so $n^* \in N_2$.

(iii) $n \in N_3$. Here $n^* > n > m$ and so $n^* \in N_3$.

Thus, for every $n \in N$,

$$n \in N_1 \cup N_2 \cup N_3 \quad \text{implies} \quad n^* \in N_1 \cup N_2 \cup N_3$$

Since $1 \in N_1 \cup N_2 \cup N_3$ we conclude that $N = N_1 \cup N_2 \cup N_3$.

Now $m \notin N_2$, since $m \nless m$; hence, $N_1 \cap N_2 = \emptyset$. Similarly, $m \ngtr m$ and so $N_1 \cap N_3 = \emptyset$. Suppose $N_2 \cap N_3 = \{p\}$ for some $p \in N$. Then $p < m$ and $p > m$ or, what is the same, $p < m$ and $m < p$. Since $<$ is transitive, we have $p < p$, a contradiction. Thus, we must conclude that $N_2 \cap N_3 = \emptyset$ and the proof is now complete for this case.

11. Prove: If $m, n \in N$ and $m < n$, then for each $p \in N$, $m + p < n + p$ and conversely.

Since $m < n$, there exists some $k \in N$ such that $m + k = n$. Then

$$n + p = (m + k) + p = m + k + p = m + p + k = (m + p) + k$$

and so

$$m + p < n + p$$

For the converse, assume $m + p < n + p$. Now either $m = n$, $m < n$, or $m > n$. If $m = n$, then $m + p = n + p$; if $m > n$, then $m + p > n + p$ (Theorem II′). Since these contradict the hypothesis, we conclude that $m < n$.

12. Prove: The set N is well-ordered.

Consider any subset $S \neq \emptyset$ of N. We are to prove that S has a least element. This is certainly true if $1 \in S$. Suppose $1 \notin S$; then $1 < s$ for every $s \in S$. Denote by K the set

$$K = \{k : k \in N, k \leqq s \text{ for each } s \in S\}$$

Since $1 \in K$, we know that $K \neq \emptyset$. Moreover $K \neq N$; hence, there must exist an $r \in K$ such that $r^* \notin K$. Now this $r \in S$ since, otherwise, $r < s$ and so $r^* \leqq s$ for every $s \in S$. But then $r^* \in K$, a contradiction of our assumption concerning r. Thus, S has a least element. Now S was any non-empty subset of N; hence, every non-empty subset of N has a least element and N is well-ordered.

Supplementary Problems

13. Prove by induction: $1 \cdot n = n$ for every $n \in N$.

14. Prove by induction: (*a*) $\mathbf{M}_1$, (*b*) $\mathbf{M}_2$, (*c*) $\mathbf{M}_3$.
Hint: Use the result of Problem 13 and $\mathbf{D}_2$ in (*b*).

15. Prove: (*a*) $\mathbf{D}_1$ by following Problem 5, (*b*) $\mathbf{D}_1$ by using $\mathbf{M}_2$.

16. Prove: (*a*) $(m + n^*)^* = m^* + n^*$
(*b*) $(m \cdot n^*)^* = m \cdot n + m^*$
(*c*) $(m^* \cdot n^*)^* = m^* + m \cdot n + n^*$
where $m, n \in N$.

17. Prove: (*a*) $(m + n) \cdot (p + q) = (m \cdot p + m \cdot q) + (n \cdot p + n \cdot q)$
(*b*) $m \cdot (n + p) \cdot q = (m \cdot n) \cdot q + m \cdot (p \cdot q)$
(*c*) $m^* + n^* = (m + n)^* + 1$
(*d*) $m^* \cdot n^* = (m \cdot n)^* + m + n$

18. Let $m, n, p, q \in N$ and define $m \cdot n \cdot p \cdot q = (m \cdot n \cdot p) \cdot q$. (*a*) Show that in $m \cdot n \cdot p \cdot q$ we may insert parentheses at will. (*b*) Prove: $m \cdot (n + p + q) = m \cdot n + m \cdot p + m \cdot q$.

19. Identify $S = \{x : x \in N, n < x < n^* \text{ for all } n \in N\}$.

20. If $m, n, p, q \in N$ and if $m < n$ and $p < q$, prove: (*a*) $m + p < n + q$, (*b*) $m \cdot p < n \cdot q$

21. Let $m, n \in N$. Prove: (*a*) If $m = n$, then $k^* \cdot m > n$ for every $k \in N$. (*b*) If $k^* \cdot m = n$ for some $k \in N$, then $m < n$.

22. Prove $\mathbf{A}_4$ and $\mathbf{M}_4$ using the Trichotomy Law and Theorems II, Page 32, and II′, Page 33.

23. For all $m \in N$ define $m^1 = m$ and $m^{p+1} = m^p \cdot m$ provided m^p is defined. When $m, n, p, q \in N$, prove: (*a*) $m^p \cdot m^q = m^{p+q}$, (*b*) $(m^p)^q = m^{p \cdot q}$, (*c*) $(m \cdot n)^p = m^p \cdot n^p$, (*d*) $(1)^p = 1$.

24. For $m, n \in N$ show that (*a*) $m^2 < m \cdot n < n^2$ if $m < n$, (*b*) $m^2 + n^2 > 2m \cdot n$ if $m \neq n$.

25. Prove, by induction, for all $n \in N$:
(*a*) $1 + 2 + 3 + \cdots + n = \frac{1}{2}n(n+1)$
(*b*) $1^2 + 2^2 + 3^2 + \cdots + n^2 = \frac{1}{6}n(n+1)(2n+1)$
(*c*) $1^3 + 2^3 + 3^3 + \cdots + n^3 = \frac{1}{4}n^2(n+1)^2$

26. For $a_1, a_2, a_3, \ldots, a_n \in N$ define $a_1 + a_2 + a_3 + \cdots + a_k = (a_1 + a_2 + a_3 + \cdots + a_{k-1}) + a_k$ for $k = 3, 4, 5, \ldots, n$. Prove:
(*a*) $a_1 + a_2 + a_3 + \cdots + a_n = (a_1 + a_2 + a_3 + \cdots + a_r) + (a_{r+1} + a_{r+2} + a_{r+3} + \cdots + a_n)$
(*b*) In any sum of n natural numbers, parentheses may be inserted at will.

27. Prove each of the following alternate forms of the Induction Principle:
(*a*) With each $n \in N$ let there be associated a proposition $P(n)$. Then $P(n)$ is true for every $n \in N$ provided:
(i) $P(1)$ is true.
(ii) For each $m \in N$ the assumption $P(k)$ is true for all $k < m$ implies $P(m)$ is true.
(*b*) Let b be some fixed natural number, and with each natural number $n \geqq b$ let there be associated a proposition $P(n)$. Then $P(n)$ is true for all values of n provided:
(i) $P(b)$ is true.
(ii) For each $m > b$ the assumption that $P(k)$ is true for all $k \in N$ such that $b \leqq k < m$ implies $P(m)$ is true.

Chapter 4

The Integers

INTRODUCTION

The system of natural numbers has an obvious defect in that, given $m, s \in N$, the equation $m + x = s$ may or may not have a solution. For example, $m + x = m$ has no solution (see Problem 7, Chapter 3) while $m + x = m^*$ has the solution $x = 1$. Everyone knows that this state of affairs is remedied by adjoining to the natural numbers (then called positive integers) the additional numbers zero and the negative integers to form the set I of all integers.

In this chapter it will be shown how the system of integers can be constructed from the system of natural numbers. For this purpose, we form the product set

$$L \;=\; N \times N \;=\; \{(s, m):\ s \in N,\ m \in N\}$$

Now we shall not say that (s, m) is a solution of $m + x = s$. However, let it be perfectly clear, we shall proceed as if this were the case. Notice that if (s, m) were a solution of $m + x = s$, then (s, m) would also be a solution of $m^* + x = s^*$ which, in turn, would have (s^*, m^*) as solution. This observation motivates the partition of L into equivalence classes such that (s, m) and (s^*, m^*) are members of the same class.

BINARY RELATION ~

Let the binary relation "~", read "wave", be defined on all $(s, m), (t, n) \in L$ by

$$(s, m) \sim (t, n) \quad \text{if and only if} \quad s + n = t + m$$

Example 1: (*a*) $(5, 2) \sim (9, 6)$ since $5 + 6 = 9 + 2$

(*b*) $(5, 2) \not\sim (8, 4)$ since $5 + 4 \neq 8 + 2$

(*c*) $(r, r) \sim (s, s)$ since $r + s = s + r$

(*d*) $(r^*, r) \sim (s^*, s)$ since $r^* + s = s^* + r$

(*e*) $(r^*, s^*) \sim (r, s)$ since $r^* + s = r + s^*$

whenever $r, s \in N$.

Now ~ is an equivalence relation (see Problem 1) and thus partitions L into a set of equivalence classes $\mathcal{J} = \{[s, m], [t, n], \ldots\}$ where

$$[s, m] \;=\; \{(a, b):\ (a, b) \in L,\ (a, b) \sim (s, m)\}$$

We recall from Chapter 2 that $(s, m) \in [s, m]$ and that, if $(c, d) \in [s, m]$, then $[c, d] = [s, m]$. Thus

$$[s, m] = [t, n] \quad \text{if and only if} \quad (s, m) \sim (t, n)$$

It will be our purpose now to show that the set $\mathcal{J}$ of equivalence classes of L relative to ~ is, except for the symbols used, the familiar set I of all integers.

ADDITION AND MULTIPLICATION ON $\mathcal{J}$

Addition and multiplication on $\mathcal{J}$ will be defined respectively by

(i) $$[s, m] + [t, n] = [(s+t), (m+n)]$$

(ii) $$[s, m] \cdot [t, n] = [(s \cdot t + m \cdot n), (s \cdot n + m \cdot t)]$$

for all $[s, m], [t, n] \in \mathcal{J}$.

An examination of the right members of (i) and (ii) shows that the Closure Laws

A₁. $$x + y \in \mathcal{J} \quad \text{for all } x, y \in \mathcal{J}$$

and

M₁. $$x \cdot y \in \mathcal{J} \quad \text{for all } x, y \in \mathcal{J}$$

are valid.

In Problem 3, we prove

Theorem I. The equivalence class to which the sum (product) of two elements, one selected from each of two equivalence classes of $\mathcal{J}$, belongs is independent of the particular elements selected.

Example 2: If $(a, b), (c, d) \in [s, m]$ and $(e, f), (g, h) \in [t, n]$, we have not only

$$[a, b] = [c, d] = [s, m] \quad \text{and} \quad [e, f] = [g, h] = [t, n]$$

but, by Theorem I, also

$$[a, b] + [e, f] = [c, d] + [g, h] = [s, m] + [t, n]$$

and $$[a, b] \cdot [e, f] = [c, d] \cdot [g, h] = [s, m] \cdot [t, n]$$

Theorem I may also be stated as follows: Addition and multiplication on $\mathcal{J}$ are *well-defined.*

By using the commutative and associate laws for addition and multiplication on N, it is not difficult to show that addition and multiplication on $\mathcal{J}$ obey these same laws. The associative law for addition and one of the distributive laws are proved in Problems 4 and 5.

THE POSITIVE INTEGERS

Let $r \in N$. From $1 + r = r^*$ it follows that r is a solution of $1 + x = r^*$. Consider now the mapping

$$[n^*, 1] \leftrightarrow n, \quad n \in N \qquad (1)$$

For this mapping, we find

$$[r^*, 1] + [s^*, 1] = [(r^* + s^*), (1+1)] = [(r+s)^*, 1] \leftrightarrow r + s$$

and

$$[r^*, 1] \cdot [s^*, 1] = [(r^* \cdot s^* + 1 \cdot 1), (r^* \cdot 1 + s^* \cdot 1)] = [(r \cdot s)^*, 1] \leftrightarrow r \cdot s$$

Thus, *(1)* is an isomorphism of the subset $\{[n^*, 1] : n \in N\}$ of $\mathcal{J}$ onto N.

Suppose now that $[s, m] = [r^*, 1]$. Then $(s, m) \sim (r^*, 1)$, $s = r + m$, and $s > m$. This suggests defining the set I^+ of positive integers by

$$I^+ = \{[s, m] : [s, m] \in \mathcal{J},\ s > m\}$$

In view of the isomorphism *(1)* the set I^+ may be replaced by the set N whenever the latter is found more convenient.

ZERO AND THE NEGATIVE INTEGERS

Let $r, s \in N$. Now $[r,r] = [s,s]$ for any choice of r and s, and $[r,r] = [s,t]$ if and only if $t = s$. We now define the integer *zero*, 0, to correspond to the equivalence class $[r,r]$, $r \in N$. Its familiar properties are

$$[s,m] + [r,r] = [s,m] \quad \text{and} \quad [s,m] \cdot [r,r] = [r,r]$$

proved in Problems 2(*b*) and 2(*c*). The first of these leads to the designation of zero as the identity element for addition.

Finally, we define the set I^- of negative integers by

$$I^- = \{[s,m] : [s,m] \in \mathcal{J},\ s < m\}$$

It follows now that for each integer $[a,b]$, $a \neq b$, there exists a unique integer $[b,a]$ such that (see Problem 2(*d*))

$$[a,b] + [b,a] = [r,r] \leftrightarrow 0 \tag{2}$$

We denote $[b,a]$ by $-[a,b]$ and call it the *negative* of $[a,b]$. The relation (*2*) suggests the designation of $[b,a]$ or $-[a,b]$ as the *additive inverse* of $[a,b]$.

THE INTEGERS

Let $p, q \in N$. By the Trichotomy Law for natural numbers, there are three possibilities:

(*a*) $p = q$, whence $[p,q] = [q,p] \leftrightarrow 0$.

(*b*) $p < q$, so that $p + a = q$ for some $a \in N$; then $p + a^* = q + 1$ and $[q,p] = [a^*, 1] \leftrightarrow a$.

(*c*) $p > q$, so that $p = q + a$ for some $a \in N$ and $[p,q] \leftrightarrow a$.

Suppose $[p,q] \leftrightarrow n \in N$. Since $[q,p] = -[p,q]$, we introduce the symbol $-n$ to denote the negative of $n \in N$ and write $[q,p] \leftrightarrow -n$. Thus, each equivalence class of $\mathcal{J}$ is now mapped onto a unique element of $I = \{0, \pm 1, \pm 2, \ldots\}$. That $\mathcal{J}$ and I are isomorphic follows readily once the familiar properties of the minus sign have been established. In proving most of the basic properties of integers, however, we shall find it expedient to use the corresponding equivalence classes of $\mathcal{J}$.

Example 3: Let $a, b \in I$. Show that $(-a) \cdot b = -(a \cdot b)$. Let $a \leftrightarrow [s,m]$ so that $-a \leftrightarrow [m,s]$ and let $b \leftrightarrow [t,n]$. Then

$$(-a) \cdot b \leftrightarrow [m,s] \cdot [t,n] = [(m \cdot t + s \cdot n),\ (m \cdot n + s \cdot t)]$$

while
$$a \cdot b \leftrightarrow [s,m] \cdot [t,n] = [(s \cdot t + m \cdot n),\ (s \cdot n + m \cdot t)]$$

Now
$$-(a \cdot b) \leftrightarrow [(s \cdot n + m \cdot t),\ (s \cdot t + m \cdot n)] \leftrightarrow [(-a) \cdot b]$$

and so
$$(-a) \cdot b = -(a \cdot b)$$

See Problems 6-7.

ORDER RELATIONS

For $a, b \in I$, let $a \leftrightarrow [s,m]$ and $b \leftrightarrow [t,n]$. The order relations "<" and ">" for integers are defined by

$$a < b \quad \text{if and only if} \quad (s+n) < (t+m)$$

and
$$a > b \quad \text{if and only if} \quad (s+n) > (t+m)$$

In Problem 8, we prove the *Trichotomy Law*: For any $a, b \in I$, one and only one of

$$(a)\ a = b, \quad (b)\ a < b, \quad (c)\ a > b$$

is true.

When $a,b,c \in I$, we have:

(1) $a+c < b+c$ if and only if $a < b$.

(1′) $a+c > b+c$ if and only if $a > b$.

(2) If $c > 0$, then $a \cdot c < b \cdot c$ if and only if $a < b$.

(2′) If $c > 0$, then $a \cdot c > b \cdot c$ if and only if $a > b$.

(3) If $c < 0$, then $a \cdot c < b \cdot c$ if and only if $a > b$.

(3′) If $c < 0$, then $a \cdot c > b \cdot c$ if and only if $a < b$.

For proofs of (1′) and (3), see Problems 9-10.

The Cancellation Law for multiplication on I,

M_4. If $z \neq 0$ and if $x \cdot z = y \cdot z$, then $x = y$

may now be established.

As an immediate consequence, we have

Theorem II. If $a,b \in I$ and if $a \cdot b = 0$, then either $a = 0$ or $b = 0$.

For a proof see Problem 11.

The order relations permit the customary listing of the integers

$$\ldots, -5, -4, -3, -2, -1, 0, 1, 2, 3, 4, 5, \ldots$$

and their representation as equally spaced points on a line. Then "$a < b$" means "a lies to the left of b", and "$a > b$" means "a lies to the right of b".

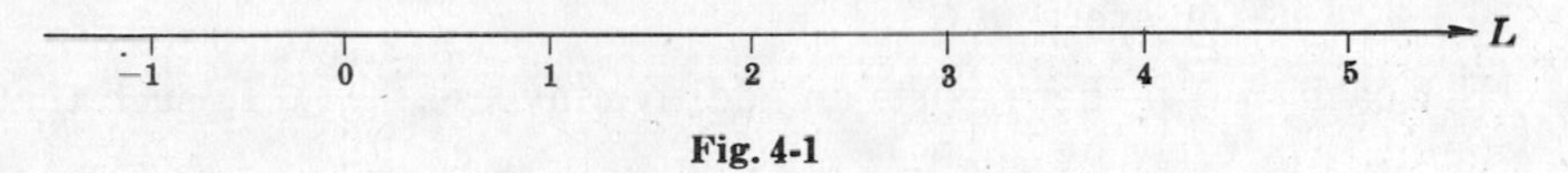

Fig. 4-1

From the above listing of the integers, we note

Theorem III. There exists no $n \in I^+$ such that $0 < n < 1$.

This theorem (see Problem 12 for a proof) is a consequence of the fact that the set I^+ of positive integers (being isomorphic to N) is well-ordered.

SUBTRACTION "−"

Subtraction, "−", on I is defined by $a - b = a + (-b)$. Subtraction is clearly a binary operation on I. It is, however, neither commutative nor associative although multiplication is distributive with respect to subtraction.

Example 4: Prove: $a - (b - c) \neq (a - b) - c$ for $a,b,c \in I$ and $c \neq 0$.

Let $a \leftrightarrow [s,m]$, $b \leftrightarrow [t,n]$, and $c \leftrightarrow [u,p]$. Then

$$b - c = b + (-c) \leftrightarrow [(t+p), (n+u)]$$

$$-(b-c) \leftrightarrow [(n+u), (t+p)]$$

and $$a - (b-c) = a + (-(b-c)) \leftrightarrow [(s+n+u), (m+t+p)]$$

while $$a - b = a + (-b) \leftrightarrow [(s+n), (m+t)]$$

and $$(a-b) - c = (a+b) + (-c) \leftrightarrow [(s+n+p), (m+t+u)]$$

Thus, when $c \neq 0$, $a - (b-c) \neq (a-b) - c$.

ABSOLUTE VALUE $|a|$

The absolute value, "$|a|$", of an integer a is defined by

$$|a| = \begin{cases} a & \text{when } a \geqq 0 \\ -a & \text{when } a < 0 \end{cases}$$

Thus, except when $a = 0$, $|a| \in I^+$.

The following laws

$$(1)\quad -|a| \leqq a \leqq |a| \qquad\qquad (2)\quad |a \cdot b| = |a| \cdot |b|$$

$$(3)\quad |a| - |b| \leqq |a+b| \qquad\qquad (3')\quad |a+b| \leqq |a| + |b|$$

$$(4)\quad |a| - |b| \leqq |a-b| \qquad\qquad (4')\quad |a-b| \leqq |a| + |b|$$

are evidently true when at least one of a, b is 0. They may be established for all $a, b \in I$ by considering the separate cases as in Problems 14 and 15.

ADDITION AND MULTIPLICATION ON I

The operations of addition and multiplication on I satisfy the laws $\mathbf{A}_1$-$\mathbf{A}_4$, $\mathbf{M}_1$-$\mathbf{M}_4$, and $\mathbf{D}_1$-$\mathbf{D}_2$ of Chapter 3 (when stated for integers) with the single modification

$\mathbf{M}_4$. **Cancellation Law:** If $m \cdot p = n \cdot p$ and if $p \neq 0 \in I$, then $m = n$ for all $m, n \in I$.

We list below two properties of I which N lacked

$\mathbf{A}_5$. There exists an identity element, $0 \in I$, relative to addition, such that $n + 0 = 0 + n = n$ for every $n \in I$.

$\mathbf{A}_6$. For each $n \in I$ there exists an additive inverse, $-n \in I$, such that $n + (-n) = (-n) + n = 0$.

and a common property of N and I

$\mathbf{M}_5$. There exists an identity element, $1 \in I$, relative to multiplication, such that $1 \cdot n = n \cdot 1 = n$ for every $n \in I$.

By Theorem III, Chapter 2, the identity elements in $\mathbf{A}_5$ and $\mathbf{M}_5$ are unique; by Theorem IV, Chapter 2, the additive inverses in $\mathbf{A}_6$ are unique.

OTHER PROPERTIES OF INTEGERS

Certain properties of the integers have been established using the equivalence classes of $\mathcal{J}$. However, once the basic laws have been established, all other properties may be obtained using the elements of I themselves.

Example 5: Prove: For all $a, b, c \in I$,

$$(a)\ a \cdot 0 = 0 \cdot a = 0 \qquad (b)\ a(-b) = -(ab) \qquad (c)\ a(b-c) = ab - ac$$

(*a*)
$$a + 0 = a \qquad (\mathbf{A}_5)$$

Then
$$a \cdot a + 0 = a \cdot a = a(a+0) = a \cdot a + a \cdot 0 \qquad (\mathbf{D}_1)$$

and
$$0 = a \cdot 0 \qquad (\mathbf{A}_4)$$

Now, by $\mathbf{M}_2$, $0 \cdot a = a \cdot 0 = 0$, as required. However, for reasons which will not be apparent until a later chapter, we will prove

$$0 \cdot a = a \cdot 0$$

without appealing to the commutative law of multiplication. We have

$$a \cdot a + 0 = a \cdot a = (a+0)a = a \cdot a + 0 \cdot a \qquad (\mathbf{D_2})$$

hence, $$0 = 0 \cdot a$$

and $$0 \cdot a = a \cdot 0$$

(b) $$0 = a \cdot 0 = a[b + (-b)] = a \cdot b + a(-b) \qquad (\mathbf{D_1})$$

thus, $a(-b)$ is an additive inverse of $a \cdot b$. But $-(a \cdot b)$ is also an additive inverse of $a \cdot b$; hence,

$$a(-b) = -(a \cdot b) \qquad \text{(Theorem IV, Chapter 2)}$$

(c) $$a(b-c) = a[b + (-c)] = ab + a(-c) \qquad (\mathbf{D_1})$$
$$= ab + (-ac) \qquad ((b) \text{ above})$$
$$= ab - ac$$

Note: In (c) we have replaced $a \cdot b$ and $-(a \cdot c)$ by the familiar ab and $-ac$ respectively.

MULTIPLES AND POWERS

The section with this title in Chapter 3 (Page 33) may be repeated here after replacing N by I^+.

It is to be noted that for the case $S = I$, we may now identify ka with $k \cdot a$. Moreover, since I contains the additive inverse of each of its elements, we have **(ix)**-**(xi)**, but not **(ix)**$'$-**(xi)**$'$, for all $m, n \in I$. This is largely trivial, of course, since when $a, b, m, n \in I$ the properties **(ix)** and **(xi)** are then the distributive laws.

Solved Problems

1. Show that $\sim$ on L is an equivalence relation.

Let $(s,m), (t,n), (u,p) \in L$. We have

(a) $(s,m) \sim (s,m)$ since $s + m = s + m$; $\sim$ is reflexive.

(b) If $(s,m) \sim (t,n)$, then $(t,n) \sim (s,m)$ since each requires $s + n = t + m$; $\sim$ is symmetric.

(c) If $(s,m) \sim (t,n)$ and $(t,n) \sim (u,p)$, then $s + n = t + m$, $t + p = u + n$, and $s + n + t + p = t + m + u + n$. Using $\mathbf{A_4}$ of Chapter 3, Page 30, the latter equality can be replaced by $s + p = m + u$; then $(s,m) \sim (u,p)$ and $\sim$ is transitive.

Thus, $\sim$, being reflexive, symmetric, and transitive, is an equivalence relation.

2. When $s, m, p, r \in N$, prove:

(a) $[(r+p), p] = [r^*, 1]$ (c) $[s,m] \cdot [r,r] = [r,r]$ (e) $[s,m] \cdot [r^*,r] = [s,m]$

(b) $[s,m] + [r,r] = [s,m]$ (d) $[s,m] + [m,s] = [r,r]$

(a) $((r+p), p) \sim (r^*, 1)$ since $r + p + 1 = r^* + p$.
Hence, $[(r+p), p] = [r^*, 1]$ as required.

(b) $[s,m] + [r,r] = [(s+r), (m+r)]$. Now $((s+r), (m+r)) \sim (s,m)$ since $(s+r) + m = s + (m+r)$.
Hence, $[(s+r), (m+r)] = [s,m] + [r,r] = [s,m]$.

(c) $[s,m] \cdot [r,r] = [(s \cdot r + m \cdot r), (s \cdot r + m \cdot r)] = [r,r]$ since $s \cdot r + m \cdot r + r = s \cdot r + m \cdot r + r$.

(d) $[s,m] + [m,s] = [(s+m), (s+m)] = [r,r]$

(e) $[s,m] \cdot [r^*,r] = [(s \cdot r^* + m \cdot r), (s \cdot r + m \cdot r^*)] = [s,m]$ since $s \cdot r^* + m \cdot r + m = s + s \cdot r + m \cdot r^* = s \cdot r^* + m \cdot r^*$.

3. Prove: The equivalence class to which the sum (product) of two elements, one selected from each of two equivalence classes of $\mathcal{J}$, belongs is independent of the elements selected.

Let $[a, b] = [s, m]$ and $[c, d] = [t, n]$. Then $(a, b) \sim (s, m)$ and $(c, d) \sim (t, n)$ so that $a + m = s + b$ and $c + n = t + d$. We shall prove:

(a) $[a, b] + [c, d] = [s, m] + [t, n]$, the first part of the theorem;

(b) $a \cdot c + b \cdot d + s \cdot n + m \cdot t = a \cdot d + b \cdot c + s \cdot t + m \cdot n$, a necessary lemma;

(c) $[a, b] \cdot [c, d] = [s, m] \cdot [t, n]$, the second part of the theorem.

(a) Since $a + m + c + n = s + b + t + d$,

$$(a + c) + (m + n) = (s + t) + (b + d)$$
$$\big((a + c), (b + d)\big) \sim \big((s + t), (m + n)\big)$$
$$[(a + c), (b + d)] = [(s + t), (m + n)]$$

and
$$[a, b] + [c, d] = [s, m] + [t, n]$$

(b) We begin with the evident equality

$$(a + m) \cdot (c + t) + (s + b) \cdot (d + n) + (c + n) \cdot (a + s) + (d + t) \cdot (b + m)$$
$$= (s + b) \cdot (c + t) + (a + m) \cdot (d + n) + (d + t) \cdot (a + s) + (c + n) \cdot (b + m)$$

which reduces readily to

$$2(a \cdot c + b \cdot d + s \cdot n + m \cdot t) + (a \cdot t + m \cdot c + s \cdot d + b \cdot n) + (s \cdot c + n \cdot a + b \cdot t + m \cdot d)$$
$$= 2(a \cdot d + b \cdot c + s \cdot t + m \cdot n) + (a \cdot t + m \cdot c + s \cdot d + b \cdot n) + (s \cdot c + n \cdot a + b \cdot t + m \cdot d)$$

and, using the Cancellation Laws of Chapter 3, to the required identity.

(c) From (b), we have

$$(a \cdot c + b \cdot d) + (s \cdot n + m \cdot t) = (s \cdot t + m \cdot n) + (a \cdot d + b \cdot c)$$

Then
$$\big((a \cdot c + b \cdot d), (a \cdot d + b \cdot c)\big) \sim \big((s \cdot t + m \cdot n), (s \cdot n + m \cdot t)\big)$$
$$[(a \cdot c + b \cdot d), (a \cdot d + b \cdot c)] = [(s \cdot t + m \cdot n), (s \cdot n + m \cdot t)]$$

and so
$$[a, b] \cdot [c, d] = [s, m] \cdot [t, n]$$

4. Prove the Associative Law for addition:

$$([s, m] + [t, n]) + [u, p] = [s, m] + ([t, n] + [u, p])$$

for all $[s, m], [t, n], [u, p] \in \mathcal{J}$.

We find

$$([s, m] + [t, n]) + [u, p] = [(s + t), (m + n)] + [u, p] = [(s + t + u), (m + n + p)]$$

while
$$[s, m] + ([t, n] + [u, p]) = [s, m] + [(t + u), (n + p)] = [(s + t + u), (m + n + p)]$$

and the law follows.

5. Prove the Distributive Law $\mathbf{D}_2$:

$$([s, m] + [t, n]) \cdot [u, p] = [s, m] \cdot [u, p] + [t, n] \cdot [u, p]$$

for all $[s, m], [t, n], [u, p] \in \mathcal{J}$.

We have

$$\begin{aligned}([s, m] + [t, n]) \cdot [u, p] &= [(s + t), (m + n)] \cdot [u, p] \\ &= \big[\big((s + t) \cdot u + (m + n) \cdot p\big), \big((s + t) \cdot p + (m + n) \cdot u\big)\big] \\ &= [(s \cdot u + t \cdot u + m \cdot p + n \cdot p), (s \cdot p + t \cdot p + m \cdot u + n \cdot u)] \\ &= \big[\big((s \cdot u + m \cdot p) + (t \cdot u + n \cdot p)\big), \big((s \cdot p + m \cdot u) + (t \cdot p + n \cdot u)\big)\big] \\ &= [(s \cdot u + m \cdot p), (s \cdot p + m \cdot u)] + [(t \cdot u + n \cdot p), (t \cdot p + n \cdot u)] \\ &= [s, m] \cdot [u, p] + [t, n] \cdot [u, p]\end{aligned}$$

6. (*a*) Show that $a+(-a) = 0$ for every $a \in I$.

Let $a \leftrightarrow [s,m]$; then $-a \leftrightarrow [m,s]$,

$$a + (-a) \leftrightarrow [s,m] + [m,s] = [(s+m), (m+s)] = [r,r] \leftrightarrow 0$$

and $a+(-a) = 0$.

(*b*) If $x+a = b$ for $a,b \in I$, show that $x = b+(-a)$.

When $x = b+(-a)$, $x+a = \big(b+(-a)\big)+a = b+\big((-a)+a\big) = b$; thus, $x = b+(-a)$ is a solution of the equation $x+a = b$. Suppose there is a second solution y. Then $y+a = b = x+a$ and, by $\mathbf{A}_4$, $y = x$. Thus the solution is unique.

7. When $a,b \in I$, prove: (1) $(-a)+(-b) = -(a+b)$, (2) $(-a)\cdot(-b) = a\cdot b$.

Let $a \leftrightarrow [s,m]$ and $b \leftrightarrow [t,n]$; then $-a \leftrightarrow [m,s]$ and $-b \leftrightarrow [n,t]$.

(1) $$(-a) + (-b) \leftrightarrow [m,s] + [n,t] = [(m+n), (s+t)]$$

and $$a + b \leftrightarrow [s,m] + [t,n] = [(s+t), (m+n)]$$

Then $$-(a+b) \leftrightarrow [(m+n), (s+t)] \leftrightarrow (-a) + (-b)$$

and $$(-a) + (-b) = -(a+b)$$

(2) $$(-a)\cdot(-b) \leftrightarrow [m,s]\cdot[n,t] = [(m\cdot n + s\cdot t), (m\cdot t + s\cdot n)]$$

and $$a\cdot b \leftrightarrow [s,m]\cdot[t,n] = [(s\cdot t + m\cdot n), (s\cdot n + m\cdot t)]$$

Now $$[(m\cdot n + s\cdot t), (m\cdot t + s\cdot n)] = [(s\cdot t + m\cdot n), (s\cdot n + m\cdot t)]$$

and $$(-a)\cdot(-b) = a\cdot b$$

8. Prove the Trichotomy Law: For any $a,b \in I$ one and only one of

$$(a)\ a = b, \quad (b)\ a < b, \quad (c)\ a > b$$

is true.

Let $a \leftrightarrow [s,m]$ and $b \leftrightarrow [t,n]$; then by the Trichotomy Law of Chapter 3, Page 32, one and only one of (*a*) $s+n = t+m$ and $a = b$, (*b*) $s+n < t+m$ and $a < b$, (*c*) $s+n > t+m$ and $a > b$ is true.

9. When $a,b,c \in I$, prove: $a+c > b+c$ if and only if $a > b$.

Take $a \leftrightarrow [s,m]$, $b \leftrightarrow [t,n]$, and $c \leftrightarrow [u,p]$. Suppose first that

$$a+c > b+c \quad \text{or} \quad ([s,m] + [u,p]) > ([t,n] + [u,p])$$

Now this implies $$[(s+u), (m+p)] > [(t+u), (n+p)]$$

which, in turn, implies $$(s+u) + (n+p) > (t+u) + (m+p)$$

Then, by Theorem II′, Chapter 3, Page 33, $(s+n) > (t+m)$ or $[s,m] > [t,n]$ and $a > b$, as required.

Suppose next that $a > b$ or $[s,m] > [t,n]$; then $(s+n) > (t+m)$. Now to compare

$$a + c \leftrightarrow [(s+u), (m+p)] \quad \text{and} \quad b + c \leftrightarrow [(t+u), (n+p)]$$

we compare $$[(s+u), (m+p)] \quad \text{and} \quad [(t+u), (n+p)]$$

or $$(s+u) + (n+p) \quad \text{and} \quad (t+u) + (m+p)$$

or $$(s+n) + (u+p) \quad \text{and} \quad (t+m) + (u+p)$$

Since $(s+n) > (t+m)$, it follows by Theorem II′, Chapter 3, Page 33, that

$$(s+n) + (u+p) > (t+m) + (u+p)$$

Then

$$(s+u) + (n+p) > (t+u) + (m+p)$$

$$[(s+u), (m+p)] > [(t+u), (n+p)]$$

and

$$a + c > b + c$$

as required.

10. When $a, b, c \in I$, prove: If $c < 0$, then $a \cdot c < b \cdot c$ if and only if $a > b$.

Take $a \leftrightarrow [s, m]$, $b \leftrightarrow [t, n]$, and $c \leftrightarrow [u, p]$ in which $u < p$ since $c < 0$.

(*a*) Suppose $a \cdot c < b \cdot c$; then

$$[(s \cdot u + m \cdot p), (s \cdot p + m \cdot u)] < [(t \cdot u + n \cdot p), (t \cdot p + n \cdot u)]$$

and

$$(s \cdot u + m \cdot p) + (t \cdot p + n \cdot u) < (t \cdot u + n \cdot p) + (s \cdot p + m \cdot u)$$

Since $u < p$, there exists some $k \in N$ such that $u + k = p$. When this replacement for p is made in the inequality immediately above, we have

$$(s \cdot u + m \cdot u + m \cdot k + t \cdot u + t \cdot k + n \cdot u) < (t \cdot u + n \cdot u + n \cdot k + s \cdot u + s \cdot k + m \cdot u)$$

whence

$$m \cdot k + t \cdot k < n \cdot k + s \cdot k$$

Then

$$(m+t) \cdot k < (n+s) \cdot k$$

$$m + t < n + s$$

$$s + n > t + m$$

and

$$a > b$$

(*b*) Suppose $a > b$. By simply reversing the steps in (*a*), we obtain $a \cdot c < b \cdot c$, as required.

11. Prove: If $a, b \in I$ and $a \cdot b = 0$, then either $a = 0$ or $b = 0$.

Suppose $a \neq 0$; then $a \cdot b = 0 = a \cdot 0$ and by $\mathbf{M}_4$, $b = 0$. Similarly, if $b \neq 0$ then $a = 0$.

12. Prove: There exists no $n \in I^+$ such that $0 < n < 1$.

Suppose the contrary and let $m \in I^+$ be the least such integer. From $0 < m < 1$, we have by the use of (*2*), Page 41, $0 < m^2 < m < 1$. Now $0 < m^2 < 1$ and $m^2 < m$ contradict the assumption that m is the least, and the theorem is established.

13. Prove: When $a, b \in I$, then $a < b$ if and only if $a - b < 0$.

Let $a \leftrightarrow [s, m]$ and $b \leftrightarrow [t, n]$. Then

$$a - b = a + (-b) \leftrightarrow [s, m] + [n, t] = [(s+n), (m+t)]$$

If $a < b$, then $s + n < m + t$ and $a - b < 0$. Conversely, if $a - b < 0$, then $s + n < m + t$ and $a < b$.

14. Prove: $|a + b| \leq |a| + |b|$ for all $a, b \in I$.

Suppose $a > 0$ and $b > 0$; then $|a + b| = a + b = |a| + |b|$.

Suppose $a < 0$ and $b < 0$; then $|a + b| = -(a + b) = -a + (-b) = |a| + |b|$.

Suppose $a > 0$ and $b < 0$ so that $|a| = a$ and $|b| = -b$. Now, either $a + b = 0$ and $|a + b| = 0 < |a| + |b|$ or

$$a + b < 0 \quad \text{and} \quad |a + b| = -(a + b) = -a + (-b) = -|a| + |b| < |a| + |b|$$

or

$$a + b > 0 \quad \text{and} \quad |a + b| = a + b = a - (-b) = |a| - |b| < |a| + |b|$$

The case $a < 0$ and $b > 0$ is left as an exercise.

15. Prove: $|a \cdot b| = |a| \cdot |b|$ for all $a, b \in I$.

Suppose $a > 0$ and $b > 0$; then $|a| = a$ and $|b| = b$. Then $|a \cdot b| = a \cdot b = |a| \cdot |b|$.

Suppose $a < 0$ and $b < 0$; then $|a| = -a$ and $|b| = -b$. Now $a \cdot b > 0$; hence, $|a \cdot b| = a \cdot b = (-a) \cdot (-b) = |a| \cdot |b|$.

Suppose $a > 0$ and $b < 0$; then $|a| = a$ and $|b| = -b$. Since $a \cdot b < 0$, $|a \cdot b| = -(a \cdot b) = a \cdot (-b) = |a| \cdot |b|$.

The case $a < 0$ and $b > 0$ is left as an exercise.

16. Prove: If a and b are integers such that $a \cdot b = 1$ then a and b are either both 1 or both -1.

First we note that neither a nor b can be zero. Now $|a \cdot b| = |a| \cdot |b| = 1$ and, by Problem 12, $|a| \geqq 1$ and $|b| \geqq 1$. If $|a| > 1$ (also, if $|b| > 1$), $|a| \cdot |b| \neq 1$. Hence $|a| = |b| = 1$ and, in view of Problem 7(*b*), the theorem follows.

Supplementary Problems

17. Prove: When $r, s \in N$,

(*a*) $(r, r) \sim (s, s) \sim (1, 1)$

(*b*) $(r^*, r) \sim (s^*, s) \sim (2, 1)$

(*c*) $(r, r^*) \sim (s, s^*) \sim (1, 2)$

(*d*) $(r^*, r) \not\sim (r, r^*)$

(*e*) $(r^*, r) \not\sim (s, s^*)$

(*f*) $(r^* \cdot s^* + 1,\ r^* + s^*) \sim \big((r \cdot s)^*,\ 1\big)$

18. State and prove: (*a*) the Associative Law for multiplication, (*b*) the Commutative Law for addition, (*c*) the Commutative Law for multiplication, (*d*) the Cancellation Law for addition on $\mathcal{J}$.

19. Prove: $[r^*, r] \leftrightarrow 1$ and $[r, r^*] \leftrightarrow -1$

20. If $a \in I$, prove: (*a*) $a \cdot 0 = 0 \cdot a = 0$, (*b*) $(-1) \cdot a = -a$, (*c*) $-0 = 0$.

21. If $a, b \in I$, prove: (*a*) $-(-a) = +a$, (*b*) $(-a)(-b) = a \cdot b$, (*c*) $(-a) + b = -\big(a + (-b)\big)$.

22. When $b \in I^+$, show that $a - b < a + b$ for all $a \in I$.

23. When $a, b \in I$, prove (1), (2), (2′) and (3′), Page 41, of the order relations.

24. When $a, b, c \in I$, prove $a \cdot (b - c) = a \cdot b - a \cdot c$.

25. Prove: If $a,b \in I$ and $a < b$, then there exists some $c \in I^+$ such that $a + c = b$.
Hint. For a and b as represented in Problem 7, take $c \leftrightarrow [(t+m), (n+s)]$.

26. Prove: When $a,b,c,d \in I$,

(a) $-a > -b$ if $a < b$.

(b) $a + c < b + d$ if $a < b$ and $c < d$.

(c) If $a < (b + c)$, then $a - b < c$.

(d) $a - b = c - d$ if and only if $a + d = b + c$.

27. Prove that the order relations are well-defined.

28. Prove the Cancellation Law for multiplication.

29. Define sums and products of $n > 2$ elements of I and show that in such sums and products parentheses may be inserted at will.

30. Prove: (a) $m^2 > 0$ for all integers $m \neq 0$.
(b) $m^3 > 0$ for all integers $m > 0$.
(c) $m^3 < 0$ for all integers $m < 0$.

31. Prove without using equivalence classes (see Example 5):

(a) $-(-a) = a$

(b) $(-a)(-b) = ab$

(c) $(b - c) = (b + a) - (c + a)$

(d) $a(b - c) = ab - ac$

(e) $(a + b)(c + d) = (ac + ad) + (bc + bd)$

(f) $(a + b)(c - d) = (ac + bc) - (ad + bd)$

(g) $(a - b)(c - d) = (ac + bd) - (ad + bc)$

Chapter 5

Some Properties of Integers

DIVISORS

An integer $a \neq 0$ is called a *divisor* (*factor*) of an integer b (written "$a \mid b$") if there exists an integer c such that $b = ac$. When $a \mid b$ we shall also say that b is an *integral multiple* of a.

Example 1: (*a*) $2 \mid 6$ since $6 = 2 \cdot 3$

(*b*) $-3 \mid 15$ since $15 = (-3)(-5)$

(*c*) $a \mid 0$, for all $a \in I$, since $0 = a \cdot 0$

In order to show that the restriction $a \neq 0$ is necessary, suppose $0 \mid b$. If $b \neq 0$, we must have $b = 0 \cdot c$ for some $c \in I$, which is impossible; while if $b = 0$, we would have $0 = 0 \cdot c$, which is true for *every* $c \in I$.

When $b, c; x, y \in I$, the integer $bx + cy$ is called a *linear* combination of b and c. In Problem 1, we prove

Theorem I. If $a \mid b$ and $a \mid c$ then $a \mid (bx + cy)$ for all $x, y \in I$.

See also Problems 2, 3.

PRIMES

Since $a \cdot 1 = (-a)(-1) = a$ for every $a \in I$, it follows that ± 1 and $\pm a$ are divisors of a. An integer $p \neq 0, \pm 1$ is called a *prime* if and only if its only divisors are ± 1 and $\pm p$.

Example 2: (*a*) The integers 2 and -5 are primes, while $6 = 2 \cdot 3$ and $-39 = 3(-13)$ are not primes.

(*b*) The first 10 positive primes are 2, 3, 5, 7, 11, 13, 17, 19, 23, 29.

It is clear that $-p$ is a prime if and only if p is a prime. Hereafter, we shall restrict our attention mainly to positive primes. In Problem 4, we prove:

The number of positive primes is infinite.

When $a = bc$ with $|b| > 1$ and $|c| > 1$, we call a *composite*. Thus every integer $a \neq 0, \pm 1$ is either a prime or a composite.

GREATEST COMMON DIVISOR

When $a \mid b$ and $a \mid c$, we call a a *common divisor* of b and c. When, in addition, every common divisor of b and c is also a divisor of a, we call a a *greatest common divisor* (*highest common factor*) of b and c.

Example 3: (*a*) $\pm 1, \pm 2, \pm 3, \pm 4, \pm 6, \pm 12$ are common divisors of 24 and 60.

(*b*) ± 12 are greatest common divisors of 24 and 60.

(*c*) The greatest common divisors of $b = 0$ and $c \neq 0$ are $\pm c$.

Suppose c and d are two different greatest common divisors of $a \neq 0$ and $b \neq 0$. Then $c \mid d$ and $d \mid c$; hence, by Problem 3, c and d differ only in sign. As a matter of convenience, we shall hereafter limit our attention to the positive greatest common divisor of two integers a and b and use either d or (a, b) to denote it. Thus, d is truly the largest (greatest) integer which divides both a and b.

Example 4: A familiar procedure for finding $(210, 510)$ consists in expressing each integer as a product of its prime factors, i.e., $210 = 2 \cdot 3 \cdot 5 \cdot 7$, $510 = 2 \cdot 3 \cdot 5 \cdot 17$, and forming the product $2 \cdot 3 \cdot 5 = 30$ of their common factors.

In Example 4 we have tacitly assumed (*a*) that every two non-zero integers have a positive greatest common divisor and (*b*) that any integer $a > 1$ has a unique factorization, except for the order of the factors, as a product of positive primes. Of course, in (*b*) it must be understood that when a is itself a prime, "a product of positive primes" consists of a single prime. We shall prove these propositions later. At the moment, we wish to exhibit another procedure for finding the greatest common divisor of two non-zero integers. We begin with:

The Division Algorithm. For any two non-zero integers a and b, there exist unique integers q and r, called respectively *quotient* and *remainder*, such that

$$a = bq + r, \quad 0 \leq r < |b| \tag{1}$$

For a proof, see Problem 5.

Example 5:
(*a*) $780 = -48(-16) + 12$
(*b*) $-2805 = 119(-24) + 51$
(*c*) $826 = 25 \cdot 33 + 1$
(*d*) $758 = 242(3) + 32$

From (*1*) it follows that $b \mid a$ and $(a, b) = b$ if and only if $r = 0$. When $r \neq 0$ it is easy to show that a common divisor of a and b also divides r and a common divisor of b and r also divides a. Then $(a, b) \mid (b, r)$ and $(b, r) \mid (a, b)$ so that, by Problem 3, $(a, b) = (b, r)$. Now either $r \mid b$ (see Examples 5(*a*) and (5*c*)) or $r \nmid b$ (see Examples 5(*b*) and 5(*d*)). In the latter case, we use the division algorithm to obtain

$$b = rq_1 + r_1, \quad 0 < r_1 < r \tag{2}$$

Again, either $r_1 \mid r$ and $(a, b) = r_1$ or, using the division algorithm,

$$r = r_1 q_2 + r_2, \quad 0 < r_2 < r_1 \tag{3}$$

and $(a, b) = (b, r) = (r, r_1) = (r_1, r_2)$.

Since the remainders $r_1, r_2, \ldots$, assuming the process to continue, constitute a set of decreasing non-negative integers there must eventually be one which is zero. Suppose the process terminates with

$$\begin{array}{llll} (k) & r_{k-3} = r_{k-2} \cdot q_{k-1} + r_{k-1} & & 0 < r_{k-1} < r_{k-2} \\ (k+1) & r_{k-2} = r_{k-1} \cdot q_k + r_k & & 0 < r_k < r_{k-1} \\ (k+2) & r_{k-1} = r_k \cdot q_{k+1} + 0 & & \end{array}$$

Then $(a, b) = (b, r) = (r, r_1) = \cdots = (r_{k-2}, r_{k-1}) = (r_{k-1}, r_k) = r_k$.

Example 6: (*a*) In Example 5(*b*), $51 \nmid 119$. Proceeding as in (*2*), we find $119 = 51(2) + 17$. Now $17 \mid 51$; hence, $(-2805, 119) = 17$.

(*b*) In Example 5(*d*), $32 \nmid 242$. From the sequence

$$\begin{aligned} 758 &= 242(3) + 32 \\ 242 &= 32(7) + 18 \\ 32 &= 18(1) + 14 \\ 18 &= 14(1) + 4 \\ 14 &= 4(3) + 2 \\ 4 &= 2(2) \end{aligned}$$

we conclude $(758, 242) = 2$.

Now solving (*1*) for $r = a - bq = a + (-q)b = m_1a + n_1b$; and

substituting in (*2*), $r_1 = b - rq_1 = b - (m_1a + n_1b)q_1$
$= -m_1q_1a + (1 - n_1q_1)b = m_2a + n_2b$

substituting in (*3*), $r_2 = r - r_1q_2 = (m_1a + n_1b) - (m_2a + n_2b)q_2$
$= (m_1 - q_2m_2)a + (n_1 - q_2n_2)b = m_3a + n_3b$

and continuing, we obtain finally

$$r_k = m_{k+1}a + n_{k+1}b$$

Thus, we have

Theorem II. When $d = (a, b)$, there exist $m, n \in I$ such that $d = (a, b) = ma + nb$.

Example 7: Find (726, 275) and express it in the form of Theorem II.

From	We obtain
$726 = 275 \cdot 2 + 176$	$11 = 77 - 22 \cdot 3 = 77 - (99 - 77) \cdot 3$
$275 = 176 \cdot 1 + 99$	$= 77 \cdot 4 - 99 \cdot 3$
$176 = 99 \cdot 1 + 77$	$= (176 - 99) \cdot 4 - 99 \cdot 3$
$99 = 77 \cdot 1 + 22$	$= 176 \cdot 4 - 99 \cdot 7$
$77 = 22 \cdot 3 + 11$	$= 176 \cdot 4 - (275 - 176) \cdot 7$
$22 = 11 \cdot 2$	$= 176 \cdot 11 - 275 \cdot 7$
	$= (726 - 275 \cdot 2) \cdot 11 - 275 \cdot 7$
	$= 11 \cdot 726 + (-29) \cdot 275$

Thus, $m = 11$ and $n = -29$.

Note 1. The procedure for obtaining m and n here is an alternate to that used in obtaining Theorem II.

Note 2. In $(a, b) = ma + nb$, the integers m and n are not unique; in fact, $(a, b) = (m + kb)a + (n - ka)b$ for every $k \in N$.

See Problem 6.

The importance of Theorem II is indicated in

Example 8: Prove: If $a \mid c$, if $b \mid c$, and if $(a, b) = d$, then $ab \mid cd$.

Since $a \mid c$ and $b \mid c$, there exist integers s and t such that $c = as = bt$. By Theorem II there exist $m, n \in I$ such that $d = ma + nb$. Then

$$cd = cma + cnb = btma + asnb = ab(tm + sn)$$

and $ab \mid cd$.

A second consequence of the Division Algorithm is

Theorem III. Any non-empty set K of integers which is closed under the binary operations addition and subtraction is either $\{0\}$ or consists of all multiples of its least positive element.

An outline of the proof when $K \neq \{0\}$ follows. Suppose K contains the integer $a \neq 0$. Since K is closed with respect to addition and subtraction, we have:

(i) $a - a = 0 \in K$

(ii) $0 - a = -a \in K$

(iii) K contains at least one positive integer.

(iv) K contains a least positive integer, say, e.

(v) By induction on n, K contains all positive multiples ne of e (show this).

(vi) K contains all integral multiples me of e.

(vii) If $b \in K$, then $b = q \cdot e + r$ where $0 \leq r < e$; hence, $r = 0$ and so every element of K is an integral multiple of e.

RELATIVELY PRIME INTEGERS

For given $a, b \in I$, suppose there exist $m, n \in I$ such that $am + bn = 1$. Now every common factor of a and b is a factor of the right member 1; hence, $(a, b) = 1$. Two integers a and b for which $(a, b) = 1$ are said to be *relatively prime*.

See Problem 7.

In Problem 8, we prove

Theorem IV. If $(a, s) = (b, s) = 1$, then $(ab, s) = 1$.

PRIME FACTORS

In Problem 9, we prove

Theorem V. If p is a prime and if $p \mid ab$, where $a, b \in I$, then $p \mid a$ or $p \mid b$.

By repeated use of Theorem V there follows

Theorem V′. If p is a prime and if p is a divisor of the product $a \cdot b \cdot c \cdot \ldots \cdot t$ of n integers, then p is a divisor of at least one of these integers.

In Problem 10, we prove

The Unique Factorization Theorem. Every integer $a > 1$ has a unique factorization, except for order,

(a) $$a = p_1 \cdot p_2 \cdot p_3 \cdot \ldots \cdot p_n$$

into a product of positive primes.

Evidently, if (a) gives the factorization of a, then

$$-a = -(p_1 \cdot p_2 \cdot p_3 \cdot \ldots \cdot p_n)$$

Moreover, since the p's of (a) are not necessarily distinct, we may write

$$a = p_1^{\alpha_1} \cdot p_2^{\alpha_2} \cdot p_3^{\alpha_3} \cdot \ldots \cdot p_s^{\alpha_s}$$

where each $\alpha_i \geqq 1$ and the primes $p_1, p_2, p_3, \ldots, p_s$ are distinct.

Example 9: Express each of 2,241,756 and 8,566,074 as a product of positive primes and obtain their greatest common divisor.

$$2{,}241{,}756 = 2^2 \cdot 3^4 \cdot 11 \cdot 17 \cdot 37 \quad \text{and} \quad 8{,}566{,}074 = 2 \cdot 3^4 \cdot 11^2 \cdot 19 \cdot 23$$

Their greatest common divisor is $2 \cdot 3^4 \cdot 11$.

CONGRUENCES

Let m be a positive integer. The relation "*congruent modulo m*", ($\equiv$ (mod m)), is defined on all $a, b \in I$ by $a \equiv b \pmod{m}$ if and only if $m \mid (a - b)$.

Example 10:

(a) $89 \equiv 25 \pmod 4$ since $4 \mid (89 - 25) = 64$

(b) $89 \equiv 1 \pmod 4$ since $4 \mid 88$

(c) $25 \equiv 1 \pmod 4$ since $4 \mid 24$

(d) $153 \equiv -7 \pmod 8$ since $8 \mid 160$

(e) $24 \not\equiv 3 \pmod 5$ since $5 \nmid 21$

(f) $243 \not\equiv 167 \pmod 7$ since $7 \nmid 76$

(g) Any integer a is congruent modulo m to the remainder obtained by dividing a by m.

An alternate definition, often more useful than the original, is $a \equiv b \pmod{m}$ if and only if a and b have the same remainder when divided by m.

As immediate consequences of these definitions, we have:

Theorem VI. If $a \equiv b \pmod{m}$ then, for any $n \in I$, $mn + a \equiv b \pmod{m}$ and conversely.

Theorem VII. If $a \equiv b(\text{mod}\ m)$, then, for all $x \in I$, $a+x \equiv b+x(\text{mod}\ m)$ and $ax \equiv bx(\text{mod}\ m)$.

Theorem VIII. If $a \equiv b(\text{mod}\ m)$ and $c \equiv e(\text{mod}\ m)$, then $a+c \equiv b+e(\text{mod}\ m)$, $a-c \equiv b-e(\text{mod}\ m)$, $ac \equiv be(\text{mod}\ m)$.

See Problem 11.

Theorem IX. Let $(c, m) = d$ and write $m = m_1 d$. If $ca \equiv cb(\text{mod}\ m)$, then $a \equiv b(\text{mod}\ m_1)$ and conversely.

For a proof, see Problem 12.

As a special case of Theorem IX, we have

Theorem X. Let $(c, m) = 1$. If $ca \equiv cb(\text{mod}\ m)$, then $a \equiv b(\text{mod}\ m)$ and conversely.

The relation $\equiv (\text{mod}\ m)$ on I is an equivalence relation and separates the integers into m equivalence classes, $[0], [1], [2], \ldots, [m-1]$, called *residue classes modulo m,* where

$$[r] = \{a\colon a \in I,\ a \equiv r(\text{mod}\ m)\}$$

Example 11: The residue classes modulo 4 are:

$$[0] = \{\ldots, -16, -12, -8, -4, 0, 4, 8, 12, 16, \ldots\}$$
$$[1] = \{\ldots, -15, -11, -7, -3, 1, 5, 9, 13, 17, \ldots\}$$
$$[2] = \{\ldots, -14, -10, -6, -2, 2, 6, 10, 14, 18, \ldots\}$$
$$[3] = \{\ldots, -13, -9, -5, -1, 3, 7, 11, 15, 19, \ldots\}$$

We will denote the set of all residue classes modulo m by $I/(m)$. For example, $I/(4) = \{[0], [1], [2], [3]\}$ and $I/(m) = \{[0], [1], [2], [3], \ldots, [m-1]\}$. Of course, $[3] \in I/(4) = [3] \in I/(m)$ if and only if $m = 4$. Two basic properties of the residue classes modulo m are:

If a and b are elements of the same residue class $[s]$, then $a \equiv b(\text{mod}\ m)$.

If $[s]$ and $[t]$ are distinct residue classes with $a \in [s]$ and $b \in [t]$, then $a \not\equiv b(\text{mod}\ m)$.

THE ALGEBRA OF RESIDUE CLASSES

Let "$\oplus$" (addition) and "$\odot$" (multiplication) be defined on the elements of $I/(m)$ as follows:

$$[a] \oplus [b] = [a+b]$$
$$[a] \odot [b] = [a \cdot b]$$

for every $[a], [b] \in I/(m)$.

Since $\oplus$ and $\odot$ on $I/(m)$ are defined respectively in terms of $+$ and $\cdot$ on I, it follows readily that $\oplus$ and $\odot$ satisfy the laws $\mathbf{A}_1$-$\mathbf{A}_4$, $\mathbf{M}_1$-$\mathbf{M}_4$, and $\mathbf{D}_1$-$\mathbf{D}_2$ as modified in Chapter 4, Page 42.

Example 12: The addition and multiplication tables for $I/(4)$ are:

$\oplus$	0	1	2	3
0	0	1	2	3
1	1	2	3	0
2	2	3	0	1
3	3	0	1	2

Table 5-1

and

$\odot$	0	1	2	3
0	0	0	0	0
1	0	1	2	3
2	0	2	0	2
3	0	3	2	1

Table 5-2

where, for convenience, $[0], [1], [2], [3]$ have been replaced by $0, 1, 2, 3$.

LINEAR CONGRUENCES

Consider the *linear congruence*

$$(b) \qquad ax \equiv b (\text{mod } m)$$

in which a, b, m are fixed integers with $m > 0$. By a *solution* of the congruence we shall mean an integer $x = x_1$ for which $m \mid (ax_1 - b)$. Now if x_1 is a solution of (b) so that $m \mid (ax_1 - b)$ then, for any $k \in I$, $m \mid (a(x_1 + km) - b)$ and $x_1 + km$ is another solution. Thus, if x_1 is a solution so also is every other element of the residue class $[x_1]$ modulo m. If then the linear congruence (b) has solutions, they consist of all the elements of one or more of the residue classes of $I/(m)$.

Example 13: (*a*) The congruence $2x \equiv 3(\text{mod } 4)$ has no solution since none of $2 \cdot 0 - 3$, $2 \cdot 1 - 3$, $2 \cdot 2 - 3$, $2 \cdot 3 - 3$ has 4 as a divisor.

(*b*) The congruence $3x \equiv 2(\text{mod } 4)$ has 6 as a solution and, hence, all elements of $[2] \in I/(4)$ as solutions. There are no others.

(*c*) The congruence $6x \equiv 2(\text{mod } 4)$ has 1 and 3 as solutions. Since $3 \not\equiv 1(\text{mod } 4)$, we shall call 1 and 3 *incongruent solutions* of the congruence. Of course, all elements of $[1], [3] \in I/(4)$ are solutions. There are no others.

Returning to (b), suppose $(a, m) = 1 = sa + tm$. Then $b = bsa + btm$ and $x_1 = bs$ is a solution. Now assume $x_2 \not\equiv x_1(\text{mod } m)$ to be another solution. Since $ax_1 \equiv b(\text{mod } m)$ and $ax_2 \equiv b(\text{mod } m)$, it follows from the transitive property of $\equiv (\text{mod } m)$ that $ax_1 \equiv ax_2(\text{mod } m)$. Then $m \mid a(x_1 - x_2)$ and $x_1 \equiv x_2(\text{mod } m)$ contrary to our assumption. Thus, (b) has just one incongruent solution, say x_1, and the residue class $[x_1] \in I/(m)$, also called a *congruence class*, includes all solutions.

Next, suppose that $(a, m) = d = sa + tm$, $d > 1$. Since $a = a_1 d$ and $m = m_1 d$, it follows that if (b) has a solution $x = x_1$ then $ax_1 - b = mq = m_1 dq$ and so $d \mid b$. Conversely, suppose that $d = (a, m)$ is a divisor of b and write $b = b_1 d$. By Theorem IX, Page 53, any solution of (b) is a solution of

$$(c) \qquad a_1 x \equiv b_1(\text{mod } m_1)$$

and any solution of (c) is a solution of (b). Now $(a_1, m_1) = 1$ so that (c) has a single incongruent solution and, hence, (b) has solutions. We have proved the first part of

Theorem XI. The congruence $ax \equiv b(\text{mod } m)$ has a solution if and only if $d = (a, m)$ is a divisor of b. When $d \mid b$, the congruence has exactly d incongruent solutions (d congruence classes of solutions).

To complete the proof, consider the subset

$$S = \{x_1,\ x_1 + m_1,\ x_1 + 2m_1,\ x_1 + 3m_1,\ \ldots,\ x_1 + (d-1)m_1\}$$

of $[x_1]$, the totality of solutions of $a_1 x \equiv b_1(\text{mod } m_1)$. We shall now show that no two distinct elements of S are congruent modulo m (thus, (b) has at least d incongruent solutions) while each element of $[x_1] - S$ is congruent modulo m to some element of S (thus, (b) has at most d incongruent solutions).

Let $x_1 + sm_1$ and $x_1 + tm_1$ be distinct elements of S. Now if $x_1 + sm_1 \equiv x_1 + tm_1(\text{mod } m)$ then $m \mid (s-t)m_1$; hence, $d \mid (s-t)$ and $s = t$, a contradiction of the assumption $s \neq t$. Thus, the elements of S are incongruent modulo m. Next, consider any element of $[x_1] - S$, say $x_1 + (qd + r)m_1$ where $q \geqq 1$ and $0 \leqq r < d$. Now $x_1 + (qd + r)m_1 = x_1 + rm_1 + qm \equiv x_1 + rm_1(\text{mod } m)$ and $x_1 + rm_1 \in S$. Thus, the congruence (b), with $(a, m) = d$ and $d \mid b$, has exactly d incongruent solutions.

See Problem 14.

POSITIONAL NOTATION FOR INTEGERS

It is well known to the reader that

$$827{,}016 = 8\cdot 10^5 + 2\cdot 10^4 + 7\cdot 10^3 + 0\cdot 10^2 + 1\cdot 10 + 6$$

What is not so well known is that this representation is an application of the congruence properties of integers. For, suppose a is a positive integer. By the division algorithm, $a = 10\cdot q_0 + r_0$, $0 \leq r_0 < 10$. If $q_0 = 0$, we write $a = r_0$; if $q_0 > 0$, then $q_0 = 10\cdot q_1 + r_1$, $0 \leq r_1 < 10$. Now if $q_1 = 0$, then $a = 10\cdot r_1 + r_2$ and we write $a = r_1 r_0$; if $q_1 > 0$, then $q_1 = 10\cdot q_2 + r_2$, $0 \leq r_2 < 10$. Again, if $q_2 = 0$, then $a = 10^2\cdot r_2 + 10\cdot r_1 + r_0$ and we write $a = r_2 r_1 r_0$; if $q_2 > 0$, we repeat the process. That it must end eventually and we have

$$a = 10^s\cdot r_s + 10^{s-1}\cdot r_{s-1} + \cdots + 10\cdot r_1 + r_0 = r_s r_{s-1} \cdots r_1 r_0$$

follows from the fact that the q's constitute a set of decreasing non-negative integers. Note that in this representation the symbols r_i used are from the set $\{0,1,2,3,\ldots,9\}$ of remainders modulo 10. (Why is the representation unique?)

In the paragraph above, we chose the particular integer 10, called the *base*, since this led to our system of representation. However, the process is independent of the base and any other positive integer may be used. Thus, if 4 is taken as base, any positive integer will be represented by a sequence of the symbols 0, 1, 2, 3. For example, the integer (base 10) $155 = 4^3\cdot 2 + 4^2\cdot 1 + 4\cdot 2 + 3 = 2123$ (base 4).

Now addition and multiplication are carried out in much the same fashion, regardless of the base, however, new tables for each operation must be memorized. These tables for base 4 are:

+	0	1	2	3
0	0	1	2	3
1	1	2	3	10
2	2	3	10	11
3	3	10	11	12

Table 5-3

and

·	0	1	2	3
0	0	0	0	0
1	0	1	2	3
2	0	2	10	12
3	0	3	12	21

Table 5-4

See Problem 15.

Solved Problems

1. Prove: If $a \mid b$ and $a \mid c$, then $a \mid (bx + cy)$ where $x, y \in I$.

Since $a \mid b$ and $a \mid c$, there exist integers s, t such that $b = as$ and $c = at$. Then $bx + cy = asx + aty = a(sx + ty)$ and $a \mid (bx + cy)$.

2. Prove: If $a \mid b$ and $b \neq 0$, then $|b| \geq |a|$.

Since $a \mid b$, we have $b = ac$ for some $c \in I$. Then $|b| = |a|\cdot|c|$ with $|c| \geq 1$. Since $|c| \geq 1$, it follows that $|a|\cdot|c| \geq |a|$, that is, $|b| \geq |a|$.

3. Prove: If $a \mid b$ and $b \mid a$, then $b = a$ or $b = -a$.

Since $a \mid b$ implies $a \neq 0$ and $b \mid a$ implies $b \neq 0$, write $b = ac$ and $a = bd$, where $c, d \in I$. Now $a \cdot b = (bd)(ac) = abcd$ and, by the Cancellation Law, $1 = cd$. Then by Problem 16, Chapter 4, $c = 1$ or -1 and $b = ac = a$ or $-a$.

4. Prove: The number of positive primes is infinite.

Suppose the contrary, i.e., suppose there are exactly n positive primes $p_1, p_2, p_3, \ldots, p_n$, written in order of magnitude. Now form $a = p_1 \cdot p_2 \cdot p_3 \cdot \ldots \cdot p_n$ and consider the integer $a+1$. Since no one of the p's is a divisor of $a+1$, it follows that $a+1$ is either a prime $> p_n$ or has a prime $> p_n$ as factor, contrary to the assumption that p_n is the largest. Thus, there is no largest positive prime and their number is infinite.

5. Prove the Division Algorithm: For any two non-zero integers a and b, there exist unique integers q and r such that

$$a = bq + r, \qquad 0 \leqq r < |b|$$

Define $S = \{a - bx : x \in I\}$. If $b < 0$, i.e., $b \leqq -1$, then $b \cdot |a| \leqq -|a| \leqq a$ and $a - b \cdot |a| \geqq 0$. If $b > 0$, i.e., $b \geqq 1$, then $b \cdot (-|a|) \leqq -|a| \leqq a$ and $a - b(-|a|) \geqq 0$. Thus, S contains non-negative integers; denote by r the smallest of these ($r \geqq 0$) and suppose $r = a - bq$. Now if $r \geqq |b|$, then $r - |b| \geqq 0$ and $r - |b| = a - bq - |b| = a - (q+1)b < r$ or $a - (q-1)b < r$, contrary to the choice of r as the smallest non-negative integer $\in S$. Hence, $r < |b|$.

Suppose we should find another pair q' and r' such that

$$a = bq' + r', \qquad 0 \leqq r' < |b|$$

Now $bq' + r' = bq + r$ or $b(q' - q) = r - r'$ implies $b \mid (r - r')$ and, since $|r - r'| < |b|$, then $r - r' = 0$; also $q' - q = 0$ since $b \neq 0$. Thus, $r' = r$, $q' = q$, and q and r are unique.

6. Find (389,167) and express it in the form $389m + 167n$.

From	We find
$389 = 167 \cdot 2 + 55$	$1 = 55 - 2 \cdot 27$
$167 = 55 \cdot 3 + 2$	$= 55 \cdot 82 - 167 \cdot 27$
$55 = 2 \cdot 27 + 1$	$= 389 \cdot 82 - 167 \cdot 191$
$2 = 1 \cdot 2$	

Thus, $(389,167) = 1 = 82 \cdot 389 + (-191)(167)$.

7. Prove: If $c \mid ab$ and if $(a, c) = 1$, then $c \mid b$.

From $1 = ma + nc$, we have $b = mab + ncb$. Since c is a divisor of $mab + ncb$, it is a divisor of b and $c \mid b$ as required.

8. Prove: If $(a, s) = (b, s) = 1$, then $(ab, s) = 1$.

Suppose the contrary, i.e., suppose $(ab, s) = d > 1$ and let $d = (ab, s) = mab + ns$. Now $d \mid ab$ and $d \mid s$. Since $(a, s) = 1$, it follows that $d \nmid a$; hence, by Problem 7, $d \mid b$. But this contradicts $(b, s) = 1$; thus, $(ab, s) = 1$.

9. Prove: If p is a prime and if $p \mid ab$, where $a, b \in I$, then $p \mid a$ or $p \mid b$.

If $p \mid a$ we have the theorem. Suppose then that $p \nmid a$. By definition, the only divisors of p are ± 1 and $\pm p$; then $(p, a) = 1 = mp + na$ for some $m, n \in I$ by Theorem II. Now $b = mpb + nab$ and, since $p \mid (mpb + nab)$, $p \mid b$ as required.

10. Prove: Every integer $a > 1$ has a unique factorization (except for order)

$$a = p_1 \cdot p_2 \cdot p_3 \cdot \ldots \cdot p_n$$

into a product of positive primes.

If a is a prime, the representation in accordance with the theorem is immediate. Suppose then that a is composite and consider the set $S = \{x: x > 1, x \mid a\}$. The least element s of S has no positive factors except 1 and s; hence s is a prime, say p_1, and

$$a = p_1 \cdot b_1, \qquad b_1 > 1$$

Now either b_1 is a prime, say p_2, and $a = p_1 \cdot p_2$ or b_1, being composite, has a prime factor p_2 and

$$a = p_1 \cdot p_2 \cdot b_2, \qquad b_2 > 1$$

A repetition of the argument leads to $a = p_1 \cdot p_2 \cdot p_3$ or

$$a = p_1 \cdot p_2 \cdot p_3 \cdot b_3, \qquad b_3 > 1$$

and so on.

Now the elements of the set $B = \{b_1, b_2, b_3, \ldots\}$ have the property $b_1 > b_2 > b_3 > \cdots$; hence, B has a least element, say b_n, which is a prime p_n and

$$a = p_1 \cdot p_2 \cdot p_3 \cdot \ldots \cdot p_n$$

as required.

To prove uniqueness, suppose we have two representations

$$a = p_1 \cdot p_2 \cdot p_3 \cdot \ldots \cdot p_n = q_1 \cdot q_2 \cdot q_3 \cdot \ldots \cdot q_m$$

Now q_1 is a divisor of $p_1 \cdot p_2 \cdot p_3 \cdot \ldots \cdot p_n$; hence, by Theorem V$'$, q_1 is a divisor of some one of the p's, say p_1. Then $q_1 = p_1$, since both are positive primes, and by $\mathbf{M}_4$ of Chapter IV,

$$p_2 \cdot p_3 \cdot \ldots \cdot p_n = q_2 \cdot q_3 \cdot \ldots \cdot q_m$$

After repeating the argument a sufficient number of times, we find that $m = n$ and the factorization is unique.

11. Find the least positive integers modulo 5 to which 19, 288, $19 \cdot 288$ and $19^3 \cdot 288^2$ are congruent.

We find

$$19 = 5 \cdot 3 + 4; \text{ hence } 19 \equiv 4 (\text{mod } 5).$$
$$288 = 5 \cdot 57 + 3; \text{ hence } 288 \equiv 3 (\text{mod } 5).$$
$$19 \cdot 288 = 5(\cdots) + 12; \text{ hence } 19 \cdot 288 \equiv 2 (\text{mod } 5).$$
$$19^3 \cdot 288^2 = 5(\cdots) + 4^3 \cdot 3^2 = 5(\cdots) + 576; \text{ hence } 19^3 \cdot 288^2 \equiv 1 (\text{mod } 5).$$

12. Prove: Let $(c, m) = d$ and write $m = m_1 d$. If $ca \equiv cb (\text{mod } m)$, then $a = b (\text{mod } m_1)$ and conversely.

Write $c = c_1 d$ so that $(c_1, m_1) = 1$. If $m \mid c(a - b)$, that is, if $m_1 d \mid c_1 d(a - b)$, then $m_1 \mid c_1(a - b)$ and, since $(c_1, m_1) = 1$, $m_1 \mid (a - b)$ and $a \equiv b (\text{mod } m_1)$.

For the converse, suppose $a \equiv b (\text{mod } m_1)$. Since $m_1 \mid (a - b)$, it follows that $m_1 \mid c_1(a - b)$ and $m_1 d \mid c_1 d(a - b)$. Thus, $m \mid c(a - b)$ and $ca \equiv cb (\text{mod } m)$.

13. Show that, when $a, b, p > 0 \in I$, $(a + b)^p \equiv a^p + b^p (\text{mod } p)$.

By the binomial theorem, $(a + b)^p = a^p + p(\cdots) + b^p$ and the theorem is immediate.

14. Find the least positive incongruent solutions of:

(*a*) $13x \equiv 9 (\text{mod } 25)$ (*c*) $259x \equiv 5 (\text{mod } 11)$ (*e*) $222x \equiv 12 (\text{mod } 18)$

(*b*) $207x \equiv 6 (\text{mod } 18)$ (*d*) $7x \equiv 5 (\text{mod } 256)$

(*a*) Since $(13, 25) = 1$, the congruence has, by Theorem XI, a single incongruent solution.

Solution I. If x_1 is the solution, then it is clear that x_1 is an integer whose unit's digit is either 3 or 8; thus $x_1 \in \{3, 8, 13, 18, 23\}$. Testing each of these in turn, we find $x_1 = 18$.

Solution II. By the greatest common divisor process we find $(13, 25) = 1 = -1 \cdot 25 + 2 \cdot 13$. Then $9 = -9 \cdot 25 + 18 \cdot 13$ and 18 is the required solution.

(*b*) Since $207 = 18 \cdot 11 + 9$, $207 \equiv 9 (\text{mod } 18)$, $207x \equiv 9x (\text{mod } 18)$ and, by transitivity, the given congruence is equivalent to $9x \equiv 6 (\text{mod } 18)$. By Theorem IX this congruence may be reduced to $3x \equiv 2 (\text{mod } 6)$. Now $(3, 6) = 3$ and $3 \nmid 2$; hence, there is no solution.

(*c*) Since $259 = 11 \cdot 23 + 6$, $259 \equiv 6 (\text{mod } 11)$ and the given congruence is equivalent to $6x \equiv 5 (\text{mod } 11)$. This congruence has a single incongruent solution which by inspection is found to be 10.

(*d*) Using the greatest common divisor process, we find $(256, 7) = 1 = 2 \cdot 256 + 7(-73)$; thus, $5 = 10 \cdot 256 + 7(-365)$. Now $-365 \equiv 147 (\text{mod } 256)$ and the required solution is 147.

(*e*) Since $222 = 18 \cdot 12 + 6$, the given congruence is equivalent to $6x \equiv 12 (\text{mod } 18)$. Since $(6, 18) = 6$ and $6 \mid 12$, there are exactly 6 incongruent solutions. As shown in the proof of Theorem XI, these 6 solutions are the first 6 positive integers in the set of all solutions of $x \equiv 2 (\text{mod } 3)$, that is, the first 6 positive integers in $[2] \in I/(\text{mod } 3)$. They are then 2, 5, 8, 11, 14, 17.

15. Write 141 and 152 with base 4. Form their sum and product, and check each result.

$141 = 4^3 \cdot 2 + 4^2 \cdot 0 + 4 \cdot 3 + 1$; the representation is 2031.

$152 = 4^3 \cdot 2 + 4^2 \cdot 1 + 4 \cdot 2 + 0$; the representation is 2120.

Sum.

$1 + 0 = 1$; $3 + 2 = 11$, we write 1 and carry 1; $1 + 1 + 0 = 2$; $2 + 2 = 10$

Thus, the sum is 10211, base 4, and 293, base 10.

Product.

Multiply by 0:	0000
Multiply by 2: $2 \cdot 1 = 2$; $2 \cdot 3 = 12$, write 2 and carry 1; etc.	10122
Multiply by 1:	2031
Multiply by 2:	10122
	11032320

The product is 11032320, base 4, and 21432, base 10.

Supplementary Problems

16. Show that the relation (|) is reflexive and transitive but not symmetric.

17. Prove: If $a \mid b$, then $-a \mid b$, $a \mid -b$, and $-a \mid -b$.

18. List all the positive primes (*a*) < 50, (*b*) < 200.
Ans. (*a*) 2, 3, 5, 7, 11, 13, 17, 19, 23, 29, 31, 37, 41, 43, 47

19. Prove: If $a = b \cdot q + r$, where $a, b, q, r \in I$, then any common divisor of a and b also divides r while any common divisor of b and r also divides a.

20. Find the greatest common divisor of each pair of integers and express it in the form of Theorem II:

(*a*) 237, 81 *Ans.* $3 = 13 \cdot 237 + (-38) \cdot 81$

(*b*) 616, 427 *Ans.* $7 = -9 \cdot 616 + 13 \cdot 427$

(*c*) 936, 666 *Ans.* $18 = 5 \cdot 936 + (-7) \cdot 666$

(*d*) 1137, 419 *Ans.* $1 = 206 \cdot 1137 + (-559) \cdot 419$

21. Prove: If $s \neq 0$, then $(sa, sb) = |s| \cdot (a, b)$.

22. Prove: (*a*) If $a \mid s$, $b \mid s$ and $(a, b) = 1$, then $ab \mid s$.
(*b*) If $m = dm_1$ and if $m \mid am_1$, then $d \mid a$.

23. Prove: If p, a prime, is a divisor of $a \cdot b \cdot c$, then $p \mid a$ or $p \mid b$ or $p \mid c$.

24. The integer $e = [a, b]$ is called the *least common multiple* of the positive integers a and b when (1) $a \mid e$ and $b \mid e$, (2) if $a \mid x$ and $b \mid x$ then $e \mid x$.

25. Find: (*a*) $[3, 7]$, (*b*) $[3, 12]$, (*c*) $[22, 715]$. *Ans.* (*a*) 21, (*b*) 12, (*c*) 1430

26. (*a*) Write the integers $a = 19{,}500$ and $b = 54{,}450$ as products of positive primes.
(*b*) Find $d = (a, b)$ and $e = [a, b]$.
(*c*) Verify $d \cdot e = a \cdot b$
(*d*) Prove the relation in (*c*) when a and b are any positive integers.
Ans. (*b*) $2 \cdot 3 \cdot 5^2$; $2^2 \cdot 3^2 \cdot 5^3 \cdot 11^2 \cdot 13$

27. Prove: If $m > 1$, $m \nmid a$, $m \nmid b$, then $m \mid (a - b)$ implies $a - mq_1 = r = b - mq_2$, $0 < r < m$, and conversely.

28. Find all solutions of:

(*a*) $4x \equiv 3 (\text{mod } 7)$
(*b*) $9x \equiv 11 (\text{mod } 26)$
(*c*) $3x + 1 \equiv 4 (\text{mod } 5)$
(*d*) $8x \equiv 6 (\text{mod } 14)$
(*e*) $153x \equiv 6 (\text{mod } 12)$
(*f*) $x + 1 \equiv 3 (\text{mod } 7)$
(*g*) $8x \equiv 6 (\text{mod } 422)$
(*h*) $363x \equiv 345 (\text{mod } 624)$

Ans. (*a*) [6], (*b*) [7], (*c*) [1], (*d*) [6], [13], (*e*) [2], [6], [10], (*f*) [2], (*g*) [159], [370], (*h*) [123], [331], [539]

29. Prove: Theorems V, VI, VII, VIII.

30. Prove: If $a \equiv b (\text{mod } m)$ and $c \equiv b (\text{mod } m)$, then $a \equiv c (\text{mod } m)$. See Examples 10(*a*), (*b*), (*c*), Page 52.

31. (*a*) Prove: If $a + x \equiv b + x (\text{mod } m)$, then $a \equiv b (\text{mod } m)$.
(*b*) Give a single numerical example to disprove: If $ax \equiv bx (\text{mod } m)$, then $ax \equiv b (\text{mod } m)$.
(*c*) Modify the false statement in (*b*) to obtain a true one.

32. (*a*) Interpret $a \equiv b (\text{mod } 0)$.
(*b*) Show that every $x \in I$ is a solution of $ax \equiv b (\text{mod } 1)$.

33. (*a*) Construct addition and multiplication tables for $I/(5)$.
(*b*) Use the multiplication table to obtain $3^2 \equiv 4 (\text{mod } 5)$, $3^4 \equiv 1 (\text{mod } 5)$, $3^8 \equiv 1 (\text{mod } 5)$.
(*c*) Obtain $3^{256} \equiv 1 (\text{mod } 5)$, $3^{514} \equiv 4 (\text{mod } 5)$, $3^{1024} \equiv 1 (\text{mod } 5)$.

34. Construct addition and multiplication tables for $I/(2)$, $I/(6)$, $I/(7)$, $I/(9)$.

35. Prove: If $[s] \in I/(m)$ and if $a, b \in [s]$, then $a \equiv b (\text{mod } m)$.

36. Prove: If $[s], [t] \in I/(m)$ and if $a \in [s]$ and $b \in [t]$, then $a \equiv b (\text{mod } m)$ if and only if $[s] = [t]$.

37. Express 212 using in turn the base (*a*) 2, (*b*) 3, (*c*) 4, (*d*) 7, and (*e*) 9.
Ans. (*a*) 11010100, (*b*) 21212, (*c*) 3110, (*d*) 422, (*e*) 255

38. Express 89 and 111 with various bases, form the sum and product, and check.

39. Prove the first part of the Unique Factorization Theorem using the induction principle stated in Problem 27, Chapter 3, Page 37.

Chapter 6

The Rational Numbers

THE RATIONAL NUMBERS

The system of integers has an obvious defect in that, given integers $m \neq 0$ and s, the equation $mx = s$ may or may not have a solution. For example, $3x = 6$ has the solution $x = 2$ but $4x = 6$ has no solution. This defect is remedied by adjoining to the integers additional numbers (common fractions) to form the system Q of *rational numbers*. The construction here is, in the main, that used in Chapter 4.

We begin with the set of ordered pairs

$$K \quad = \quad I \times (I - \{0\}) \quad = \quad \{(s,m): \ s \in I, \ m \in I - \{0\}\}$$

and define the binary relation $\sim$ on all $(s,m),(t,n) \in K$ by

$$(s,m) \sim (t,n) \quad \text{if and only if} \quad sn = mt$$

(Note carefully that 0 may appear as first component but *never* as second component in any (s,m).)

Now $\sim$ is an equivalence relation (prove it) and thus partitions K into a set of equivalence classes

$$\mathcal{J} \quad = \quad \{[s,m],[t,n], \ldots\}$$

where $$[s,m] \quad = \quad \{(a,b): \ (a,b) \in K, \ (a,b) \sim (s,m)\}$$

We shall call the equivalence classes of $\mathcal{J}$ rational numbers and in the following sections indicate that $\mathcal{J}$ is isomorphic to the system Q as we know it.

ADDITION AND MULTIPLICATION

Addition and Multiplication on $\mathcal{J}$ will be defined respectively by

(i) $$[s,m] + [t,n] \quad = \quad [(sn+mt), mn]$$

and

(ii) $$[s,m] \cdot [t,n] \quad = \quad [st, mn]$$

These operations, being defined in terms of well-defined operations on integers, are (see Problem 1) themselves well-defined.

We now define two special rational numbers

$$zero: \quad [0,m] \leftrightarrow 0 \qquad\qquad one: \quad [m,m] \leftrightarrow 1$$

and the inverses

(additive): $\ -[s,m] = [-s,m]$ for each $[s,m] \in \mathcal{J}$

(multiplicative): $\ [s,m]^{-1} = [m,s]$ for each $[s,m] \in \mathcal{J}$ when $s \neq 0$.

By paralleling the procedures in Chapter 4, it is easily shown that addition and multiplication obey the laws $\mathbf{A_1}$-$\mathbf{A_6}$, $\mathbf{M_1}$-$\mathbf{M_5}$, $\mathbf{D_1}$-$\mathbf{D_2}$ as stated for integers.

A property of $\mathcal{J}$, but not of I, is:

$\mathbf{M_6}$: For every $x \neq 0 \in \mathcal{J}$ there exists a multiplicative inverse $x^{-1} \in \mathcal{J}$ such that $x \cdot x^{-1} = x^{-1} \cdot x = 1$.

By Theorem IV, Chapter 2, the inverses defined in $\mathbf{M}_6$ are unique.

In Problem 2, we prove

Theorem I. If x and y are non-zero elements of $\mathcal{J}$ then $(x \cdot y)^{-1} = y^{-1} \cdot x^{-1}$.

SUBTRACTION AND DIVISION

Subtraction and division are defined on $\mathcal{J}$ by

(iii) $\quad x - y = x + (-y) \quad$ for all $x, y \in \mathcal{J}$

and

(iv) $\quad x \div y = x \cdot y^{-1} \quad$ for all $x \in \mathcal{J}$, $y \neq 0 \in \mathcal{J}$

respectively.

These operations are neither associative nor commutative (prove this). However, as on I, multiplication is distributive with respect to subtraction.

REPLACEMENT

The mapping
$$[t, 1] \in \mathcal{J} \quad \leftrightarrow \quad t \in I$$
is an isomorphism of a certain subset of $\mathcal{J}$ onto the set of integers. We may then, whenever more convenient, replace the subset $\mathcal{J}^* = \{[t,1] : [t,1] \in \mathcal{J}\}$ by I. To complete the identification of $\mathcal{J}$ with Q, we have merely to replace
$$x \cdot y^{-1} \quad \text{by} \quad x/y$$
and, in particular, $[s, m]$ by s/m.

ORDER RELATIONS

An element $x \in Q$, i.e., $x \leftrightarrow [s, m] \in \mathcal{J}$, is called *positive* if and only if $s \cdot m > 0$. The subset of all positive elements of Q will be denoted by Q^+ and the corresponding subset of $\mathcal{J}$ by $\mathcal{J}^+$. Similarly, $[s, m]$ is called *negative* if and only if $s \cdot m < 0$. The subset of all negative elements of Q will be denoted by Q^- and the corresponding subset of $\mathcal{J}$ by $\mathcal{J}^-$. Since, by the Trichotomy Law of Chapter 4, either $s \cdot m > 0$, $s \cdot m < 0$ or $s \cdot m = 0$, it follows that each element of $\mathcal{J}$ is either positive, negative or zero.

The order relations $<$ and $>$ on Q are defined as follows:

For each $x, y \in Q$,
$$x < y \quad \text{if and only if} \quad x - y < 0$$
$$x > y \quad \text{if and only if} \quad x - y > 0$$

These relations are transitive but neither reflexive nor symmetric.

Q also satisfies

The Trichotomy Law: If $x, y \in Q$, one and only one of
$$(a)\ x = y \qquad (b)\ x < y \qquad (c)\ x > y$$
holds.

REDUCTION TO LOWEST TERMS

Consider any arbitrary $[s, m] \in \mathcal{J}$ with $s \neq 0$. Let the (positive) greatest common divisor of s and m be d and write $s = ds_1$, $m = dm_1$. Since $(s, m) \sim (s_1, m_1)$, it follows that $[s, m] = [s_1, m_1]$, i.e., $s/m = s_1/m_1$. Thus, any rational number $\neq 0$ can be written uniquely in the form a/b where a and b are relatively prime integers. Whenever s/m has been replaced by a/b, we shall say that s/m has been reduced to lowest terms. Hereafter, any arbitrary rational number introduced in any discussion is to be assumed reduced to lowest terms.

In Problem 3 we prove:

Theorem II. If x and y are positive rationals with $x < y$, then $1/x > 1/y$.

In Problems 4 and 5, we prove:

The Density Property. If x and y, with $x < y$, are two rational numbers, there exists a rational number z such that $x < z < y$.

and

The Archimedean Property. If x and y are positive rational numbers, there exists a positive integer p such that $px > y$.

DECIMAL REPRESENTATION

Consider the positive rational number a/b in which $b > 1$. Now

$$a = q_0 b + r_0 \qquad 0 \leq r_0 < b$$

and

$$10r_0 = q_1 b + r_1 \qquad 0 \leq r_1 < b$$

Since $r_0 < b$ and, hence, $q_1 b + r_1 = 10r_0 < 10b$, it follows that $q_1 < 10$. If $r_1 = 0$, then $r_0 = \frac{q_1}{10}b$, $a = q_0 b + \frac{q_1}{10}b$, and $\frac{a}{b} = q_0 + q_1/10$. We write $a/b = q_0.q_1$ and call $q_0.q_1$ the decimal representation of a/b. If $r_1 \neq 0$, we have

$$10r_1 = q_2 b + r_2 \qquad 0 \leq r_2 < b$$

in which $q_2 < 10$. If $r_2 = 0$, then $r_1 = \frac{q_2}{10}b$ so that $r_0 = \frac{q_1}{10}b + \frac{q_2}{10^2}b$ and the decimal representation of a/b is $q_0.q_1q_2$; if $r_2 = r_1$, the decimal representation of a/b is the repeating decimal $q_0.q_1q_2q_2q_2\ldots\ldots$; if $r_2 \neq 0, r_1$, we repeat the process.

Now the distinct remainders $r_0, r_1, r_2, \ldots$ are elements of the set $\{0, 1, 2, 3, \ldots, b-1\}$ of residues modulo b so that, in the extreme case, r_b must be identical with some one of $r_0, r_1, r_2, \ldots, r_{b-1}$, say r_c, and the decimal representation of a/b is the repeating decimal

$$q_0.q_1q_2q_3\ldots.q_{b-1}\,q_{c+1}\,q_{c+2}\ldots q_{b-1}\,q_{c+1}\,q_{c+2}\ldots.q_{b-1}\ldots.$$

Thus, every rational number can be expressed as either a terminating or a repeating decimal.

Example 1: (*a*) $5/4 = 1.25$

(*b*) $3/8 = 0.375$

(*c*) For 11/6, we find

$$\begin{aligned} 11 &= 1 \cdot 6 + 5; \quad q_0 = 1,\ r_0 = 5 \\ 10 \cdot 5 &= 8 \cdot 6 + 2; \quad q_1 = 8,\ r_1 = 2 \\ 10 \cdot 2 &= 3 \cdot 6 + 2; \quad q_2 = 3,\ r_2 = 2 = r_1 \end{aligned}$$

and $11/6 = 1.833333\ldots.$

(*d*) For 25/7, we find

$$\begin{aligned} 25 &= 3 \cdot 7 + 4; \quad q_0 = 3,\ r_0 = 4 \\ 10 \cdot 4 &= 5 \cdot 7 + 5; \quad q_1 = 5,\ r_1 = 5 \\ 10 \cdot 5 &= 7 \cdot 7 + 1; \quad q_2 = 7,\ r_2 = 1 \\ 10 \cdot 1 &= 1 \cdot 7 + 3; \quad q_3 = 1,\ r_3 = 3 \\ 10 \cdot 3 &= 4 \cdot 7 + 2; \quad q_4 = 4,\ r_4 = 2 \\ 10 \cdot 2 &= 2 \cdot 7 + 6; \quad q_5 = 2,\ r_5 = 6 \\ 10 \cdot 6 &= 8 \cdot 7 + 4; \quad q_6 = 8,\ r_6 = 4 = r_0 \end{aligned}$$

and $25/7 = 3.571428\,571428\ldots.$

Conversely, it is clear that every terminating decimal is a rational number. For example, $0.17 = 17/100$ and $0.175 = 175/1000 = 7/40$.

In Problem 6, we prove

Theorem III. Every repeating decimal is a rational number.

The proof makes use of two preliminary theorems:

(i) Every repeating decimal may be written as the sum of an infinite geometric progression.

(ii) The sum of an infinite geometric progression whose common ratio r satisfies $|r| < 1$ is a finite number.

A discussion of these theorems can be found in any College Algebra book.

Solved Problems

1. Show that addition and multiplication on $\mathcal{J}$ are well-defined.

Let $[a,b] = [s,m]$ and $[c,d] = [t,n]$. Then $(a,b) \sim (s,m)$ and $(c,d) \sim (t,n)$, so that $am = bs$ and $cn = dt$. Now

$$\begin{aligned}[a,b] + [c,d] &= [(ad + bc), bd] = [(ad+bc)mn, bd \cdot mn] \\ &= [(am \cdot dn + cn \cdot bm), bd \cdot mn] \\ &= [(bs \cdot dn + dt \cdot bm), bd \cdot mn] \\ &= [bd(sn + tm), bd \cdot mn] \\ &= [sn + tm, mn] = [s,m] + [t,n]\end{aligned}$$

and addition is well-defined.

Also

$$\begin{aligned}[a,b] \cdot [c,d] &= [ac, bd] = [ac \cdot mn, bd \cdot mn] \\ &= [am \cdot cn, bd \cdot mn] = [bs \cdot dt, bd \cdot mn] \\ &= [bd \cdot st, bd \cdot mn] = [st, mn] \\ &= [s,m] \cdot [t,n]\end{aligned}$$

and multiplication is well-defined.

2. Prove: If x, y are non-zero rational numbers then $(x \cdot y)^{-1} = y^{-1} \cdot x^{-1}$.

Let $x \leftrightarrow [s,m]$ and $y \leftrightarrow [t,n]$, so that $x^{-1} \leftrightarrow [m,s]$ and $y^{-1} \leftrightarrow [n,t]$. Then $x \cdot y \leftrightarrow [s,m] \cdot [t,n] = [st, mn]$ and $(x \cdot y)^{-1} \leftrightarrow [mn, st] = [n,t] \cdot [m,s] \leftrightarrow y^{-1} \cdot x^{-1}$.

3. Prove: If x and y are positive rationals with $x < y$, then $1/x > 1/y$.

Let $x \leftrightarrow [s,m]$ and $y \leftrightarrow [t,n]$; then $sm > 0$, $tn > 0$, and $sn < mt$. Now, for $1/x = x^{-1} \leftrightarrow [m,s]$ and $1/y \leftrightarrow [n,t]$, the inequality $mt > sn$ implies $1/x > 1/y$, as required.

4. Prove: If x and y, with $x < y$, are two rational numbers, there exists a rational number z such that $x < z < y$.

Since $x < y$, we have

$$2x = x + x < x + y \quad \text{and} \quad x + y < y + y = 2y$$

Then

$$2x < x + y < 2y$$

and, multiplying by $\frac{1}{2}$, $x < \frac{1}{2}(x+y) < y$. Thus, $\frac{1}{2}(x+y)$ meets the requirement for z.

5. Prove: If x and y are positive rational numbers, there exists a positive integer p such that $px > y$.

Let $x \leftrightarrow [s, m]$ and $y \leftrightarrow [t, n]$, where s, m, t, n are positive integers. Now $px > y$ if and only if $psn > mt$. Since $sn \geq 1$ and $2sn > 1$, the inequality is certainly satisfied if we take $p = 2mt$.

6. Prove: Every repeating decimal represents a rational number.

Consider the repeating decimal

$$x.yzdefdef.... \quad = \quad x.yz \; + \; 0.00def \; + \; 0.00000def \; + \; \cdots$$

Now $x.yz$ is a rational fraction since it is a terminating decimal, while $0.00def + 0.00000def + \cdots$ is an infinite geometric progression with first term $a = 0.00def$, common ratio $r = 0.001$, and sum

$$S \; = \; \frac{a}{1-r} \; = \; \frac{0.00def}{0.999} \; = \; \frac{def}{99900}, \quad \text{a rational fraction}$$

Thus the repeating decimal, being the sum of two rational numbers, is a rational number.

7. Express (*a*) $\frac{27}{32}$ with base 4, (*b*) $\frac{1}{3}$ with base 5.

(*a*) $27/32 \; = \; 3(\frac{1}{4}) + 3/32 \; = \; 3(\frac{1}{4}) + 1(\frac{1}{4})^2 + 1/32 \; = \; 3(\frac{1}{4}) + 1(\frac{1}{4})^2 + 2(\frac{1}{4})^3$

The required representation is 0.312.

(*b*)
$$\begin{aligned} 1/3 &= 1(\tfrac{1}{5}) + \tfrac{2}{15} = 1(\tfrac{1}{5}) + 3(\tfrac{1}{5})^2 + 1/75 \\ &= 1(\tfrac{1}{5}) + 3(\tfrac{1}{5})^2 + 1(\tfrac{1}{5})^3 + 2/375 \\ &= 1(\tfrac{1}{5}) + 3(\tfrac{1}{5})^2 + 1(\tfrac{1}{5})^3 + 3(\tfrac{1}{5})^4 + 1/1875 \end{aligned}$$

The required representation is 0.131313...... .

Supplementary Problems

8. Verify: (*a*) $[s, m] + [0, n] \; = \; [s, m]$ (*c*) $[s, m] + [-s, m] \; = \; [0, n] \; = \; [s, m] + [s, -m]$
(*b*) $[s, m] \cdot [0, n] \; = \; [0, n]$ (*d*) $[s, m] \cdot [m, s] \; = \; [n, n]$

9. Restate the laws $\mathbf{A}_1$-$\mathbf{A}_6$, $\mathbf{M}_1$-$\mathbf{M}_5$, $\mathbf{D}_1$-$\mathbf{D}_2$ of Chapter 4 for rational numbers and prove them.

10. Prove: (*a*) $\mathcal{Q}^+$ is closed with respect to addition and multiplication.
(*b*) If $[s, m] \in \mathcal{Q}^+$, so also does $[s, m]^{-1}$.

11. Prove: (*a*) $\mathcal{Q}^-$ is closed with respect to addition but not with respect to multiplication.
(*b*) If $[s, m] \in \mathcal{Q}^-$, so also does $[s, m]^{-1}$.

12. Prove: If $x, y \in Q$ and $x \cdot y = 0$, then $x = 0$ or $y = 0$.

13. Prove: If $x, y \in Q$, then (*a*) $-(x+y) = -x-y$ and (*b*) $-(-x) = x$.

14. Prove: The Trichotomy Law.

15. If $x, y, z \in Q$, prove:
(*a*) $x+z < y+z$ if and only if $x < y$;
(*b*) when $z > 0$, $xz < yz$ if and only if $x < y$;
(*c*) when $z < 0$, $xz < yz$ if and only if $x > y$.

16. If $w, x, y, z \in Q$ with $xz \neq 0$ in (*a*) and (*b*), and $xyz \neq 0$ in (*c*), prove:
(*a*) $(w \div x) \pm (y \div z) \; = \; (wz \pm xy) \div xz$
(*b*) $(w \div x) \cdot (y \div z) \; = \; wy \div xz$
(*c*) $(w \div x) \div (y \div z) \; = \; wz \div xy$

17. Prove: If $a, b \in Q^+$ and $a < b$, then $a^2 < ab < b^2$. What is the corresponding inequality if $a, b \in Q^-$?

Chapter 7

The Real Numbers

INTRODUCTION

Chapters 4 and 6 began with the observation that the system X of numbers previously studied had an obvious defect. This defect was remedied by enlarging the system X. In doing so, we defined on the set of ordered pairs of elements of X an equivalence relation, etc., etc. In this way we developed from N the systems I and Q satisfying $N \subset I \subset Q$. For what is to follow, it is important to realize that each of the new systems I and Q has a single simple characteristic, namely,

I is the smallest set in which, for arbitrary $m, s \in N$, the equation $m + x = s$ always has a solution.

Q is the smallest set in which, for arbitrary integers $m \neq 0$ and s, the equation $mx = s$ always has a solution.

Now the situation here is not that the system Q has a single defect; rather, it is there are many defects and these so diverse that the procedure of the previous chapters will not remedy all of them. We mention only two:

(1) The equation $x^2 = 3$ has no solution in Q. For, suppose the contrary, and assume that the rational a/b, reduced to lowest terms, be such that $(a/b)^2 = 3$. Since $a^2 = 3b^2$, it follows that $3 \mid a^2$ and, by Theorem V, Chapter 5, that $3 \mid a$. Write $a = 3a_1$; then $3a_1^2 = b^2$ so that $3 \mid b^2$ and, hence, $3 \mid b$. But this contradicts the assumption that a/b was expressed in lowest terms.

(2) The circumference c of a circle of diameter $d \in Q$ is not an element of Q, i.e., in $c = \pi d$, $\pi \notin Q$. Moreover, $\pi^2 \notin Q$ so that π is not a solution of $x^2 = q$ for any $q \in Q$. (In fact, π satisfies *no* equation of the form $ax^n + bx^{n-1} + \cdots + sx + t = 0$ with $a, b, \ldots, s, t \in Q$.)

The method, introduced in the next section, of extending the rational numbers to the real numbers is due to the German mathematician R. Dedekind. In an effort to motivate the definition of the basic concept of a *Dedekind cut* or, more simply, a *cut* of the rational numbers, we shall first discuss it in non-algebraic terms.

Consider the rational scale of Fig. 7-1, that is, a line L on which the non-zero elements of Q are attached to points at proper (scaled) distances from the origin which bears the label 0. For convenience, call each point of L to which a rational number is attached a rational point. (Not every point of L is a rational point. For, let P be an intersection of the circle with center at 0 and radius 2 units with the line parallel to and 1 unit above L. Then let the perpendicular to L through P meet L in T; by (1) above, T is not a rational point.) Suppose the line L to be cut into two pieces at some one of its points. There are two possibilities:

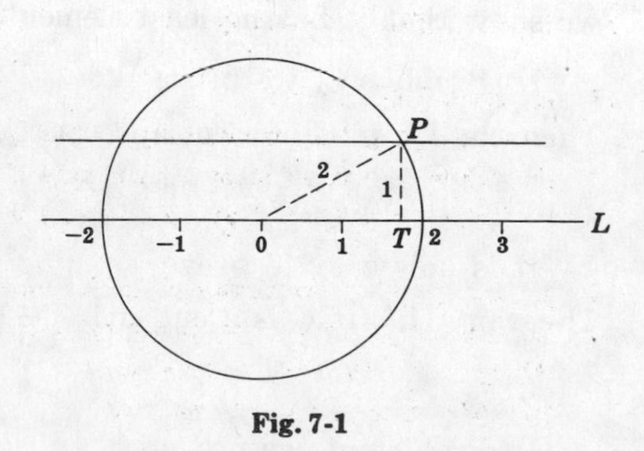

Fig. 7-1

(*a*) The point at which L is cut is not a rational point. Then every rational point of L is on one but not both of the pieces.

(*b*) The point at which L is cut is a rational point. Then, with the exception of this point, every other rational point is on one but not both of the pieces. Let us agree to place the exceptional point always on the right-hand piece.

In either case, then, the effect of cutting L at one of its points is to determine two non-empty proper subsets of Q. Since these subsets are disjoint while their union is Q, either defines the other and we may limit our attention to the left-hand subset. This left-hand subset is called a cut and we now proceed to define it algebraically, that is, without reference to any line.

DEDEKIND CUTS

By a *cut* C in Q we shall mean a non-empty proper subset of Q having the additional properties:

(i) if $c \in C$ and $a \in Q$ with $a < c$, then $a \in C$.

(ii) for every $c \in C$ there exists $b \in C$ such that $b > c$.

The gist of these properties is that a cut has neither a least (first) element nor a greatest (last) element. There is, however, a sharp difference in the reasons for this state of affairs. A cut C has no least element because, if $c \in C$, every rational number $a < c$ is an element of C. On the other hand, while there always exist elements of C which are greater than any selected element $c \in C$, there also exist rational numbers greater than c which *do not* belong to C, i.e., are greater than every element of C.

Example 1: Let r be an arbitrary rational number. Show that $C(r) = \{a: a \in Q, a < r\}$ is a cut.

Since Q has neither a first nor last element, it follows that there exists $r_1 \in Q$ such that $r_1 < r$ (thus, $C(r) \neq \emptyset$) and $r_2 \in Q$ such that $r_2 > r$ (thus, $C(r) \neq Q$). Hence, $C(r)$ is a non-empty proper subset of Q. Let $c \in C(r)$, that is, $c < r$. Now for any $a \in Q$ such that $a < c$, we have $a < c < r$; thus, $a \in C(r)$ as required in (i). By the Density Property of Q there exists $d \in Q$ such that $c < d < r$; then $d > c$ and $d \in C(r)$, as required in (ii). Thus, $C(r)$ is a cut.

The cut defined in Example 1 will be called a *rational cut* or, more precisely, *the cut at the rational number r.* For an example of a non-rational cut, see Problem 1.

When C is a cut, we shall denote by C' the complement of C in Q. For example, if $C = C(r)$ of Example 1, then $C' = C'(r) = \{a': a' \in Q, a' \geq r\}$. Thus, the complement of a rational cut is a proper subset of Q having a least but no greatest element. Clearly, the complement of the non-rational cut of Problem 1 has no greatest element; in Problem 2, we show that it has no least element.

In Problem 3, we prove

Theorem I. If C is a cut and $r \in Q$, then

$$(a)\ D = \{r + a: a \in C\} \text{ is a cut} \quad \text{and} \quad (b)\ D' = \{r + a': a' \in C'\}$$

It is now easy to prove

Theorem II. If C is a cut and $r \in Q^+$, then

$$(a)\ E = \{ra: a \in C\} \text{ is a cut} \quad \text{and} \quad (b)\ E' = \{ra': a' \in C'\}$$

In Problem 4, we prove

Theorem III. If C is a cut and $r \in Q^+$, there exists $b \in C$ such that $r + b \in C'$.

POSITIVE CUTS

Denote by $\mathcal{K}$ the set of all cuts of the rational numbers and by $\mathcal{K}^+$ the set of all cuts (called *positive cuts*) which contain one or more elements of Q^+. Let the remaining cuts in $\mathcal{K}$ be partitioned into the cut 0, i.e., $0 = C(0) = \{a: a \in Q^-\}$, and the set $\mathcal{K}^-$ of all cuts containing some but not all of the elements of Q^-. For example, $C(2) \in \mathcal{K}^+$ while $C(-5) \in \mathcal{K}^-$. We shall, for the present, restrict our attention solely to the cuts of $\mathcal{K}^+$ for which it is easy to prove

Theorem IV. If $C \in \mathcal{K}^+$ and $r > 1 \in Q^+$, there exists $c \in C$ such that $rc \in C'$.

Each $C \in \mathcal{K}^+$ consists of all elements of Q^-, 0, and see (Problem 5) an infinitude of elements of Q^+. For each $C \in \mathcal{K}^+$, define $\mathcal{C} = \{a: a \in C, a > 0\}$ and denote the set of all $\mathcal{C}$'s by $\mathcal{H}$. For example, if $C = C(3)$ then $\mathcal{C}(3) = \{a: a \in Q, 0 < a < 3\}$ and C may be written as $C = C(3) = Q^- \cup \{0\} \cup \mathcal{C}(3)$. Note that each $C \in \mathcal{K}^+$ defines a unique $\mathcal{C} \in \mathcal{H}$ and, conversely, each $\mathcal{C} \in \mathcal{H}$ defines a unique $C \in \mathcal{K}^+$. Let us agree on the convention: when $C_i \in \mathcal{K}^+$, then $C_i = Q^- \cup \{0\} \cup \mathcal{C}_i$.

We define addition (+) and multiplication (·) on $\mathcal{K}^+$ as follows:

$$\left.\begin{aligned} C_1 + C_2 &= Q^- \cup \{0\} \cup (\mathcal{C}_1 + \mathcal{C}_2) \\ C_1 \cdot C_2 &= Q^- \cup \{0\} \cup (\mathcal{C}_1 \cdot \mathcal{C}_2) \end{aligned}\right\} \quad \text{for each } C_1, C_2 \in \mathcal{K}^+$$

where (i)

$$\begin{cases} \mathcal{C}_1 + \mathcal{C}_2 = \{c_1 + c_2 : c_1 \in \mathcal{C}_1, c_2 \in \mathcal{C}_2\} \\ \mathcal{C}_1 \cdot \mathcal{C}_2 = \{c_1 \cdot c_2 : c_1 \in \mathcal{C}_1, c_2 \in \mathcal{C}_2\} \end{cases}$$

It is easy to see that both $C_1 + C_2$ and $C_1 \cdot C_2$ are elements of $\mathcal{K}^+$. Moreover, since

$$\mathcal{C}_1 + \mathcal{C}_2 = \{a: a \in C_1 + C_2,\ a > 0\}$$

and

$$\mathcal{C}_1 \cdot \mathcal{C}_2 = \{a: a \in C_1 \cdot C_2,\ a > 0\}$$

it follows that $\mathcal{H}$ is closed under both addition and multiplication as defined in (i).

Example 2: Verify: (*a*) $C(3) + C(7) = C(10)$, (*b*) $C(3)\,C(7) = C(21)$.

Denote by $\mathcal{C}(3)$ and $\mathcal{C}(7)$ respectively the subsets of all positive rational numbers of $C(3)$ and $C(7)$. We need then only verify

$$\mathcal{C}(3) + \mathcal{C}(7) = \mathcal{C}(10) \qquad \text{and} \qquad \mathcal{C}(3) \cdot \mathcal{C}(7) = \mathcal{C}(21)$$

(*a*) Let $c_1 \in \mathcal{C}(3)$ and $c_2 \in \mathcal{C}(7)$. Since $0 < c_1 < 3$ and $0 < c_2 < 7$, we have $0 < c_1 + c_2 < 10$. Then $c_1 + c_2 \in \mathcal{C}(10)$ and $\mathcal{C}(3) + \mathcal{C}(7) \subseteq \mathcal{C}(10)$. Next, suppose $c_3 \in \mathcal{C}(10)$. Then, since $0 < c_3 < 10$,

$$0 < \frac{3}{10}c_3 < 3 \qquad \text{and} \qquad 0 < \frac{7}{10}c_3 < 7$$

that is, $\frac{3}{10}c_3 \in \mathcal{C}(3)$ and $\frac{7}{10}c_3 \in \mathcal{C}(7)$. Now $c_3 = \frac{3}{10}c_3 + \frac{7}{10}c_3$; hence, $\mathcal{C}(10) \subseteq \mathcal{C}(3) + \mathcal{C}(7)$. Thus, $\mathcal{C}(3) + \mathcal{C}(7) = \mathcal{C}(10)$ as required.

(*b*) For c_1 and c_2 as in (*a*), we have $0 < c_1 \cdot c_2 < 21$. Then $c_1 \cdot c_2 \in \mathcal{C}(21)$ and $\mathcal{C}(3) \cdot \mathcal{C}(7) \subseteq \mathcal{C}(21)$. Now suppose $c_3 \in \mathcal{C}(21)$ so that $0 < c_3 < 21$ and $0 < \frac{c_3}{21} = q < 1$. Write $q = q_1 \cdot q_2$ with $0 < q_1 < 1$ and $0 < q_2 < 1$. Then $c_3 = 21q = (3q_1)(7q_2)$ with $0 < 3q_1 < 3$ and $0 < 7q_2 < 7$, i.e., $3q_1 \in \mathcal{C}(3)$ and $7q_2 \in \mathcal{C}(7)$. Then $\mathcal{C}(21) \subseteq \mathcal{C}(3) \cdot \mathcal{C}(7)$ and $\mathcal{C}(3) \cdot \mathcal{C}(7) = \mathcal{C}(21)$ as required.

The laws $\mathbf{A}_1$-$\mathbf{A}_4$, $\mathbf{M}_1$-$\mathbf{M}_4$, $\mathbf{D}_1$-$\mathbf{D}_2$ in $\mathcal{K}^+$ follow immediately from the definitions of addition and multiplication in $\mathcal{K}^+$ and the fact that these laws hold in Q^+ and, hence, in $\mathcal{H}$. Moreover, it is easily shown that $C(1)$ is the multiplicative identity, so that $\mathbf{M}_5$ also holds.

MULTIPLICATIVE INVERSES

Consider now an arbitrary $C = Q^- \cup \{0\} \cup \mathcal{C} \in \mathcal{K}^+$ and define

$$\mathcal{C}^{-1} = \{b: b \in Q^+,\ b < a^{-1} \text{ for some } a \in C'\}$$

In Problem 6 we prove:

If $C = Q^- \cup \{0\} \cup \mathcal{C}$ then $C^{-1} = Q^- \cup \{0\} \cup \mathcal{C}^{-1}$ is a positive cut.

In Problem 7 we prove:

For each $C \in \mathcal{K}^+$, its multiplicative inverse is $C^{-1} \in \mathcal{K}^+$.

At this point we may move in either of two ways to expand $\mathcal{K}^+$ into a system in which each element has an additive inverse:

(1) Repeat Chapter 4 using $\mathcal{K}^+$ instead of N.

(2) Identify each element of $\mathcal{K}^-$ as the additive inverse of a unique element of $\mathcal{K}^+$. We shall proceed in this fashion.

ADDITIVE INVERSES

The definition of the sum of two positive cuts is equivalent to

$$C_1 + C_2 = \{c_1 + c_2: c_1 \in C_1,\ c_2 \in C_2\}, \qquad C_1, C_2 \in \mathcal{K}^+$$

We now extend the definition to embrace all cuts by

$$C_1 + C_2 = \{c_1 + c_2: c_1 \in C_1,\ c_2 \in C_2\}, \qquad C_1, C_2 \in \mathcal{K} \tag{1}$$

Example 3: Verify $C(3) + C(-7) = C(-4)$.

Let $c_1 + c_2 \in C(3) + C(-7)$, where $c_1 \in C(3)$ and $c_2 \in C(-7)$. Since $c_1 < 3$ and $c_2 < -7$, it follows that $c_1 + c_2 < -4$ so that $c_1 + c_2 \in C(-4)$. Thus, $C(3) + C(-7) \subseteq C(-4)$.

Conversely, let $c_3 \in C(-4)$. Then $c_3 < -4$ and $-4 - c_3 = d \in Q^+$. Now $c_3 = -4 - d = (3 - \frac{1}{2}d) + (-7 - \frac{1}{2}d)$; then since $3 - \frac{1}{2}d \in C(3)$ and $-7 - \frac{1}{2}d \in C(-7)$, it follows that $c_3 \in C(3) + C(-7)$ and $C(-4) \subseteq C(3) + C(-7)$. Thus, $C(3) + C(-7) = C(-4)$ as required.

Now, for each $C \in \mathcal{K}$, define

$$-C = \{x: x \in Q,\ x < -a \text{ for some } a \in C'\}$$

For $C = C(-3)$, we have $-C = \{x: x \in Q,\ x < 3\}$ since -3 is the least element of C'. But this is precisely $C(3)$; hence, in general,

$$-C(r) = C(-r) \quad \text{when } r \in Q$$

In Problem 8 we show that $-C$ is truly a cut, and in Problem 9 we show that $-C$ is the additive inverse of C. Now the laws $\mathbf{A}_1$-$\mathbf{A}_4$ hold in $\mathcal{K}$.

In Problem 10 we prove

The Trichotomy Law. For any $C \in \mathcal{K}$, one and only one of

$$C = C(0) \qquad C \in \mathcal{K}^+ \qquad -C \in \mathcal{K}^+$$

holds.

MULTIPLICATION ON $\mathcal{K}$

For all $C \in \mathcal{K}$, we define

$$C > C(0) \quad \text{if and only if} \quad C \in \mathcal{K}^+$$
$$C < C(0) \quad \text{if and only if} \quad -C \in \mathcal{K}^+$$

and
$$|C| = C \quad \text{when} \quad C \geqq C(0)$$
$$|C| = -C \quad \text{when} \quad C < C(0)$$
Thus, $|C| \geqq C(0)$, that is, $|C| = C(0)$ or $|C| \in \mathcal{K}^+$.

For all $C_1, C_2 \in \mathcal{K}$, we define

$$\begin{cases} C_1 \cdot C_2 = C(0) & \text{when } C_1 = C(0) \text{ or } C_2 = C(0) \\ C_1 \cdot C_2 = |C_1| \cdot |C_2| & \text{when } C_1 > C(0) \text{ and } C_2 > C(0) \text{ or when } C_1 < C(0) \text{ and } C_2 < C(0) \\ C_1 \cdot C_2 = -(|C_1| \cdot |C_2|) & \text{when } C_1 > C(0) \text{ and } C_2 < C(0) \text{ or when } C_1 < C(0) \text{ and } C_2 > C(0) \end{cases} \tag{2}$$

Finally, for all $C \neq C(0)$, we define
$$C^{-1} = |C|^{-1} \text{ when } C > C(0) \qquad \text{and} \qquad C^{-1} = -(|C|^{-1}) \text{ when } C < C(0)$$

It now follows easily that the laws $\mathbf{A}_1$-$\mathbf{A}_6$, $\mathbf{M}_1$-$\mathbf{M}_6$, $\mathbf{D}_1$-$\mathbf{D}_2$ (see Page 71) hold in $\mathcal{K}$.

SUBTRACTION AND DIVISION

Paralleling the corresponding section in Chapter 6, we define for all $C_1, C_2 \in \mathcal{K}$
$$C_1 - C_2 \quad = \quad C_1 + (-C_2) \tag{3}$$
and, when $C_2 \neq C(0)$,
$$C_1 \div C_2 \quad = \quad C_1 \cdot C_2^{-1} \tag{4}$$

Note. We now find ourselves in the uncomfortable position of having two completely different meanings for $C_1 - C_2$. Throughout this chapter, then, we shall agree to consider $C_1 - C_2$ and $C_1 \cap C_2'$ as having different meanings.

ORDER RELATIONS

For any two distinct cuts $C_1, C_2 \in \mathcal{K}$, we define
$$C_1 < C_2, \text{ also } C_2 > C_1, \quad \text{to mean} \quad C_1 - C_2 < C(0)$$
In Problem 11, we show
$$C_1 < C_2, \text{ also } C_2 > C_1, \quad \text{if and only if} \quad C_1 \subset C_2$$

There follows easily

The Trichotomy Law. For any $C_1, C_2 \in \mathcal{K}$, one and only one of the following holds:
$$(a)\ C_1 = C_2 \qquad (b)\ C_1 < C_2 \qquad (c)\ C_1 > C_2$$

PROPERTIES OF THE REAL NUMBERS

Define $\mathcal{K}^* = \{C(r) : C(r) \in \mathcal{K},\ r \in Q\}$. We leave for the reader to prove

Theorem V. The mapping $C(r) \in \mathcal{K}^* \to r \in Q$ is an isomorphism of $\mathcal{K}^*$ onto Q.

The elements of $\mathcal{K}$ are called *real numbers* and, whenever more convenient, $\mathcal{K}$ will be replaced by the familiar R while $A, B, \ldots$ will denote arbitrary elements of R. Now $Q \subset R$; the elements of the complement of Q in R are called *irrational numbers.*

In Problems 12 and 13, we prove

The Density Property. If $A, B \in R$ with $A < B$, there exists a rational number $C(r)$ such that $A < C(r) < B$.

and

The Archimedean Property. If $A, B \in R^+$, there exists a positive integer $C(n)$ such that $C(n) \cdot A > B$.

In order to state below an important property of the real numbers which does not hold for the rational numbers, we make the following definition:

Let $S \neq \emptyset$ be a set on which an order relation $<$ is well-defined, and let T be any proper subset of S. An element $s \in S$, if one exists, such that $s \geqq t$ for every $t \in T$ ($s \leqq t$ for every $t \in T$) is called an *upper bound* (*lower bound*) of T.

Example 4: (*a*) If $S = Q$ and $T = \{-5, -1, 0, 1, 3/2\}$, then $3/2, 2, 102, \ldots$ are upper bounds of T while $-5, -17/3, -100, \ldots$ are lower bounds of T.

(*b*) When $S = Q$ and $T = C \in \mathcal{K}$, then T has no lower bound while any $t' \in T' = C'$ is an upper bound. On the other hand, T' has no upper bound while any $t \in T$ is a lower bound.

If the set of all upper bounds (lower bounds) of a subset T of a set S contains a least element (greatest element) e, then e is called the *least upper bound* (*greatest lower bound*) of T.

Let the universal set be Q and consider the rational cut $C(r) \in \mathcal{K}$. Since r is the least element of $C'(r)$, every $s \geqq r$ in Q is an upper bound of $C(r)$ and every $t \leqq r$ in Q is a lower bound of $C'(r)$. Thus, r is both the least upper bound (l.u.b.) of $C(r)$ and the greatest lower bound (g.l.b.) of $C'(r)$.

Example 5: (*a*) The set T of Example 4(*a*) has $3/2$ as l.u.b. and -5 as g.l.b.

(*b*) Let the universal set be Q. The cut C of Problem 1 has no lower bounds and, hence, no g.l.b. Also, it has upper bounds but no l.u.b. since C has no greatest element and C' has no least element.

(*c*) Let the universal set be R. Any cut $C \in \mathcal{K}$, being a subset of Q, is a subset of R. The cut $C(r)$ then has upper bounds in R and $r \in R$ as l.u.b. *Also*, the cut C of Problem 1 has upper bounds in R and $\sqrt{3} \in R$ as l.u.b.

Example 5(*c*) illustrates

Theorem VI. If $\mathcal{S}$ is a non-empty subset of $\mathcal{K}$ and if $\mathcal{S}$ has an upper bound in $\mathcal{K}$, it has a l.u.b. in $\mathcal{K}$.

For a proof, see Problem 14.

Similarly, we have

Theorem VI′. If $\mathcal{S}$ is a non-empty subset of $\mathcal{K}$ and if $\mathcal{S}$ has a lower bound in $\mathcal{K}$, it has a g.l.b. in $\mathcal{K}$.

Thus, the set R of real numbers has the

Completeness Property. Every non-empty subset of R having a lower bound (upper bound) has a greatest lower bound (least upper bound).

Suppose $\theta^n = \alpha$, where $\alpha, \theta \in R^+$ and $n \in I^+$. We call θ the principal nth root of α and write $\theta = \alpha^{1/n}$. Then for $r = m/n \in Q$, there follows $\alpha^r = \theta^m$.

Other properties of R are:

(1) For each $\alpha \in R^+$ and each $n \in I^+$, there exists a unique $\theta \in R^+$ such that $\theta^n = \alpha$.

(2) For real numbers $\alpha > 1$ and β, define α^β as the l.u.b. of $\{\alpha^r : r \in Q, r < \beta\}$. Then α^β is defined for all $\alpha > 0$, $\beta \in R$ by setting $\alpha^\beta = (1/\alpha)^{-\beta}$ when $0 < \alpha < 1$.

SUMMARY

As a partial summary to date, there follows a listing of the basic properties of the system R of all real numbers. At the right, in parentheses, is indicated other systems N, I, Q for which each property holds.

Addition

$\mathbf{A_1}$.	Closure Law	$r+s \in R$, for all $r,s \in R$	(N,I,Q)
$\mathbf{A_2}$.	Commutative Law	$r+s = s+r$, for all $r,s \in R$	(N,I,Q)
$\mathbf{A_3}$.	Associative Law	$r+(s+t) = (r+s)+t$, for all $r,s,t \in R$	(N,I,Q)
$\mathbf{A_4}$.	Cancellation Law	If $r+t = s+t$, then $r = s$ for all $r,s,t \in R$.	(N,I,Q)
$\mathbf{A_5}$.	Additive Identity	There exists a unique additive identity element $0 \in R$ such that $r+0 = 0+r = r$, for every $r \in R$.	(I,Q)
$\mathbf{A_6}$.	Additive Inverses	For each $r \in R$, there exists a unique additive inverse $-r \in R$ such that $r+(-r) = (-r)+r = 0$.	(I,Q)

Multiplication

$\mathbf{M_1}$.	Closure Law	$r \cdot s \in R$, for all $r,s \in R$	(N,I,Q)
$\mathbf{M_2}$.	Commutative Law	$r \cdot s = s \cdot r$, for all $r,s \in R$	(N,I,Q)
$\mathbf{M_3}$.	Associative Law	$r \cdot (s \cdot t) = (r \cdot s) \cdot t$, for all $r,s,t \in R$	(N,I,Q)
$\mathbf{M_4}$.	Cancellation Law	If $m \cdot p = n \cdot p$, then $m = n$, for all $m,n \in R$ and $p \neq 0 \in R$.	(N,I,Q)
$\mathbf{M_5}$.	Multiplicative Identity	There exists a unique multiplicative identity element $1 \in R$ such that $1 \cdot r = r \cdot 1 = r$ for every $r \in R$.	(N,I,Q)
$\mathbf{M_6}$	Multiplicative Inverses	For each $r \neq 0 \in R$, there exists a unique multiplicative inverse $r^{-1} \in R$ such that $r \cdot r^{-1} = r^{-1} \cdot r = 1$.	(Q)

Distributive Laws	For every $r,s,t \in R$,	
$\mathbf{D_1}$.	$r \cdot (s+t) = r \cdot s + r \cdot t$	
$\mathbf{D_2}$.	$(s+t) \cdot r = s \cdot r + t \cdot r$	(N,I,Q)
Density Property	For each $r,s \in R$, with $r < s$, there exists $t \in Q$ such that $r < t < s$.	(Q)
Archimedean Property	For each $r,s \in R^+$, with $r < s$, there exists $n \in I^+$ such that $n \cdot r > s$.	(Q)
Completeness Property	Every non-empty subset of R having a lower bound (upper bound) has a greatest lower bound (least upper bound).	(N,I)

Solved Problems

1. Show that the set S consisting of Q^-, zero, and all $s \in Q^+$ such that $s^2 < 3$ is a cut.

First, S is a proper subset of Q since $1 \in S$ and $2 \notin S$. Next, let $c \in S$ and $a \in Q$ with $a < c$. Clearly, $a \in S$ when $a \leq 0$ and also when $c \leq 0$.

For the remaining case $(0 < a < c)$, $a^2 < ac < c^2 < 3$; then $a^2 < 3$ and $a \in S$ as required in (i), Page 66. Property (ii), Page 66, is satisfied by $b = 1$ when $c \leq 0$; then S will be a cut provided that for each $c > 0$ with $c^2 < 3$ an $m \in Q^+$ can always be found such that $(c+m)^2 < 3$. We can simplify matters somewhat by noting that if p/q, where $p,q \in I^+$, should be such an m so also is $1/q$. Now $q \geq 1$; hence $\left(c+\frac{1}{q}\right)^2 = c^2 + \frac{2c}{q} + \frac{1}{q^2} \leq c^2 + \frac{2c+1}{q}$; thus $\left(c+\frac{1}{q}\right)^2 < 3$ provided $\frac{2c+1}{q} < 3 - c^2$, that is, provided $(3-c^2)q > 2c+1$. Since $(3-c^2) \in Q^+$ and $(2c+1) \in Q^+$, the existence of $q \in I^+$ satisfying the latter inequality is assured by the Archimedean Property of Q^+. Thus, S is a cut.

2. Show that the complement S' of the set S of Problem 1 has no smallest element.

For any $r \in S' = \{a: a \in Q^+, a^2 \geq 3\}$, we are to show that a positive rational number m can always be found such that $(r-m)^2 > 3$, that is, the choice r will never be the smallest element in S'. As in Problem 1, matters will be simplified by seeking an m of the form $1/q$ where $q \in I^+$. Now $\left(r - \frac{1}{q}\right)^2 = r^2 - \frac{2r}{q} + \frac{1}{q^2} > r^2 - \frac{2r}{q}$; hence, $\left(r - \frac{1}{q}\right)^2 > 3$ whenever $\frac{2r}{q} < r^2 - 3$, that is, provided $(r^2-3)q > 2r$. As in Problem 1, the Archimedean Property of Q^+ assures the existence of $q \in I^+$ satisfying the latter inequality. Thus S' has no smallest element.

3. Prove: If C is a cut and $r \in Q$, then (a) $D = \{r+a: a \in C\}$ is a cut and (b) $D' = \{r+a': a' \in C'\}$.

(a) $D \neq \emptyset$ since $C \neq \emptyset$; moreover, for any $c' \in C'$, $r+c' \notin D$ and $D \neq Q$. Thus D is a proper subset of Q.

Let $b \in C$. For any $s \in Q$ such that $s < r+b$, we have $s-r < b$ so that $s-r \in C$ and then $s = r+(s-r) \in D$ as required in (i), Page 66. Also, for $b \in C$ there exists an element $c \in C$ such that $c > b$; then $r+b, r+c \in D$ and $r+c > r+b$ as required in (ii), Page 66. Thus D is a cut.

(b) Let $b' \in C'$. Then $r+b' \notin D$ since $b' \notin C$; hence, $r+b' \in D'$. On the other hand, if $q' = r+p' \in D'$ then $p' \notin C$ since if it did we would have $D \cap D' \neq \emptyset$. Thus D' is as defined.

4. Prove: If C is a cut and $r \in Q^+$, there exists $b \in C$ such that $r+b \in C'$.

From Problem 3, $D = \{r+a: a \in C\}$ is a cut. Since $r > 0$, it follows that $C \subset D$. Let $q \in Q$ such that $p = r+q \in D$ but not in C. Then $q \in C$ but $r+q \in C'$. Thus q satisfies the requirement of b in the theorem.

5. Prove: If $C \in \mathcal{K}^+$, then C contains an infinitude of elements of Q^+.

Since $C \in \mathcal{K}^+$ there exists at least one $r \in Q^+$ such that $r \in C$. Then for all $q \in N$ we have $r/q \in C$. Thus, no $C \in \mathcal{K}^+$ contains only a finite number of positive rational numbers.

6. Prove: If $C = Q^- \cup \{0\} \cup \mathcal{C} \in \mathcal{K}^+$, then $C^{-1} = Q^- \cup \{0\} \cup \mathcal{C}^{-1}$ is a positive cut.

Since $\mathcal{C} \neq Q^+$, it follows that $C' \neq \emptyset$; and since $\mathcal{C} \neq \emptyset$, it follows that $\mathcal{C}' \neq Q^+$. Let $d \in C'$. Then $(d+1)^{-1} \in Q^+$ and $(d+1)^{-1} < d^{-1}$ so that $(d+1)^{-1} \in \mathcal{C}^{-1}$ and $\mathcal{C}^{-1} \neq \emptyset$. Also, if $c \in C$ then for every $a \in C'$ we have $c < a$ and $c^{-1} > a^{-1}$; hence, $c^{-1} \notin \mathcal{C}^{-1}$ and $\mathcal{C}^{-1} \neq Q^+$. Thus C^{-1} is a proper subset of Q.

Let $c \in \mathcal{C}^{-1}$ and $r \in Q^+$ such that $r < c$. Then $r < c < d^{-1}$ for some $d \in C'$ and $r \in \mathcal{C}^{-1}$ as required in (i), Page 66. Also, since $c \neq d^{-1}$ there exists $s \in Q^+$ such that $c < s < d^{-1}$ and $s \in \mathcal{C}^{-1}$ as required in (ii). Thus C^{-1} is a positive cut.

7. Prove: For each $C \in \mathcal{K}^+$, its multiplicative inverse is $C^{-1} \in \mathcal{K}^+$.

Let $C = Q^- \cup \{0\} \cup \mathcal{C}$ so that $C^{-1} = Q^- \cup \{0\} \cup \mathcal{C}^{-1}$. Then $\mathcal{C} \cdot \mathcal{C}^{-1} = \{c \cdot b: c \in \mathcal{C}, b \in \mathcal{C}^{-1}\}$. Now $b < d^{-1}$ for some $d \in C'$, and so $bd < 1$; also, $c < d$ so that $bc < 1$. Thus, $\mathcal{C} \cdot \mathcal{C}^{-1} \subseteq \mathcal{C}(1)$.

Let $n \in \mathcal{C}(1)$ so that $n^{-1} > 1$. By Theorem IV there exists $c \in C$ such that $c \cdot n^{-1} \in C'$. For each $a \in C$ such that $a > c$, we have $n \cdot a^{-1} < n \cdot c^{-1} = (c \cdot n^{-1})^{-1}$; thus, $n \cdot a^{-1} = e \in \mathcal{C}^{-1}$. Then $n = ae \in \mathcal{C} \cdot \mathcal{C}^{-1}$ and $\mathcal{C}(1) \subseteq \mathcal{C} \cdot \mathcal{C}^{-1}$. Hence, $\mathcal{C} \cdot \mathcal{C}^{-1} = \mathcal{C}(1)$ and $C \cdot C^{-1} = C(1)$. By Problem 6, $C^{-1} \in \mathcal{K}^+$.

8. When $C \in \mathcal{K}$, show that $-C$ is a cut.

Note first that $-C \neq \emptyset$ since $C' \neq \emptyset$. Now let $c \in C$; then $-c \notin -C$, for if it were we would have $-c < -c'$ (for some $c' \in C'$) so that $c' < c$, a contradiction. Thus, $-C$ is a proper subset of Q. Property (i), Page 66, is immediate. To prove property (ii), let $x \in -C$, i.e., $x < -c'$ for some $c' \in C'$. Now $x < \frac{1}{2}(x - c') < -c'$. Thus, $\frac{1}{2}(x-c') > x$ and $\frac{1}{2}(x-c') \in -C$.

9. Show that $-C$ of Problem 8 is the additive inverse of C, i.e., $C + (-C) = -C + C = C(0)$.

Let $c + x \in C + (-C)$, where $c \in C$ and $x \in -C$. Now if $x < -c'$ for $c' \in C'$, we have $c + x < c - c' < 0$ since $c < c'$. Then $C + (-C) \subseteq C(0)$. Conversely, let $y, z \in C(0)$ with $z > y$. Then, by Theorem III, there exist $c \in C$ and $c' \in C'$ such that $c + (z - y) = c'$. Since $z - c' < -c'$, it follows that $z - c' \in -C$. Thus, $y = c + (z - c') \in C + (-C)$ and $C(0) \subseteq C + (-C)$. Hence, $C + (-C) = -C + C = C(0)$.

10. Prove the Trichotomy Law: For $C \in \mathcal{K}$, one and only one of

$$C = C(0) \qquad C \in \mathcal{K}^+ \qquad -C \in \mathcal{K}^+$$

is true.

Clearly, neither $C(0)$ nor $-C(0) \in \mathcal{K}^+$. Suppose now that $C \neq C(0)$ and $C \notin \mathcal{K}^+$. Since every $c \in C$ is a negative rational number but $C \neq Q^-$, there exists $c' \in Q^-$ such that $c' \in C'$. Since $c' < \frac{1}{2}c' < 0$, it follows that $0 < -\frac{1}{2}c' < -c'$. Then $-\frac{1}{2}c' \in -C$ and so $-C \in \mathcal{K}^+$. On the contrary, if $C \in \mathcal{K}^+$, every $c' \in C'$ is also $\in Q^+$. Then every element of $-C$ is negative and $-C \notin \mathcal{K}^+$.

11. Prove: For any two cuts $C_1, C_2 \in \mathcal{K}$, we have $C_1 < C_2$ if and only if $C_1 \subset C_2$.

Suppose $C_1 \subset C_2$. Choose $a' \in C_2 \cap C_1'$ and then choose $b \in C_2$ such that $b > a'$. Since $-b < -a'$, it follows that $-b \in -C_1$. Now $-C_1$ is a cut; hence there exists an element $c \in -C_1$ such that $c > -b$. Then $b + c > 0$ and $b + c \in C_2 + (-C_1) = C_2 - C_1$ so that $C_2 - C_1 \in \mathcal{K}^+$. Thus $C_2 - C_1 > C(0)$ and $C_1 < C_2$.

For the converse, suppose $C_1 < C_2$. Then $C_2 - C_1 > C(0)$ and $C_2 - C_1 \in \mathcal{K}^+$. Choose $d \in Q^+$ such that $d \in C_2 - C_1$ and write $d = b + a$ where $b \in C_2$ and $a \in -C_1$. Then $-b < a$ since $d > 0$; also, since $a \in -C_1$, we may choose an $a' \in C_1'$ such that $a < -a'$. Now $-b < -a'$; then $a' < b$ so that $b \notin C_1$ and, thus, $C_2 \not\subseteq C_1$. Next, consider any $x \in C_1$. Then $x < b$ so that $x \in C_2$ and, hence, $C_1 \subset C_2$.

12. Prove: If $A, B \in R$ with $A < B$, there exists a rational number $C(r)$ such that $A < C(r) < B$.

Since $A < B$, there exist rational numbers r and s with $s < r$ such that $r, s \in B$ but not in A. Then $A \leqq C(s) < C(r) < B$, as required.

13. Prove: If $A, B \in R^+$, there exists a positive integer $C(n)$ such that $C(n) \cdot A > B$.

Since this is trivial for $A \geqq B$, suppose $A < B$. Let r, s be positive rational numbers such that $r \in A$ and $s \in B'$; then $C(r) < A$ and $C(s) > B$. By the Archimedean Property of Q, there exists a positive integer n such that $nr > s$, i.e., $C(n) \cdot C(r) > C(s)$. Then

$$C(n) \cdot A \;\geqq\; C(n) \cdot C(r) \;>\; C(s) \;>\; B$$

as required.

14. Prove: If $\mathcal{S}$ is a non-empty subset of $\mathcal{K}$ and if $\mathcal{S}$ has an upper bound (in $\mathcal{K}$), it has a l.u.b. in $\mathcal{K}$.

Let $\mathcal{S} = \{C_1, C_2, C_3, \ldots\}$ be the subset and C be an upper bound. The union $U = C_1 \cup C_2 \cup C_3 \cup \ldots$ of the cuts of $\mathcal{S}$ is itself a cut $\in \mathcal{K}$; also, since $C_1 \subseteq U$, $C_2 \subseteq U$, $C_3 \subseteq U$, $\ldots$, U is an upper bound of $\mathcal{S}$. But $C_1 \subseteq C$, $C_2 \subseteq C$, $C_3 \subseteq C$, $\ldots$; hence $U \subseteq C$ and U is the l.u.b. of $\mathcal{S}$.

Supplementary Problems

15. (*a*) Define $C(3)$ and $C(-7)$. Prove that each is a cut.
(*b*) Define $C'(3)$ and $C'(-7)$.
(*c*) Locate $-10, -5, 0, 1, 4$ as $\in$ or $\notin$ each of $C(3), C(-7), C'(3), C'(-7)$.
(*d*) Find 5 rational numbers in $C(3)$ but not in $C(-7)$.

16. Prove: $C(r) \subset C(s)$ if and only if $r < s$.

17. Prove: If A and B are cuts, then $A \subset B$ implies $A \neq B$.

18. Prove: Theorem II, Page 66.

19. Prove: If C is a cut and $r \in Q^+$, then $C \leq D = \{a + r : a \in C\}$.

20. Prove: Theorem IV, Page 67.

21. Let $r \in Q$ but not in $C \in \mathcal{K}$. Prove: $C \leq C(r)$.

22. Prove: (*a*) $C(2) + C(5) = C(7)$ (*c*) $C(r) + C(0) = C(r)$
(*b*) $C(2) \cdot C(5) = C(10)$ (*d*) $C(r) \cdot C(1) = C(r)$

23. Prove: (*a*) If $C \in \mathcal{K}^+$, then $-C \in \mathcal{K}^-$.
(*b*) If $C \in \mathcal{K}^-$, then $-C \in \mathcal{K}^+$.

24. Prove: $-(-C) = C$

25. Prove: (*a*) If $C_1, C_2 \in \mathcal{K}$, then $C_1 + C_2$ and $|C_1| \cdot |C_2|$ are cuts.
(*b*) If $C_1 \neq C(0)$, then $|C_1|^{-1}$ is a cut.
(*c*) $(C^{-1})^{-1} = C$ for all $C \neq C(0)$.

26. Prove: (*a*) If $C \in \mathcal{K}^+$, then $C^{-1} \in \mathcal{K}^+$.
(*b*) If $C \in \mathcal{K}^-$, then $C^{-1} \in \mathcal{K}^-$.

27. Prove: If $r, s \in Q$ with $r < s$, there exists an irrational number α such that $r < \alpha < s$.
Hint. Consider $\alpha = r + \dfrac{s - r}{\sqrt{2}}$.

28. Prove: If A and B are real numbers with $A < B$, there exists an irrational number α such that $A < \alpha < B$.

Hint. Use Problem 12 to prove: If A and B are real numbers with $A < B$, there exist rational numbers t and r such that $A < C(t) < C(r) < B$.

29. Prove: Theorem V, Page 69.

Chapter 8

The Complex Numbers

THE SYSTEM C OF COMPLEX NUMBERS

The system C of complex numbers is the number system of ordinary algebra. It is the smallest set in which, for example, the equation $x^2 = a$ can be solved when a is any element of R. In our development of the set C, we begin with the product set $R \times R$. The binary relation "$=$" requires

$$(a, b) = (c, d) \quad \text{if and only if} \quad a = c \text{ and } b = d$$

Now each of the resulting equivalence classes contains but a single element. Hence, we shall denote a class as (a, b) rather than as $[a, b]$ and so, hereafter, denote $R \times R$ by C.

ADDITION AND MULTIPLICATION ON C

Addition and multiplication on C are defined respectively by

$$\text{(i)} \qquad (a, b) + (c, d) = (a + c, b + d)$$

$$\text{(ii)} \qquad (a, b) \cdot (c, d) = (ac - bd, ad + bc)$$

for all $(a, b), (c, d) \in C$.

The calculations necessary to show that these operations obey $\mathbf{A_1}$-$\mathbf{A_4}$, $\mathbf{M_1}$-$\mathbf{M_4}$, $\mathbf{D_1}$-$\mathbf{D_2}$ of Chapter 7, when restated in terms of C, are routine and will be left for the reader. It is easy to verify that $(0, 0)$ is the identity element for addition and $(1, 0)$ is the identity element for multiplication; also, that the additive inverse of (a, b) is $-(a, b) = (-a, -b)$ and the multiplicative inverse of $(a, b) \neq (0, 0)$ is $(a, b)^{-1} = \left(\frac{a}{a^2 + b^2}, \frac{-b}{a^2 + b^2}\right)$. Hence, the set of complex numbers have the properties $\mathbf{A_5}$-$\mathbf{A_6}$ and $\mathbf{M_5}$-$\mathbf{M_6}$ of Chapter 7, restated in terms of C.

We shall show in the next section that $R \subset C$, and one might expect then that C has all of the basic properties of R. But this is false since it is not possible to extend (redefine) the order relation "$<$" of R to include all elements of C.

See Problem 1.

PROPERTIES OF COMPLEX NUMBERS

The real numbers are a proper subset of the complex numbers C. For, if in (i) and (ii) we take $b = d = 0$, we see that the first components combine exactly as do the real numbers a and c. Thus, the mapping $a \leftrightarrow (a, 0)$ is an isomorphism of R onto a certain subset $\{(a, b):\ a \in R,\ b = 0\}$ of C.

The elements $(a, b) \in C$ in which $b \neq 0$, are called *imaginary numbers* and those imaginary numbers (a, b) in which $a = 0$ are called *pure imaginary numbers*.

For each complex number $z = (a, b)$, we define the complex number $\bar{z} = \overline{(a, b)} = (a, -b)$ to be the *conjugate* of z. Clearly, every real number is its own conjugate while the conjugate of each pure imaginary is its negative.

There follow easily

Theorem I. The sum (product) of any complex number and its conjugate is a real number.

Theorem II. The square of every pure imaginary number is a negative real number.

See also Problem 2.

The special role of the complex number $(1, 0)$ suggests an investigation of another, $(0, 1)$. We find

$$(x, y) \cdot (0, 1) = (-y, x) \quad \text{for every} \quad (x, y) \in C$$

and, in particular,

$$(y, 0) \cdot (0, 1) = (0, 1) \cdot (y, 0) = (0, y)$$

Moreover, $(0, 1)^2 = (0, 1) \cdot (0, 1) = (-1, 0) \leftrightarrow -1$ in the mapping above so that $(0, 1)$ is a solution of $z^2 = -1$.

Defining $(0, 1)$ as the *pure imaginary unit* and denoting it by i, we have

$$i^2 = -1$$

and, for every $(x, y) \in C$,

$$(x, y) = (x, 0) + (0, y) = (x, 0) + (y, 0) \cdot (0, 1) = x + yi$$

In this familiar notation, x is called the *real part* and y is called the *imaginary part* of the complex number. We summarize:

the negative of $z = x + yi$ is $-z = -(x + yi) = -x - yi$

the conjugate of $z = x + yi$ is $\bar{z} = \overline{x + yi} = x - yi$

for each $z = x + yi$, $z \cdot \bar{z} = x^2 + y^2 \in R$

for each $z \neq 0 + 0 \cdot i = 0$, $z^{-1} = \dfrac{1}{z} = \dfrac{\bar{z}}{z \cdot \bar{z}} = \dfrac{x}{x^2 + y^2} - \dfrac{y}{x^2 + y^2} i.$

SUBTRACTION AND DIVISION ON C

Subtraction and division on C are defined by

(iii) $\quad z - w = z + (-w), \quad$ for all $z, w \in C$

(iv) $\quad z \div w = z \cdot w^{-1}, \quad$ for all $w \neq 0$, $z \in C$

TRIGONOMETRIC REPRESENTATION

The representation of a complex number z by (x, y) and by $x + yi$ suggests the mapping (isomorphism)

$$x + yi \leftrightarrow (x, y)$$

of the set C of all complex numbers onto the points (x, y) of the real plane. We may therefore speak of the point $P(x, y)$ or of $P(x + yi)$ as best suits our purpose at the moment. The use of a single coordinate, surprisingly, often simplifies many otherwise tedious computations. One example will be discussed below; another will be outlined briefly in Problem 20.

Consider in Fig. 8-1 the point $P(x + yi) \neq 0$ whose distance r from O is given by $r = \sqrt{x^2 + y^2}$. If θ is the positive angle which OP makes with the positive x-axis, we have

$$x = r \cos \theta, \quad y = r \sin \theta$$

whence $\quad z = x + yi = r(\cos \theta + i \sin \theta)$

The right member of this equality is called the *trigonometric form* (*polar form*) of z. The non-negative real number

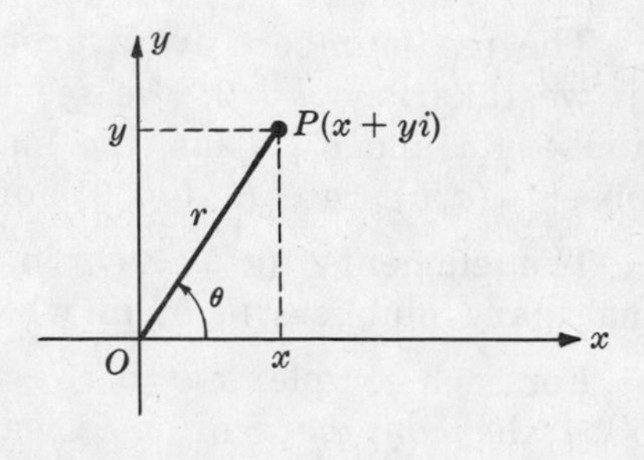

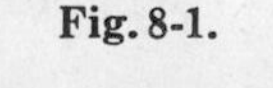
Fig. 8-1.

$$r = |z| = \sqrt{z \cdot \bar{z}} = \sqrt{x^2+y^2}$$

is called the *modulus* (*absolute value*) of z, and θ is called the *angle* (*amplitude* or *argument*) of z. Now θ satisfies $x = r\cos\theta$, $y = r\sin\theta$, $\tan\theta = y/x$ and any two of these determine θ up to an additive multiple of 2π. Usually we shall choose as θ the smallest positive angle. (*Note.* When P is at O, we have $r = 0$ and θ arbitrary.)

Example 1: Express (*a*) $1 + i$, (*b*) $-\sqrt{3} + i$ in trigonometric form.

(*a*) We have $r = \sqrt{1+1} = \sqrt{2}$. Since $\tan\theta = 1$ and $\cos\theta = 1/\sqrt{2}$, we take θ to be the first quadrant angle $45° = \pi/4$. Thus, $1 + i = \sqrt{2}(\cos\pi/4 + i\sin\pi/4)$.

(*b*) Here $r = \sqrt{3+1} = 2$, $\tan\theta = -1/\sqrt{3}$ and $\cos\theta = -\frac{1}{2}\sqrt{3}$. Taking θ to be the second quadrant angle $5\pi/6$, we have

$$-\sqrt{3} + i = 2(\cos 5\pi/6 + i\sin 5\pi/6)$$

It follows that two complex numbers are equal if and only if their absolute values are equal and their angles differ by an integral multiple of 2π, i.e., are congruent modulo 2π.

In Problems 3 and 4, we prove

Theorem III. The absolute value of the product of two complex numbers is the product of their absolute values, and the angle of the product is the sum of their angles.

and

Theorem IV. The absolute value of the quotient of two complex numbers is the quotient of their absolute values, and the angle of the quotient is the angle of the numerator minus the angle of the denominator.

Example 2: (*a*) When $z_1 = 2(\cos\frac{1}{4}\pi + i\sin\frac{1}{4}\pi)$ and $z_2 = 4(\cos\frac{3}{4}\pi + i\sin\frac{3}{4}\pi)$, we have

$$\begin{aligned} z_1 \cdot z_2 &= 2(\cos\tfrac{1}{4}\pi + i\sin\tfrac{1}{4}\pi)\cdot 4(\cos\tfrac{3}{4}\pi + i\sin\tfrac{3}{4}\pi) \\ &= 8(\cos\pi + i\sin\pi) = -8 \end{aligned}$$

$$\begin{aligned} z_2/z_1 &= 4(\cos\tfrac{3}{4}\pi + i\sin\tfrac{3}{4}\pi) \div 2(\cos\tfrac{1}{4}\pi + i\sin\tfrac{1}{4}\pi) \\ &= 2(\cos\tfrac{1}{2}\pi + i\sin\tfrac{1}{2}\pi) = 2i \end{aligned}$$

$$z_1/z_2 = \tfrac{1}{2}\big(\cos(-\tfrac{1}{2}\pi) + i\sin(-\tfrac{1}{2}\pi)\big) = \tfrac{1}{2}(\cos 3\pi/2 + i\sin 3\pi/2) = -\tfrac{1}{2}i$$

(*b*) When $z = 2(\cos\pi/6 + i\sin\pi/6)$,

$$z^2 = z \cdot z = 4(\cos\pi/3 + i\sin\pi/3) = 2(1 + i\sqrt{3})$$

and $$z^3 = z^2 \cdot z = 8(\cos\tfrac{1}{2}\pi + i\sin\tfrac{1}{2}\pi) = 8i$$

As a consequence of Theorem IV, we have

Theorem V. If n is a positive integer,

$$[r(\cos\theta + i\sin\theta)]^n = r^n(\cos n\theta + i\sin n\theta)$$

ROOTS

The equation $$z^n = A = \rho(\cos\phi + i\sin\phi)$$

where n is a positive integer and A is any complex number, will now be shown to have exactly n roots. If $z = r(\cos\theta + i\sin\theta)$ is one of these, we have by Theorem V,

$$r^n(\cos n\theta + i\sin n\theta) = \rho(\cos\phi + i\sin\phi)$$

Then $$r^n = \rho \quad \text{and} \quad n\theta = \phi + 2k\pi \quad (k\text{, an integer})$$

so that $$r = \rho^{1/n} \quad \text{and} \quad \theta = \phi/n + 2k\pi/n$$

The number of distinct roots are the number of non-coterminal angles of the set $\{\phi/n + 2k\pi/n\}$. For any positive integer $k = nq + m$, $0 \leq m < n$, it is clear that $\phi/n + 2k\pi/n$ and $\phi/n + 2m\pi/n$ are coterminal. Thus, there are exactly n distinct roots, given by

$$\rho^{1/n}[\cos(\phi/n + 2k\pi/n) + i\sin(\phi/n + 2k\pi/n)], \quad k = 0, 1, 2, 3, \ldots, n-1$$

These n roots are coordinates of n equispaced points on the circle, centered at the origin, of radius $\sqrt[n]{|A|}$. If then $z = \sqrt[n]{|A|}(\cos\theta + i\sin\theta)$ is any one of the nth roots of A, the remaining roots may be obtained by successively increasing the angle θ by $2\pi/n$ and reducing modulo 2π whenever the angle is greater than 2π.

Example 3: (*a*) One root of $z^4 = 1$ is $r_1 = 1 = \cos 0 + i\sin 0$. Increasing the angle successively by $2\pi/4 = \pi/2$, we find $r_2 = \cos\frac{1}{2}\pi + i\sin\frac{1}{2}\pi$, $r_3 = \cos\pi + i\sin\pi$, and $r_4 = \cos 3\pi/2 + i\sin 3\pi/2$. Note that had we begun with another root, say $-1 = \cos\pi + i\sin\pi$, we would obtain $\cos 3\pi/2 + i\sin 3\pi/2$, $\cos 2\pi + i\sin 2\pi = \cos 0 + i\sin 0$, and $\cos\frac{1}{2}\pi + i\sin\frac{1}{2}\pi$. These are, of course, the roots obtained above in a different order.

(*b*) One of the roots of $z^6 = -4\sqrt{3} - 4i = 8(\cos 7\pi/6 + i\sin 7\pi/6)$ is

$$r_1 = \sqrt{2}(\cos 7\pi/36 + i\sin 7\pi/36)$$

Increasing the angle successively by $2\pi/6$, we have

$$r_2 = \sqrt{2}(\cos 19\pi/36 + i\sin 19\pi/36)$$

$$r_3 = \sqrt{2}(\cos 31\pi/36 + i\sin 31\pi/36)$$

$$r_4 = \sqrt{2}(\cos 43\pi/36 + i\sin 43\pi/36)$$

$$r_5 = \sqrt{2}(\cos 55\pi/36 + i\sin 55\pi/36)$$

$$r_6 = \sqrt{2}(\cos 67\pi/36 + i\sin 67\pi/36)$$

As a consequence of Theorem V, we have

Theorem VI. The n nth roots of unity are

$$\rho = \cos 2\pi/n + i\sin 2\pi/n, \quad \rho^2, \rho^3, \rho^4, \ldots, \rho^{n-1}, \rho^n = 1$$

PRIMITIVE ROOTS OF UNITY

An nth root z of 1 is called *primitive* if and only if $z^m \neq 1$ when $0 < m < n$. Using the results of Problem 5, it is easy to show that ρ and ρ^5 are primitive sixth roots of 1 while $\rho^2, \rho^3, \rho^4, \rho^6$ are not. This illustrates

Theorem VII. Let $\rho = \cos 2\pi/n + i\sin 2\pi/n$. If $(m, n) = d > 1$, then ρ^m is an n/dth root of 1.

For a proof, see Problem 6.

There follows

Corollary. The primitive nth roots of 1 are those and only those nth roots $\rho, \rho^2, \rho^3, \ldots, \rho^n$ of 1 whose exponents are relatively prime to n.

Example 4: The primitive 12th roots of 1 are

$$\rho = \cos 2\pi/12 + i\sin 2\pi/12 = \tfrac{1}{2}\sqrt{3} + \tfrac{1}{2}i$$

$$\rho^5 = \cos 5\pi/6 + i\sin 5\pi/6 = -\tfrac{1}{2}\sqrt{3} + \tfrac{1}{2}i$$

$$\rho^7 = \cos 7\pi/6 + i\sin 7\pi/6 = -\tfrac{1}{2}\sqrt{3} - \tfrac{1}{2}i$$

$$\rho^{11} = \cos 11\pi/6 + i\sin 11\pi/6 = \tfrac{1}{2}\sqrt{3} - \tfrac{1}{2}i$$

Solved Problems

1. Express in the form $x + yi$:

(a) $3 - 2\sqrt{-1}$, (b) $3 + \sqrt{-4}$, (c) 5, (d) $\frac{1}{3+4i}$, (e) $\frac{5-i}{2-3i}$, (f) i^3.

(a) $3 - 2\sqrt{-1} = 3 - 2i$

(b) $3 + \sqrt{-4} = 3 + 2i$

(c) $5 = 5 + 0 \cdot i$

(d) $\frac{1}{3+4i} = \frac{3-4i}{(3+4i)(3-4i)} = \frac{3-4i}{25} = \frac{3}{25} - \frac{4}{25}i$

(e) $\frac{5-i}{2-3i} = \frac{(5-i)(2+3i)}{(2-3i)(2+3i)} = \frac{13+13i}{13} = 1 + i$

(f) $i^3 = i^2 \cdot i = -i = 0 - i$

2. Prove: The mapping $z \leftrightarrow \bar{z}$, $z \in C$ is an isomorphism of C onto C.

We are to show that addition and multiplication are preserved under the mapping. This follows since for $z_1 = x_1 + y_1 i$, $z_2 = x_2 + y_2 i \in C$,

$$\begin{aligned} \overline{z_1 + z_2} &= \overline{(x_1 + y_1 i) + (x_2 + y_2 i)} = \overline{(x_1 + x_2) + (y_1 + y_2)i} \\ &= (x_1 + x_2) - (y_1 + y_2)i = (x_1 - y_1 i) + (x_2 - y_2 i) = \bar{z}_1 + \bar{z}_2 \end{aligned}$$

and

$$\begin{aligned} \overline{z_1 \cdot z_2} &= \overline{(x_1x_2 - y_1y_2) + (x_1y_2 + x_2y_1)i} = (x_1x_2 - y_1y_2) - (x_1y_2 + x_2y_1)i \\ &= (x_1 - y_1 i) \cdot (x_2 - y_2 i) = \bar{z}_1 \cdot \bar{z}_2 \end{aligned}$$

3. Prove: The absolute value of the product of two complex numbers is the product of their absolute values, and the angle of the product is the sum of their angles.

Let $z_1 = r_1(\cos\theta_1 + i\sin\theta_1)$ and $z_2 = r_2(\cos\theta_2 + i\sin\theta_2)$. Then

$$\begin{aligned} z_1 \cdot z_2 &= r_1r_2[(\cos\theta_1 \cdot \cos\theta_2 - \sin\theta_1 \cdot \sin\theta_2) + i(\sin\theta_1 \cdot \cos\theta_2 + \sin\theta_2 \cdot \cos\theta_1)] \\ &= r_1r_2[\cos(\theta_1 + \theta_2) + i\sin(\theta_1 + \theta_2)] \end{aligned}$$

4. Prove: The absolute value of the quotient of two complex numbers is the quotient of their absolute values, and the angle of the quotient is the angle of the numerator minus the angle of the denominator.

For the complex numbers z_1 and z_2 of Problem 3,

$$\begin{aligned} \frac{z_1}{z_2} &= \frac{r_1(\cos\theta_1 + i\sin\theta_1)}{r_2(\cos\theta_2 + i\sin\theta_2)} = \frac{r_1(\cos\theta_1 + i\sin\theta_1)(\cos\theta_2 - i\sin\theta_2)}{r_2(\cos\theta_2 + i\sin\theta_2)(\cos\theta_2 - i\sin\theta_2)} \\ &= \frac{r_1}{r_2}[(\cos\theta_1\cos\theta_2 + \sin\theta_1\sin\theta_2) + i(\sin\theta_1\cos\theta_2 - \sin\theta_2\cos\theta_1)] \\ &= \frac{r_1}{r_2}[\cos(\theta_1 - \theta_2) + i\sin(\theta_1 - \theta_2)] \end{aligned}$$

5. Find the 6 sixth roots of 1 and show that they include both the square roots and the cube roots of 1.

The sixth roots of 1 are

$$\rho = \cos\pi/3 + i\sin\pi/3 = \tfrac{1}{2} + \tfrac{1}{2}\sqrt{3}\,i \qquad \rho^4 = \cos 4\pi/3 + i\sin 4\pi/3 = -\tfrac{1}{2} - \tfrac{1}{2}\sqrt{3}\,i$$

$$\rho^2 = \cos 2\pi/3 + i\sin 2\pi/3 = -\tfrac{1}{2} + \tfrac{1}{2}\sqrt{3}\,i \qquad \rho^5 = \cos 5\pi/3 + i\sin 5\pi/3 = \tfrac{1}{2} - \tfrac{1}{2}\sqrt{3}\,i$$

$$\rho^3 = \cos\pi + i\sin\pi = -1 \qquad \rho^6 = \cos 2\pi + i\sin 2\pi = 1$$

Of these, $\rho^3 = -1$ and $\rho^6 = 1$ are square roots of 1 while $\rho^2 = -\frac{1}{2} + \frac{1}{2}\sqrt{3}\,i$, $\rho^4 = -\frac{1}{2} - \frac{1}{2}\sqrt{3}\,i$, and $\rho^6 = 1$ are cube roots of 1.

6. Prove: Let $\rho = \cos 2\pi/n + i \sin 2\pi/n$. If $(m, n) = d > 1$, then ρ^m is an n/dth root of 1.

Let $m = m_1 d$ and $n = n_1 d$. Since $\rho^m = \cos 2m\pi/n + i \sin 2m\pi/n = \cos 2m_1\pi/n_1 + i \sin 2m_1\pi/n_1$ and $(\rho^m)^{n_1} = \cos 2m_1\pi + i \sin 2m_1\pi = 1$, it follows that ρ^m is an $n_1 = n/d$th root of 1.

Supplementary Problems

7. Express each of the following in the form $x + yi$: (a) $2 + \sqrt{-5}$, (b) $(4 + \sqrt{-5}) + (3 - 2\sqrt{-5})$, (c) $(4 + \sqrt{-5}) - (3 - 2\sqrt{-5})$, (d) $(3 + 4i) \cdot (4 - 5i)$, (e) $\dfrac{1}{2 - 3i}$, (f) $\dfrac{2 + 3i}{5 - 2i}$, (g) $\dfrac{5 - 2i}{2 + 3i}$, (h) i^4, (i) i^5, (j) i^6, (k) i^8.

Ans. (a) $2 + \sqrt{5}\,i$, (b) $7 - \sqrt{5}\,i$, (c) $1 + 3\sqrt{5}\,i$, (d) $32 + i$, (e) $2/13 + 3i/13$, (f) $4/29 + 19i/29$, (g) $4/13 - 19i/13$, (h) $1 + 0 \cdot i$, (i) $0 + i$, (j) $-1 + 0 \cdot i$, (k) $1 + 0 \cdot i$

8. Write the conjugate of each of the following: (a) $2 + 3i$, (b) $2 - 3i$, (c) 5, (d) $2i$.

Ans. (a) $2 - 3i$, (b) $2 + 3i$, (c) 5, (d) $-2i$

9. Prove: The conjugate of the conjugate of z is z itself.

10. Prove: For every $z \neq 0 \in C$, $\overline{(z^{-1})} = (\bar{z})^{-1}$.

11. Locate all points whose coordinates have the form (a) $(a + 0 \cdot i)$, (b) $(0 + bi)$, where $a, b \in R$. Show that any point z and its conjugate are located symmetrically with respect to the x-axis.

12. Express each of the following in trigonometric form:

(a) 5 (c) $-1 - \sqrt{3}\,i$ (e) -6 (g) $-3 + \sqrt{3}\,i$ (i) $1/i$

(b) $4 - 4i$ (d) $-3i$ (f) $\sqrt{2} + \sqrt{2}\,i$ (h) $-1/(1 + i)$

Ans. (a) 5 cis 0 (d) 3 cis $3\pi/2$ (g) $2\sqrt{3}$ cis $5\pi/6$

(b) $4\sqrt{2}$ cis $7\pi/4$ (e) 6 cis π (h) $\sqrt{2}/2$ cis $3\pi/4$

(c) 2 cis $4\pi/3$ (f) 2 cis $\pi/4$ (i) cis $3\pi/2$

where $\text{cis}\,\theta = \cos\theta + i \sin\theta$.

13. Express each of the following in the form $a + bi$:

(a) 5 cis 60° (e) (2 cis 25°) · (3 cis 335°)

(b) 2 cis 90° (f) (10 cis 100°) · (cis 140°)

(c) cis 150° (g) (6 cis 170°) ÷ (3 cis 50°)

(d) 2 cis 210° (h) (4 cis 20°) ÷ (8 cis 80°)

Ans. (a) $5/2 + 5\sqrt{3}\,i/2$ (c) $-\frac{1}{2}\sqrt{3} + \frac{1}{2}i$ (e) 6 (g) $-1 + \sqrt{3}\,i$

(b) $2i$ (d) $-\sqrt{3} - i$ (f) $-5 - 5\sqrt{3}\,i$ (h) $\frac{1}{4} - \frac{1}{4}\sqrt{3}\,i$

14. Find the cube roots of: (a) 1, (b) 8, (c) $27i$, (d) $-8i$, (e) $-4\sqrt{3} - 4i$.

Ans. (a) $-\frac{1}{2} \pm \frac{1}{2}\sqrt{3}\,i$, 1; (b) $-1 \pm \sqrt{3}\,i$, 2; (c) $\pm\dfrac{3\sqrt{3}}{2} + \dfrac{3}{2}i$, $-3i$; (d) $2i$, $\pm\sqrt{3} - i$; (e) 2 cis $7\pi/18$, 2 cis $19\pi/18$, 2 cis $31\pi/18$

15. Find (a) the primitive fifth roots of 1, (b) the primitive eighth roots of 1.

16. Prove: The sum of the n distinct nth roots of 1 is zero.

17. Use Fig. 8-2 to prove:

(a) $|z_1 + z_2| \leqq |z_1| + |z_2|$

(b) $|z_1 - z_2| \geqq |z_1| - |z_2|$

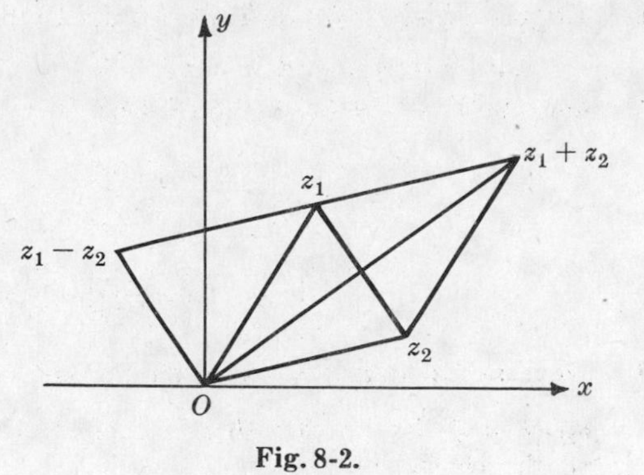

Fig. 8-2.

18. If r is any cube root of $a \in C$, then $r, \omega r, \omega^2 r$, where $\omega = -\frac{1}{2} + \frac{1}{2}\sqrt{3}\,i$ and ω^2 are the imaginary cube roots of 1, are the three cube roots of a.

19. Describe geometrically the mappings

(a) $z \to \bar{z}$ (b) $z \to zi$ (c) $z \to \bar{z}i$

20. In the real plane let K be the circle with center at 0 and radius 1 and let $A_1A_2A_3$, where $A_j(x_j, y_j) = A_j(z_j) = A_j(x_j + y_j i)$, $j = 1, 2, 3$, be an arbitrary inscribed triangle. Denote by $P(z) = P(x + yi)$ an arbitrary (variable) point in the plane.

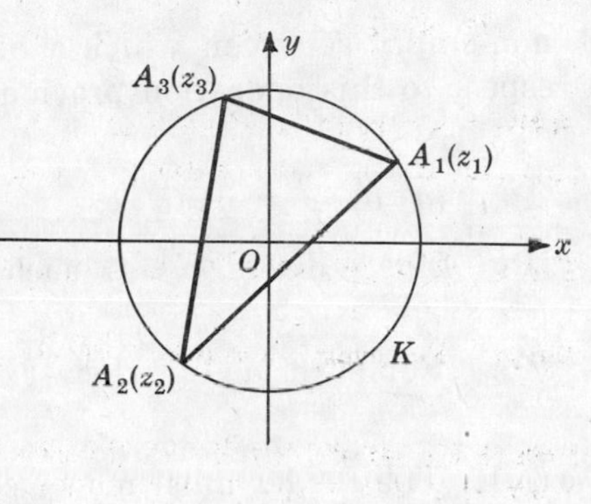

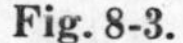
Fig. 8-3.

(a) Show that the equation of K is $z \cdot \bar{z} = 1$.

(b) Show that $P_r(x_r, y_r)$, where $x_r = \dfrac{ax_j + bx_k}{a + b}$ and $y_r = \dfrac{ay_j + by_k}{a + b}$, divides the line segment A_jA_k in the ratio $b:a$. Then, since A_j, A_k and $P_r\left(\dfrac{az_j + bz_k}{a + b}\right)$ lie on the line A_jA_k, verify that its equation is $z + z_jz_k\bar{z} = z_j + z_k$.

(c) Verify: The equation of any line parallel to A_jA_k has the form $z + z_jz_k\bar{z} = \eta$ by showing that the midpoints B_j and B_k of A_iA_j and A_iA_k lie on the line $z + z_jz_k\bar{z} = \frac{1}{2}(z_i + z_j + z_k + \bar{z}_iz_jz_k)$.

(d) Verify: The equation of any line perpendicular to A_jA_k has the form $z - z_jz_k\bar{z} = \mu$ by showing that 0 and the midpoint of A_jA_k lie on the line $z - z_jz_k\bar{z} = 0$.

(e) Use $z = z_i$ in $z - z_jz_k\bar{z} = \mu$ to obtain the equation $z - z_jz_k\bar{z} = z_i - \bar{z}_iz_jz_k$ of the altitude of $A_1A_2A_3$ through A_i. Then eliminate $\bar{z}$ between the equations of any two altitudes to obtain their common point $H(z_1 + z_2 + z_3)$. Show that H also lies on the third altitude.

Chapter 9

Groups

GROUPS

A non-empty set $\mathcal{G}$ on which a binary operation $\circ$ is defined is said to form a group with respect to this operation provided, for arbitrary $a, b, c \in \mathcal{G}$, the following properties hold:

P₁: $(a \circ b) \circ c = a \circ (b \circ c)$ (associative law)

P₂: There exists $u \in \mathcal{G}$ such that $a \circ u = u \circ a = a$ (existence of identity element)

P₃: For each $a \in \mathcal{G}$ there exists $a^{-1} \in \mathcal{G}$ such that $a \circ a^{-1} = a^{-1} \circ a = u$ (existence of inverses)

Note 1. The reader must not be confused by the use in **P₃** of a^{-1} to denote the inverse of a under the operation $\circ$. The notation is merely borrowed from that previously used in connection with multiplication. Whenever the group operation is addition, a^{-1} is to be interpreted as the additive inverse $-a$.

Note 2. The preceding chapters contain many examples of groups for most of which the group operation is commutative. We therefore call attention here to the fact that the commutative law is not one of the requisite properties listed above. A group is called *abelian* or *non-abelian* according as the group operation is or is not commutative. For the present, however, we shall not make this distinction.

Example 1: (*a*) The set I of all integers forms a group with respect to addition; the identity element is 0 and the inverse of $a \in I$ is $-a$. Thus, we may hereafter speak of the additive group I. On the other hand, I is not a multiplicative group since, for example, neither 0 nor 2 has a multiplicative inverse.

(*b*) The set A of Example 12(*c*), Chapter 2, forms a group with respect to $\circ$. The identity element is a. Find the inverse of each element.

(*c*) The set $A = \{-3, -2, -1, 0, 1, 2, 3\}$ is not a group with respect to addition on I although 0 is the identity element, each element of A has an inverse, and addition is associative. The reason is, of course, that addition is not a binary operation on A, that is, the set A is not closed with respect to addition.

(*d*) The set $A = \{\omega_1 = -\frac{1}{2} + \frac{1}{2}\sqrt{3}\,i,\ \omega_2 = -\frac{1}{2} - \frac{1}{2}\sqrt{3}\,i,\ \omega_3 = 1\}$ of the cube roots of 1 forms a group with respect to multiplication on the set of complex numbers C since (i) the product of any two elements of the set is an element of the set, (ii) the associative law holds in C and, hence, in A, (iii) ω_3 is the identity element, and (iv) the inverses of $\omega_1, \omega_2, \omega_3$ are $\omega_2, \omega_1, \omega_3$ respectively.

This is also evident from (ii) and the adjacent operation table.

$\cdot$	ω_1	ω_2	ω_3
ω_1	ω_2	ω_3	ω_1
ω_2	ω_3	ω_1	ω_2
ω_3	ω_1	ω_2	ω_3

Table 9-1

See also Problems 1-2.

SIMPLE PROPERTIES OF GROUPS

The uniqueness of the identity element and of the inverses of the elements of a group were established in Theorems III and IV, Chapter 2, Page 20. There follow readily

Theorem I (Cancellation Law). If $a, b, c \in \mathcal{G}$, then $a \circ b = a \circ c$, (also, $b \circ a = c \circ a$), implies $b = c$.

For a proof, see Problem 3.

Theorem II. For $a, b \in \mathcal{G}$, each of the equations $a \circ x = b$ and $y \circ a = b$ has a unique solution.

For a proof, see Problem 4.

Theorem III. For every $a \in \mathcal{G}$, the inverse of the inverse of a is a, i.e., $(a^{-1})^{-1} = a$.

Theorem IV. For every $a, b \in \mathcal{G}$, $(a \circ b)^{-1} = b^{-1} \circ a^{-1}$.

Theorem V. For every $a, b, \ldots, p, q \in \mathcal{G}$, $(a \circ b \circ \cdots \circ p \circ q)^{-1} = q^{-1} \circ p^{-1} \circ \cdots \circ b^{-1} \circ a^{-1}$.

For any $a \in \mathcal{G}$ and $m \in I^+$, we define

$$a^m = a \circ a \circ a \circ \cdots \circ a \qquad \text{to } m \text{ factors}$$

$$a^0 = u, \text{ the identity element of } \mathcal{G}$$

$$a^{-m} = (a^{-1})^m = a^{-1} \circ a^{-1} \circ a^{-1} \circ \cdots \circ a^{-1} \qquad \text{to } m \text{ factors}$$

Once again the notation has been borrowed from that used when multiplication is the operation. Whenever the operation is addition, a^n when $n > 0$ is to be interpreted as $na = a + a + a + \cdots + a$ to n terms, a^0 as u, and a^{-n} as $n(-a) = -a + (-a) + (-a) + \cdots + (-a)$ to n terms. Note that na is also shorthand and is not to be considered as the product of $n \in I$ and $a \in \mathcal{G}$.

In Problem 5 we prove the first part of

Theorem VI. For any $a \in \mathcal{G}$, (i) $a^m \circ a^n = a^{m+n}$ and (ii) $(a^m)^n = a^{mn}$, where $m, n \in I$.

By the *order of a group* $\mathcal{G}$ is meant the number of elements in the set $\mathcal{G}$. The additive group I of Example 1(*a*) is of infinite order; the groups of Example 1(*b*) and 1(*d*) are finite groups of orders 5 and 3 respectively.

By the *order of an element* $a \in \mathcal{G}$ is meant the least positive integer n, if one exists, for which $a^n = u$, the identity element of $\mathcal{G}$. If $a \neq 0$ is an element of the additive group I, then $na \neq 0$ for all $n > 0$ and a is defined to be of *infinite* order. The element ω_1 of Example 1(*d*) is of order 3 since ω_1 and ω_1^2 are different from 1 while $\omega_1^3 = 1$, the identity element.

SUBGROUPS

Let $\mathcal{G} = \{a, b, c, \ldots\}$ be a group with respect to $\circ$. Any non-empty subset $\mathcal{G}'$ of $\mathcal{G}$ is called a *subgroup* of $\mathcal{G}$ if $\mathcal{G}'$ is itself a group with respect to $\circ$. Clearly $\mathcal{G}' = \{u\}$, where u is the identity element of $\mathcal{G}$, and $\mathcal{G}$ itself are subgroups of any group $\mathcal{G}$. They will be called *improper* subgroups; other subgroups of $\mathcal{G}$, if any, will be called *proper*. We note in passing that every subgroup of a group $\mathcal{G}$ contains u as its identity element.

Example 2: (*a*) A proper subgroup of the multiplicative group $\mathcal{G} = \{1, -1, i, -i\}$ is $\mathcal{G}' = \{1, -1\}$. (Are there others?)

(*b*) Consider the multiplicative group $\mathcal{G} = \{\rho, \rho^2, \rho^3, \rho^4, \rho^5, \rho^6 = 1\}$ of the sixth roots of unity (see Problem 5, Chapter 8, Page 79). It has $\mathcal{G}' = \{\rho^3, \rho^6\}$ and $\mathcal{G}'' = \{\rho^2, \rho^4, \rho^6\}$ as proper subgroups.

The next two theorems are useful in determining whether a subset of a group $\mathcal{G}$ with group operation $\circ$ is a subgroup.

Theorem VII. A non-empty subset $\mathcal{G}'$ of a group $\mathcal{G}$ is a subgroup of $\mathcal{G}$ if and only if (i) $\mathcal{G}'$ is closed with respect to $\circ$, (ii) $\mathcal{G}'$ contains the inverse of each of its elements.

Theorem VIII. A non-empty subset $\mathcal{G}'$ of a group $\mathcal{G}$ is a subgroup of $\mathcal{G}$ if and only if for all $a, b \in \mathcal{G}'$, $a^{-1} \circ b \in \mathcal{G}'$.

For a proof, see Problem 6.

There follow

Theorem IX. Let a be an element of a group $\mathcal{G}$. The set $\mathcal{G}' = \{a^n : n \in I\}$ of all integral powers of a is a subgroup of $\mathcal{G}$.

Theorem X. If S is any set of subgroups of a group $\mathcal{G}$, the intersection of these subgroups is also a subgroup of $\mathcal{G}$.

For a proof, see Problem 7.

CYCLIC GROUPS

A group $\mathcal{G}$ is called *cyclic* if, for some $a \in \mathcal{G}$, every $x \in \mathcal{G}$ is of the form a^m, where $m \in I$. The element a is then called a *generator* of $\mathcal{G}$. Clearly, every cyclic group is abelian.

Example 3: (*a*) The additive group I is cyclic with generator $a = 1$ since, for every $m \in I$, $a^m = ma = m$.

(*b*) The multiplicative group of fifth roots of 1 is cyclic. Unlike the group of (*a*) which has only 1 and -1 as generators, this group may be generated by any of its elements except 1.

(*c*) The group $\mathcal{G}$ of Example 2(*b*) is cyclic. Its generators are ρ and ρ^5.

Examples 3(*b*) and 3(*c*) illustrate

Theorem XI. An element a^t of a finite cyclic group $\mathcal{G}$ of order n is a generator of $\mathcal{G}$ if and only if $(n, t) = 1$.

In Problem 8, we prove

Theorem XII. Every subgroup of a cyclic group is itself a cyclic group.

PERMUTATION GROUPS

The set S_n of the $n!$ permutations of n symbols was considered in Chapter 2. A review of this material shows that S_n is a group with respect to the permutation operation $\circ$. Since $\circ$ is not commutative, this is our first example of a non-abelian group.

It is customary to call the group S_n the *symmetric group* on n symbols and to call any subgroup of S_n a *permutation group* on n symbols.

Example 4: (*a*) $S_4 = \{(1), (12), (13), (14), (23), (24), (34), \alpha = (123), \alpha^2 = (132), \beta = (124), \beta^2 = (142), \gamma = (134), \gamma^2 = (143), \delta = (234), \delta^2 = (243), \rho = (1234), \rho^2 = (13)(24), \rho^3 = (1432), \sigma = (1243), \sigma^2 = (14)(23), \sigma^3 = (1342), \tau = (1324), \tau^2 = (12)(34), \tau^3 = (1423)\}$.

(*b*) The subgroups of S_4: (i) $\{(1), (12)\}$, (ii) $\{(1), \alpha, \alpha^2\}$, (iii) $\{(1), (12), (34), (12)(34)\}$, and (iv) $A_4 = \{(1), \alpha, \alpha^2\ \beta, \beta^2, \gamma, \gamma^2, \delta, \delta^2, \rho^2, \sigma^2, \tau^2\}$ are examples of permutation groups on 4 symbols. (A_4 consists of all even permutations in S_4 and is known as the *alternating group* on 4 symbols.) Which of the above subgroups are cyclic? which abelian? List other subgroups of S_4.

See Problems 9-10.

HOMOMORPHISMS

Let $\mathcal{G}$, with operation $\circ$, and $\mathcal{G}'$ with operation $\square$, be two groups. By a *homomorphism* of $\mathcal{G}$ into $\mathcal{G}'$ is meant a mapping

$$\mathcal{G} \rightarrow \mathcal{G}' : g \rightarrow g'$$

such that

(i) every $g \in \mathcal{G}$ has a unique image $g' \in \mathcal{G}'$

(ii) if $a \to a'$ and $b \to b'$, then $a \circ b \to a' \square b'$

If, in addition, the mapping satisfies

(iii) every $g' \in \mathcal{G}'$ is an image

we have a homomorphism of $\mathcal{G}$ *onto* $\mathcal{G}'$ and we then call $\mathcal{G}'$ a *homomorphic image* of $\mathcal{G}$.

Example 5: (*a*) Consider the mapping $n \to i^n$ of the additive group I onto the multiplicative group of the fourth roots of 1. This is a homomorphism since

$$m + n \to i^{m+n} = i^m \cdot i^n$$

and the group operations are preserved.

(*b*) Consider the cyclic group $\mathcal{G} = \{a, a^2, a^3, \ldots, a^{12} = u\}$ and its subgroup $\mathcal{G}' = \{a^2, a^4, a^6, \ldots, a^{12}\}$. It follows readily that the mapping

$$a^n \to a^{2n}$$

is a homomorphism of $\mathcal{G}$ onto $\mathcal{G}'$ while the mapping

$$a^n \to a^n$$

is a homomorphism of $\mathcal{G}'$ into $\mathcal{G}$.

See Problem 11.

In Problem 12 we prove

Theorem XIII. In any homomorphism between two groups $\mathcal{G}$ and $\mathcal{G}'$, their identity elements correspond; and if $x \in \mathcal{G}$ and $x' \in \mathcal{G}'$ correspond, so also do their inverses.

There follows

Theorem XIV. The homomorphic image of any cyclic group is cyclic.

ISOMORPHISMS

If the mapping of the section above is one-to-one, i.e.,

$$g \leftrightarrow g'$$

such that (iii) is satisfied, we say that $\mathcal{G}$ and $\mathcal{G}'$ are *isomorphic* and call the mapping an *isomorphism*.

Example 6: Show that $\mathcal{G}$, the additive group $I/(4)$, is isomorphic to $\mathcal{G}'$, the multiplicative group of the non-zero elements of $I/(5)$.

Consider the operation tables

$\mathcal{G}$

+	0	1	2	3
0	0	1	2	3
1	1	2	3	0
2	2	3	0	1
3	3	0	1	2

Table 9-2

$\mathcal{G}'$

·	1	3	4	2
1	1	3	4	2
3	3	4	2	1
4	4	2	1	3
2	2	1	3	4

Table 9-3

in which, for convenience, [0], [1], ... have been replaced by 0, 1, ... It follows readily that the mapping

$$\mathcal{G} \to \mathcal{G}' : 0 \to 1,\ 1 \to 3,\ 2 \to 4,\ 3 \to 2$$

is an isomorphism. For example, $1 = 2 + 3 \to 4 \cdot 2 = 3$, etc.

Rewrite the operation table for $\mathcal{G}'$ to show that

$$\mathcal{G} \to \mathcal{G}' : 0 \to 1,\ 1 \to 2,\ 2 \to 4,\ 3 \to 3$$

is another isomorphism of $\mathcal{G}$ onto $\mathcal{G}'$. Can you find still another?

In Problem 13 we prove the first part of

Theorem XV. (*a*) Every cyclic group of infinite order is isomorphic to the additive group I.

(*b*) Every cyclic group of finite order n is isomorphic to the additive group $I/(n)$.

The most remarkable result of this section is

Theorem XVI (Cayley). Every finite group of order n is isomorphic to a permutation group on n symbols.

Since the proof, given in Problem 14, consists in showing how to construct an effective permutation group, the reader may wish to examine first

Example 7: Consider the adjacent operation table for the group $\mathcal{G} = \{g_1, g_2, g_3, g_4, g_5, g_6\}$ with group operation $\square$.

The elements of any column of the table, say the fifth: $g_1 \square g_5 = g_5$, $g_2 \square g_5 = g_3$, $g_3 \square g_5 = g_4$, $g_4 \square g_5 = g_2$, $g_5 \square g_5 = g_6$, $g_6 \square g_5 = g_1$ are the elements of the row labels (i.e., the elements of $\mathcal{G}$) rearranged. This permutation will be indicated by

$$\begin{pmatrix} g_1 & g_2 & g_3 & g_4 & g_5 & g_6 \\ g_5 & g_3 & g_4 & g_2 & g_6 & g_1 \end{pmatrix} = (156)(234) = p_5$$

$\square$	g_1	g_2	g_3	g_4	g_5	g_6
g_1	g_1	g_2	g_3	g_4	g_5	g_6
g_2	g_2	g_1	g_5	g_6	g_3	g_4
g_3	g_3	g_6	g_1	g_5	g_4	g_2
g_4	g_4	g_5	g_6	g_1	g_2	g_3
g_5	g_5	g_4	g_2	g_3	g_6	g_1
g_6	g_6	g_3	g_4	g_2	g_1	g_5

Table 9-4

It follows readily that $\mathcal{G}$ is isomorphic to $P = \{p_1, p_2, p_3, p_4, p_5, p_6\}$ where $p_1 = (1)$, $p_2 = (12)(36)(45)$, $p_3 = (13)(25)(46)$, $p_4 = (14)(26)(35)$, $p_5 = (156)(234)$, $p_6 = (165)(243)$ under the mapping

$$g_i \leftrightarrow p_i \qquad (i = 1, 2, 3, \ldots, 6)$$

COSETS

Let $\mathcal{G}$ be a finite group with group operation $\circ$, H be a subgroup of $\mathcal{G}$, and a be an arbitrary element of $\mathcal{G}$. We define as the *right coset* Ha of H in $\mathcal{G}$, generated by a, the subset of $\mathcal{G}$

$$Ha = \{h \circ a : h \in H\}$$

and as the *left coset* aH of H in $\mathcal{G}$, generated by a, the subset of $\mathcal{G}$

$$aH = \{a \circ h : h \in H\}$$

Example 8: The subgroup $H = \{(1), (12), (34), (12)(34)\}$ and the element $a = (1432)$ of S_4 generate the right coset

$$\begin{aligned} Ha &= \{(1) \circ (1432),\ (12) \circ (1432),\ (34) \circ (1432),\ (12)(34) \circ (1432)\} \\ &= \{(1432),\ (243),\ (142),\ (24)\} \end{aligned}$$

and the left coset

$$\begin{aligned} aH &= \{(1432) \circ (1),\ (1432) \circ (12),\ (1432) \circ (34),\ (1432) \circ (12)(34)\} \\ &= \{(1432),\ (143),\ (132),\ (13)\} \end{aligned}$$

In investigating the properties of cosets, we shall usually limit our attention to right cosets and leave for the reader the formulation of the corresponding properties of left cosets. First, note that $a \in Ha$ since u, the identity element of $\mathcal{G}$, is also the identity element of H. If H contains m elements so also does Ha, for Ha contains at most m elements and $h_1 \circ a = h_2 \circ a$ for any $h_1, h_2 \in H$ implies $h_1 = h_2$. Finally, if C_r denotes the set of all distinct right cosets of H in $\mathcal{G}$, then $H \in C_r$ since $Ha = H$ when $a \in H$.

Consider now two right cosets Ha and Hb, $a \neq b$, of H in $\mathcal{G}$. Suppose that c is a common element of these cosets so that for some $h_1, h_2 \in H$ we have $c = h_1 \circ a = h_2 \circ b$. Then $a = h_1^{-1} \circ (h_2 \circ b) = (h_1^{-1} \circ h_2) \circ b$ and, since $h_1^{-1} \circ h_2 \in H$ (Theorem VIII), it follows that $a \in Hb$ and $Ha = Hb$. Thus C_r consists of mutually disjoint right cosets of $\mathcal{G}$ and so is a partition of $\mathcal{G}$. We shall call C_r a *decomposition* of $\mathcal{G}$ into right cosets with respect to H.

Example 9: (*a*) Take $\mathcal{G}$ as the additive group of integers and H as the subgroup of all integers divisible by 5. The decomposition of $\mathcal{G}$ into right cosets with respect to H consists of the five residue classes modulo 5, i.e., $H = \{x : 5 \mid x\}$, $H1 = \{x : 5 \mid (x-1)\}$, $H2 = \{x : 5 \mid (x-2)\}$, $H3 = \{x : 5 \mid (x-3)\}$, and $H4 = \{x : 5 \mid (x-4)\}$. There is no distinction here between right and left cosets since $\mathcal{G}$ is an abelian group.

(*b*) Let $\mathcal{G} = S_4$ and $H = A_4$, the subgroup of all even permutations of S_4. Then there are only two right (left) cosets of $\mathcal{G}$ generated by H, namely, A_4 and the subset of odd permutations of S_4. Here again there is no distinction between right and left cosets but note that S_4 is not abelian.

(*c*) Let $\mathcal{G} = S_4$ and H be the octic group of Problem 9. The distinct right cosets generated by H are

$$H = \{(1), (1234), (13)(24), (1432), (12)(34), (14)(23), (13), (24)\}$$
$$H(12) = \{(12), (234), (1324), (143), (34), (1423), (132), (124)\}$$
$$H(23) = \{(23), (134), (1243), (142), (1342), (14), (123), (243)\}$$

and the distinct left cosets generated by H are

$$H = \{(1), (1234), (13)(24), (1432), (12)(34), (14)(23), (13), (24)\}$$
$$(12)H = \{(12), (134), (1423), (243), (34), (1324), (123), (142)\}$$
$$(23)H = \{(23), (124), (1342), (143), (1243), (14), (132), (234)\}$$

Thus, $\mathcal{G} = H \cup H(12) \cup H(23) = H \cup (12)H \cup (23)H$. Here, the decomposition of $\mathcal{G}$ obtained by using the right and left cosets of H are distinct.

Let $\mathcal{G}$ be a finite group of order n and H be a subgroup of order m of $\mathcal{G}$. The number of distinct right cosets of H in $\mathcal{G}$ (called the *index* of H in $\mathcal{G}$) is r where $n = mr$; hence,

Theorem XVII (Lagrange). The order of each subgroup of a finite group $\mathcal{G}$ is a divisor of the order of $\mathcal{G}$.

As consequences, we have

Theorem XVIII. If $\mathcal{G}$ is a finite group of order n, then the order of any element $a \in \mathcal{G}$ (i.e., the order of the cyclic subgroup generated by a) is a divisor of n.

and

Theorem XIX. Every group of prime order is cyclic.

INVARIANT SUBGROUPS

A subgroup H of a group $\mathcal{G}$ is called an *invariant subgroup* (*normal subgroup* or *normal divisor*) of $\mathcal{G}$ if

(i) $$gH = Hg \qquad \text{for every } g \in \mathcal{G}$$

Since $g^{-1} \in \mathcal{G}$ whenever $g \in \mathcal{G}$, (i) may be replaced by

(i′) $$g^{-1}Hg = H \qquad \text{for every } g \in \mathcal{G}$$

Now (i′) requires

(i'_1) for any $g \in \mathcal{G}$ and any $h \in H$, then $g^{-1} \circ h \circ g \in H$

and

(i'_2) for any $g \in \mathcal{G}$ and each $h \in H$, there exists some $k \in H$ such that $g^{-1} \circ k \circ g = h$ or $k \circ g = g \circ h$.

We shall show that (i'_1) implies (i'_2). Consider any $h \in H$. By (i'_1), $(g^{-1})^{-1} \circ h \circ g^{-1} = g \circ h \circ g^{-1} = k \in H$ since $g^{-1} \in \mathcal{G}$; then $g^{-1} \circ k \circ g = h$ as required. We have proved

Theorem XX. If H is a subgroup of a group $\mathcal{G}$ and if $g^{-1} \circ h \circ g \in H$ for all $g \in \mathcal{G}$ and all $h \in H$, then H is an invariant subgroup of $\mathcal{G}$.

Example 10: (a) Every subgroup H of an abelian group $\mathcal{G}$ is an invariant subgroup of $\mathcal{G}$ since $g \circ h = h \circ g$, for any $g \in \mathcal{G}$ and every $h \in H$.

(b) Every group $\mathcal{G}$ has at least two invariant subgroups: $\{u\}$, since $u \circ g = g \circ u$ for every $g \in \mathcal{G}$, and $\mathcal{G}$ itself, since for any $g, h \in \mathcal{G}$ we have

$$g \circ h = g \circ h \circ (g^{-1} \circ g) = (g \circ h \circ g^{-1}) \circ g = k \circ g \quad \text{and} \quad k = g \circ h \circ g^{-1} \in \mathcal{G}$$

(c) If H is a subgroup of index 2 of $\mathcal{G}$ [see Example 9(b)] the cosets generated by H consist of H and $\mathcal{G} - H$. Hence, H is an invariant subgroup of $\mathcal{G}$.

(d) For $\mathcal{G} = \{a, a^2, a^3, \ldots, a^{12} = u\}$, its subgroups $\{u, a^2, a^4, \ldots, a^{10}\}$, $\{u, a^3, a^6, a^9\}$, $\{u, a^4, a^8\}$, and $\{u, a^6\}$ are invariant.

(e) For the octic group (Problem 9), $\{u, \rho^2, \sigma^2, \tau^2\}$, $\{u, \rho^2, b, e\}$ and $\{u, \rho, \rho^2, \rho^3\}$ are invariant subgroups of order 4 while $\{u, \rho^2\}$ is an invariant subgroup of order 2. (Use Table 9-7 to check this.)

(f) The octic group P is not an invariant subgroup of S_4 since for $\rho = (1234) \in P$ and $(12) \in S_4$, we have $(12)^{-1} \rho\, (12) = (1342) \notin P$.

In Problem 15, we prove

Theorem XXI. Under any homomorphism of a group $\mathcal{G}$ with group operation $\circ$ and identity element u into a group $\mathcal{G}'$ with group operation $\square$ and identity element u', the subset S of all elements of $\mathcal{G}$ which are mapped onto u' is an invariant subgroup of $\mathcal{G}$.

The invariant subgroup of $\mathcal{G}$ defined in Theorem XXI is called the *kernel* of the homomorphism.

In Example 10(b) it was shown that any group $\mathcal{G}$ has $\{u\}$ and $\mathcal{G}$ itself as invariant subgroups. They are called *improper* while other invariant subgroups, if any, of $\mathcal{G}$ are called *proper*. A group $\mathcal{G}$ having no proper invariant subgroups is called *simple*.

QUOTIENT GROUPS

Let H be an invariant subgroup of a group $\mathcal{G}$ with group operation $\circ$ and denote by $\mathcal{G}/H$ the set of (distinct) cosets of H in $\mathcal{G}$, i.e.,

$$\mathcal{G}/H = \{Ha, Hb, Hc, \ldots\}$$

We define the "product" of pairs of these cosets by

$$(Ha)(Hb) = \{(h_1 \circ a) \circ (h_2 \circ b) : h_1, h_2 \in H\} \qquad \text{for all } Ha, Hb \in \mathcal{G}/H$$

and, in Problem 16, prove that this operation is well-defined.

Now $\mathcal{G}/H$ is a group with respect to the operation just defined. To prove this, we note first that

$$\begin{aligned}(h_1 \circ a) \circ (h_2 \circ b) &= h_1 \circ (a \circ h_2) \circ b = h_1 \circ (h_3 \circ a) \circ b \\ &= (h_1 \circ h_3) \circ (a \circ b) = h_4 \circ (a \circ b), \qquad h_3, h_4 \in H\end{aligned}$$

Then
$$(Ha)(Hb) = H(a \circ b) \in \mathcal{G}/H$$

and
$$[(Ha)(Hb)](Hc) = H[(a \circ b) \circ c] = H[a \circ (b \circ c)] = (Ha)[(Hb)(Hc)]$$

Next, for u the identity element of $\mathcal{G}$, $(Hu)(Ha) = (Ha)(Hu) = Ha$ so that $Hu = H$ is the identity element of $\mathcal{G}/H$. Finally, since $(Ha)(Ha^{-1}) = (Ha^{-1})(Ha) = Hu = H$, it follows that $\mathcal{G}/H$ contains the inverse Ha^{-1} of each $Ha \in \mathcal{G}/H$.

The group $\mathcal{G}/H$ is called the *quotient group* (*factor group*) of $\mathcal{G}$ by H.

Example 11: (a) When $\mathcal{G}$ is the octic group of Problem 9 and $H = \{u, \rho^2, b, e\}$, then $\mathcal{G}/H = \{H, H\rho\}$. This representation of $\mathcal{G}/H$ is, of course, not unique. The reader will show that $\mathcal{G}/H = \{H, H\rho^3\} = \{H, H\sigma^2\} = \{H, H\tau^2\} = \{H, H\rho\}$.

(b) For the same $\mathcal{G}$ and $H = \{u, \rho^2\}$, we have

$$\mathcal{G}/H = \{H, H\rho, H\sigma^2, Hb\} = \{H, H\rho^3, H\tau^2, He\}$$

The examples above illustrate

Theorem XXII. If H, of order m, is an invariant subgroup of $\mathcal{G}$, of order n, then the quotient group $\mathcal{G}/H$ is of order n/m.

From $(Ha)(Hb) = H(a \circ b) \in \mathcal{G}/H$, obtained above, there follows

Theorem XXIII. If H is an invariant subgroup of a group $\mathcal{G}$, the mapping

$$\mathcal{G} \to \mathcal{G}/H: \ g \to Hg$$

is a homomorphism of $\mathcal{G}$ onto $\mathcal{G}/H$.

In Problem 17, we prove

Theorem XXIV. Any quotient group of a cyclic group is cyclic.

We leave as an exercise the proof of

Theorem XXV. If H is an invariant subgroup of a group $\mathcal{G}$ and if H is also a subgroup of a subgroup K of $\mathcal{G}$, then H is an invariant subgroup of K.

PRODUCT OF SUBGROUPS

Let $H = \{h_1, h_2, \ldots, h_r\}$ and $K = \{b_1, b_2, \ldots, b_p\}$ be subgroups of a group $\mathcal{G}$ and define the "product"

$$HK = \{h_i \circ b_j : \ h_i \in H, \ b_j \in K\}$$

In Problems 65-67, the reader is asked to examine such products and, in particular, to prove

Theorem XXVI. If H and K are invariant subgroups of a group $\mathcal{G}$, so also is HK.

COMPOSITION SERIES

An invariant subgroup H of a group $\mathcal{G}$ is called *maximal* provided there exists no proper invariant subgroup K of $\mathcal{G}$ having H as a proper subgroup.

Example 12: (a) A_4 of Example 4(b) is a maximal invariant subgroup of S_4 since it is a subgroup of index 2 in S_4. Also, $\{u, \rho^2, \sigma^2, \tau^2\}$ is a maximal invariant subgroup of A_4. (Show this.)

(b) The cyclic group $\mathcal{G} = \{u, a, a^2, \ldots, a^{11}\}$ has $H = \{u, a^2, a^4, \ldots, a^{10}\}$ and $K = \{u, a^3, a^6, a^9\}$ as maximal invariant subgroups. Also, $J = \{u, a^4, a^8\}$ is a maximal invariant subgroup of H while $L = \{u, a^6\}$ is a maximal invariant subgroup of both H and K.

For any group $\mathcal{G}$ a sequence of its subgroups

$$\mathcal{G}, H, J, K, \ldots, U = \{u\}$$

will be called a *composition series for* $\mathcal{G}$ if each element except the first is a maximal invariant subgroup of its predecessor. The groups $\mathcal{G}/H, H/J, J/K, \ldots$ are then called the *quotient groups of the composition series.*

In Problem 18 we prove

Theorem XXVII. Every finite group has at least one composition series.

Example 13: (*a*) The cyclic group $\mathcal{G} = \{u, a, a^2, a^3, a^4\}$ has only one composition series: $\mathcal{G}, U = \{u\}$.

(*b*) A composition series for $\mathcal{G} = S_4$ is

$$S_4,\ A_4,\ \{(1), \rho^2, \sigma^2, \tau^2\},\ \{(1), \rho^2\},\ U \ = \ \{(1)\}$$

Is every element of the composition series an invariant subgroup of $\mathcal{G}$?

(*c*) For the cyclic group of Example 12(*b*), there are three composition series: (i) $\mathcal{G}, H, J, U$, (ii) $\mathcal{G}, K, L, U$, (iii) $\mathcal{G}, H, L, U$. Is every element of each composition series an invariant subgroup of $\mathcal{G}$?

In Problem 19, we illustrate

Theorem XXVIII (The Jordan-Hölder Theorem). For any finite group with distinct composition series, all series are of the same length, i.e., have the same number of elements. Moreover, the quotient groups for any pair of composition series may be put into one-to-one correspondence so that corresponding quotient groups are isomorphic.

Before attempting a proof of Theorem XXVIII (see Problem 23) it will be necessary to examine certain relations which exist between the subgroups of a group $\mathcal{G}$ and the subgroups of its quotient groups. Let then H, of order r, be an invariant subgroup of a group $\mathcal{G}$ of order n and write

$$S \ = \ \mathcal{G}/H \ = \ \{Ha_1, Ha_2, Ha_3, \ldots, Ha_s\}, \qquad a_i \in \mathcal{G} \tag{1}$$

where, for convenience, $a_1 = u$. Further, let

$$P \ = \ \{Hb_1, Hb_2, Hb_3, \cdots, Hb_p\} \tag{2}$$

be any subset of S and denote by

$$K \ = \ Hb_1 \cup Hb_2 \cup Hb_3 \cup \cdots \cup Hb_p \tag{3}$$

the subset of $\mathcal{G}$ whose elements are the pr distinct elements (of $\mathcal{G}$) which belong to the cosets in P.

Suppose now that P is a subgroup of index t of S. Then $n = prt$ and some one of the b's, say b_1, is the identity element u of $\mathcal{G}$. It follows that K is a subgroup of index t of $\mathcal{G}$ and $P = K/H$ since

(i) P is closed with respect to coset multiplication; hence, K is closed with respect to the group operation on $\mathcal{G}$.

(ii) The associative law holds for $\mathcal{G}$ and thus for K.

(iii) $H \in P$; hence, $u \in K$.

(iv) P contains the inverse Hb_i^{-1} of each coset $Hb_i \in P$; hence, K contains the inverse of each of its elements.

(v) K is of order pr; hence K is of index t in $\mathcal{G}$.

Conversely, suppose K is a subgroup of index t of $\mathcal{G}$ which contains H, an invariant subgroup of $\mathcal{G}$. Then, by Theorem XXV, H is an invariant subgroup of K and so $P = K/H$ is of index t in $S = \mathcal{G}/H$.

We have proved

Theorem XXIX. Let H be an invariant subgroup of a finite group $\mathcal{G}$. A set P of the cosets of $S = \mathcal{G}/H$ is a subgroup of index t of S if and only if K, the set of group elements which belong to the cosets in P, is a subgroup of index t of $\mathcal{G}$.

We now assume $b_1 = u$ in (*2*) and (*3*) above and state

Theorem XXX. Let $\mathcal{G}$ be a group of order $n = rpt$, K be a subgroup of order rp of $\mathcal{G}$, and H be an invariant subgroup of order r of both K and $\mathcal{G}$. Then K is an invariant subgroup of $\mathcal{G}$ if and only if $P = K/H$ is an invariant subgroup of $S = \mathcal{G}/H$.

For a proof, see Problem 20.

Theorem XXXI. Let H and K be invariant subgroups of $\mathcal{G}$ with H an invariant subgroup of K, and let $P = K/H$ and $S = \mathcal{G}/H$. Then the quotient groups S/P and $\mathcal{G}/K$ are isomorphic.

For a proof, see Problem 21.

Theorem XXXII. If H is a maximal invariant subgroup of a group $\mathcal{G}$ then $\mathcal{G}/H$ is simple, and conversely.

Theorem XXXIII. Let H and K be distinct maximal invariant subgroups of a group $\mathcal{G}$. Then

(a) $D = H \cap K$ is an invariant subgroup of $\mathcal{G}$, and

(b) H/D is isomorphic to $\mathcal{G}/K$ and K/D is isomorphic to $\mathcal{G}/H$.

For a proof, see Problem 22.

Solved Problems

1. Does $I/(3)$, the set of residue classes modulo three, form a group with respect to addition? with respect to multiplication?

From the addition and multiplication tables for $I/(3)$ in which $[0], [1], [2]$ have been replaced by $0, 1, 2,$

+	0	1	2
0	0	1	2
1	1	2	0
2	2	0	1

Table 9-5

·	0	1	2
0	0	0	0
1	0	1	2
2	0	2	1

Table 9-6

it is clear that $I/(3)$ forms a group with respect to addition. The identity element is 0 and the inverses of $0, 1, 2$ are respectively $0, 2, 1$. It is equally clear that while these residue classes do not form a group with respect to multiplication, the non-zero residue classes do. Here the identity element is 1 and each of the elements $1, 2$ is its own inverse.

2. Do the non-zero residue classes modulo four form a group with respect to multiplication?

From Table 5-2 of Example 12, Chapter 5, Page 53, it is clear that these residue classes do not form a group with respect to multiplication.

3. Prove: If $a, b, c \in \mathcal{G}$, then $a \circ b = a \circ c$ (also, $b \circ a = c \circ a$) implies $b = c$.

Consider $a \circ b = a \circ c$. Operating on the left with $a^{-1} \in \mathcal{G}$, we have $a^{-1} \circ (a \circ b) = a^{-1} \circ (a \circ c)$. Using the associative law, $(a^{-1} \circ a) \circ b = (a^{-1} \circ a) \circ c$; hence, $u \circ b = u \circ c$ and so $b = c$. Similarly, $(b \circ a) \circ a^{-1} = (c \circ a) \circ a^{-1}$ reduces to $b = c$.

4. Prove: When $a, b \in \mathcal{G}$, each of the equations $a \circ x = b$ and $y \circ a = b$ has a unique solution.

We obtain readily $x = a^{-1} \circ b$ and $y = b \circ a^{-1}$ as solutions. To prove uniqueness, assume x' and y' to be a second set of solutions. Then $a \circ x = a \circ x'$ and $y \circ a = y' \circ a$ whence, by Theorem I, $x = x'$ and $y = y'$.

5. Prove: For any $a \in \mathcal{G}$, $a^m \circ a^n = a^{m+n}$ when $m, n \in I$.

We consider in turn all cases resulting when each of m and n are positive, zero, or negative. When m and n are positive,

$$a^m \circ a^n = (\overbrace{a \circ a \circ \cdots \circ a}^{m \text{ factors}}) \circ (\overbrace{a \circ a \circ \cdots \circ a}^{n \text{ factors}}) = \overbrace{a \circ a \circ \cdots \circ a}^{m+n \text{ factors}} = a^{m+n}$$

When $m = -r$ and $n = s$, where r and s are positive integers,

$$a^m \circ a^n = a^{-r} \circ a^s = (a^{-1})^r \circ a^s = (a^{-1} \circ a^{-1} \circ \cdots \circ a^{-1}) \circ (a \circ a \circ \cdots \circ a)$$

$$= \begin{cases} a^{s-r} = a^{m+n} & \text{when } s \geqq r \\ (a^{-1})^{r-s} = a^{s-r} = a^{m+n} & \text{when } s < r \end{cases}$$

The remaining cases are left for the reader.

6. Prove: A non-empty subset $\mathcal{G}'$ of a group $\mathcal{G}$ is a subgroup of $\mathcal{G}$ if and only if, for all $a, b \in \mathcal{G}'$, $a^{-1} \circ b \in \mathcal{G}'$.

Suppose $\mathcal{G}'$ is a subgroup of $\mathcal{G}$. If $a, b \in \mathcal{G}'$, then $a^{-1} \in \mathcal{G}'$ and, by the Closure Law, so also does $a^{-1} \circ b$.

Conversely, suppose $\mathcal{G}'$ is a non-empty subset of $\mathcal{G}$ for which $a^{-1} \circ b \in \mathcal{G}'$ whenever $a, b \in \mathcal{G}'$. Now $a^{-1} \circ a = u \in \mathcal{G}'$. Then $u \circ a^{-1} = a^{-1} \in \mathcal{G}'$ and every element of $\mathcal{G}'$ has an inverse in $\mathcal{G}'$. Finally, for every $a, b \in \mathcal{G}'$, $(a^{-1})^{-1} \circ b = a \circ b \in \mathcal{G}'$, and the Closure Law holds. Thus, $\mathcal{G}'$ is a group and, hence, a subgroup of $\mathcal{G}$.

7. Prove: If S is any set of subgroups of a group $\mathcal{G}$, the intersection of these subgroups is also a subgroup of $\mathcal{G}$.

Let a and b be elements of the intersection and, hence, elements of each of the subgroups which make up S. By Theorem VIII, $a^{-1} \circ b$ belongs to each subgroup and, hence, to the intersection. Thus, the intersection is a subgroup of $\mathcal{G}$.

8. Prove: Every subgroup of a cyclic group is itself a cyclic group.

Let $\mathcal{G}'$ be a subgroup of a cyclic group $\mathcal{G}$ whose generator is a. Suppose that m is the least positive integer such that $a^m \in \mathcal{G}'$. Now every element of $\mathcal{G}'$, being an element of $\mathcal{G}$, is of the form a^k, $k \in I$. Writing

$$k = mq + r, \qquad 0 \leqq r < m$$

we have

$$a^k = a^{mq+r} = (a^m)^q \circ a^r$$

and, hence,

$$a^r = (a^m)^{-q} \circ a^k$$

Since both a^m and $a^k \in \mathcal{G}'$, it follows that $a^r \in \mathcal{G}'$. But since $r < m$, $r = 0$. Thus $k = mq$, every element of $\mathcal{G}'$ is of the form $(a^m)^q$, and $\mathcal{G}$ is the cyclic group generated by a^m.

9. The subset $\{u = (1), \rho, \rho^2, \rho^3, \sigma^2, \tau^2, b = (13), e = (24)\}$ of S_4 is a group (see the operation table below), called the *octic group of a square* or the *dihedral group*. We shall now show how this permutation group may be obtained using properties of symmetry of a square.

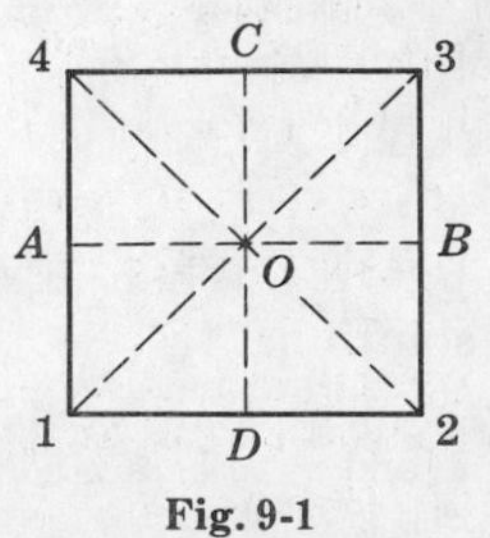

Fig. 9-1

Consider the square (Fig. 9-1) with vertices denoted by 1, 2, 3, 4; locate its center O, the bisectors AOB and COD of its parallel sides, and the diagonals $1O3$ and $2O4$. We shall be concerned with all rigid motions (rotations in the plane about O and in space about the bisectors and diagonals) such that the square will look the same after the motion as before.

Denote by ρ the counterclockwise rotation of the square about O through 90°. Its effect is to carry 1 into 2, 2 into 3, 3 into 4, and 4 into 1; thus, $\rho = (1234)$. Now $\rho^2 = \rho \circ \rho = (13)(24)$ is a rotation about O of $180°$, $\rho^3 = (1432)$ is a rotation of $270°$, and $\rho^4 = (1) = u$ is a rotation about O of $360°$ or $0°$. The rotations through $180°$ about the bisectors AOB and COD give rise respectively to $\sigma^2 = (14)(23)$ and $\tau^2 = (12)(34)$ while the rotations through $180°$ about the diagonals $1O3$ and $2O4$ give rise to $e = (24)$ and $b = (13)$.

The operation table for this group is

	u	ρ	ρ^2	ρ^3	σ^2	τ^2	b	e
u	u	ρ	ρ^2	ρ^3	σ^2	τ^2	b	e
ρ	ρ	ρ^2	ρ^3	u	b	e	τ^2	σ^2
ρ^2	ρ^2	ρ^3	u	ρ	τ^2	σ^2	e	b
ρ^3	ρ^3	u	ρ	ρ^2	e	b	σ^2	τ^2
σ^2	σ^2	e	τ^2	b	u	ρ^2	ρ^3	ρ
τ^2	τ^2	b	σ^2	e	ρ^2	u	ρ	ρ^3
b	b	σ^2	e	τ^2	ρ	ρ^3	u	ρ^2
e	e	τ^2	b	σ^2	ρ^3	ρ	ρ^2	u

Table 9-7

In forming the table

(1) fill in the first row and first column and complete the upper left 4×4 block,

(2) complete the second row, $\left(\rho \circ \sigma^2 = (1234) \circ (14)(23) = (13) = b, \ldots\right)$

and then the third and fourth rows,

$$\left(\rho^2 \circ \sigma^2 = \rho \circ (\rho \circ \sigma^2) = \rho \circ b = \tau^2, \ldots\right)$$

(3) complete the second column and then the third and fourth columns,

$$\left(\sigma^2 \circ \rho^2 = (\sigma^2 \circ \rho) \circ \rho = e \circ \rho = \tau^2, \ldots\right)$$

(4) complete the table, $\left(\sigma^2 \circ \tau^2 = \sigma^2 \circ (\sigma^2 \circ \rho^2) = \rho^2, \ldots\right)$

10. A permutation group on n symbols is called *regular* if each of its elements except the identity moves all n symbols. Find the regular permutation groups on four symbols.

Using Example 4, the required groups are:

$$\{\rho, \rho^2, \rho^3, \rho^4 = (1)\}, \quad \{\sigma, \sigma^2, \sigma^3, \sigma^4 = (1)\}, \quad \text{and} \quad \{\tau, \tau^2, \tau^3, \tau^4 = (1)\}$$

11. Prove: The mapping $I \to I/(n) : m \to [m]$ is a homomorphism of the additive group I onto the additive group $I/(n)$ of integers modulo n.

Since $[m] = [r]$ whenever $m = nq + r$, $0 \leq r < n$, it is evident that the mapping is not one-to-one. However, every $m \in I$ has a unique image in the set $\{[0], [1], [2], \ldots, [n-1]\}$ of residue classes modulo n, and every element of this latter set is an image. Also, if $a \to [r]$ and $b \to [s]$, then $a + b \to [r] + [s] = [t]$ the residue class modulo n of $c = a + b$. Thus the group operations are preserved and the mapping is a homomorphism of I onto $I/(n)$.

12. Prove: In a homomorphism between two groups $\mathcal{G}$ and $\mathcal{G}'$, their identity elements correspond, and if $x \in \mathcal{G}$ and $x' \in \mathcal{G}'$ correspond so also do their inverses.

Denote the identity elements of $\mathcal{G}$ and $\mathcal{G}'$ by u and u' respectively. Suppose now that $u \to v'$ and, for $x \neq u$, $x \to x'$. Then $x = u \circ x \to v' \,\square\, x' = x' = u' \,\square\, x'$ whence, by the Cancellation Law, $v' = u'$ and we have the first part of the theorem.

For the second part, suppose $x \to x'$ and $x^{-1} \to y'$. Then $u = x \circ x^{-1} \to x' \,\square\, y' = u' = x' \,\square\, (x')^{-1}$ so that $y' = (x')^{-1}$.

13. Prove: Every cyclic group of infinite order is isomorphic to the additive group I.

Consider the infinite cyclic group $\mathcal{G}$ generated by a and the mapping

$$n \to a^n, \quad n \in I$$

of I into $\mathcal{G}$. Now this mapping is clearly onto; moreover, it is one-to-one since, if for $s > t$ we had $s \leftrightarrow a^s$ and $t \leftrightarrow a^t$ with $a^s = a^t$, then $a^{s-t} = u$ and $\mathcal{G}$ would be finite. Finally, $s + t \leftrightarrow a^{s+t} = a^s \cdot a^t$ and the mapping is an isomorphism.

14. Prove: Every finite group of order n is isomorphic to a permutation group on n symbols.

Let $\mathcal{G} = \{g_1, g_2, g_3, \ldots, g_n\}$ with group operation $\square$ and define

$$p_j = \begin{pmatrix} g_i \\ g_i \square g_j \end{pmatrix} = \begin{pmatrix} g_1 & g_2 & g_3 & \cdots & g_n \\ g_1 \square g_j & g_2 \square g_j & g_3 \square g_j & \cdots & g_n \square g_j \end{pmatrix}, \qquad (j = 1, 2, 3, \ldots, n)$$

The elements in the second row of p_j are those in the column of the operation table of $\mathcal{G}$ labeled g_j and, hence, are a permutation of the elements of the row labels. Thus $P = \{p_1, p_2, p_3, \ldots, p_n\}$ is a subset of the elements of the symmetric group S_n on n symbols. It is left for the reader to show that P satisfies the conditions of Theorem VII for a group. Now consider the one-to-one correspondence

$$(a) \qquad g_i \leftrightarrow p_i, \quad i = 1, 2, 3, \ldots, n$$

If $g_t = g_r \square g_s$, then $g_t \leftrightarrow p_r \circ p_s$ so that

$$g_t \leftrightarrow \begin{pmatrix} g_i \\ g_i \square g_r \end{pmatrix} \circ \begin{pmatrix} g_i \\ g_i \square g_s \end{pmatrix} = \begin{pmatrix} g_i \\ g_i \square g_r \end{pmatrix} \circ \begin{pmatrix} g_i \square g_r \\ g_i \square g_r \square g_s \end{pmatrix} = \begin{pmatrix} g_i \\ g_i \square g_t \end{pmatrix}$$

and (a) is an isomorphism of $\mathcal{G}$ onto P. Note that P is regular.

15. Prove: Under any homomorphism of a group $\mathcal{G}$ with group operation $\circ$ and identity element u into a group $\mathcal{G}'$ with group operation $\square$ and identity element u', the subset S of all elements of $\mathcal{G}$ which are mapped onto u' is an invariant subgroup of $\mathcal{G}$.

As consequences of Theorem XIII, we have

(a) $u \to u'$; hence, S is non-empty.

(b) if $a \in S$, then $a^{-1} \to (u')^{-1} = u'$; hence, $a^{-1} \in S$.

(c) if $a, b \in S$, then $a^{-1} \circ b \to u' \square u' = u'$; hence, $a^{-1} \circ b \in S$.

Thus S is a subgroup of $\mathcal{G}$.

For arbitrary $a \in S$ and $g \in \mathcal{G}$,

$$g^{-1} \circ a \circ g \to (g')^{-1} \square u' \square g' = u'$$

so that $g^{-1} \circ a \circ g \in S$. Then by Theorem XX, S is an invariant subgroup of $\mathcal{G}$ as required.

16. Prove: The product of cosets

$$(Ha)(Hb) = \{(h_1 \circ a) \circ (h_2 \circ b) : h_1, h_2 \in H\} \qquad \text{for all } Ha, Hb \in \mathcal{G}/H$$

where H is an invariant subgroup of $\mathcal{G}$, is well defined.

First, we show: For any $x, x' \in \mathcal{G}$, $Hx' = Hx$ if and only if $x' = v \circ x$ for some $v \in H$. Suppose $Hx' = Hx$. Then $x' \in Hx$ requires $x' = v \circ x$ for some $v \in H$. Conversely, if $x' = v \circ x$ with $v \in H$, then $Hx' = H(v \circ x) = (Hv)x = Hx$.

Now let Ha' and Hb' be other representations of Ha and Hb respectively with $a' = a \circ r$, $b' = b \circ s$, and $r, s \in H$. In $(Ha')(Hb') = \{[h_1 \circ (a \circ r)] \circ [h_2 \circ (b \circ s)] : h_1, h_2 \in H\}$ we have, using (i_2'), Page 87,

$$\begin{aligned} [h_1 \circ (a \circ r)] \circ [h_2 \circ (b \circ s)] &= (h_1 \circ a) \circ (r \circ h_2) \circ (b \circ s) \\ &= (h_1 \circ a) \circ h_3 \circ (t \circ b) = (h_1 \circ a) \circ (h_3 \circ t) \circ b \\ &= (h_1 \circ a) \circ (h_4 \circ b) \qquad \text{where } h_3, t, h_4 \in H \end{aligned}$$

Then

$$(Ha')(Hb') = (Ha)(Hb)$$

and the product $(Ha)(Hb)$ is well-defined.

17. Prove: Any quotient group of a cyclic group $\mathcal{G}$ is cyclic.

Let H be any (invariant) subgroup of the cyclic group $\mathcal{G} = \{u, a, a^2, \ldots, a^r\}$ and consider the homomorphism

$$\mathcal{G} \to \mathcal{G}/H : a^i \to Ha^i$$

Since every element of $\mathcal{G}/H$ has the form Ha^i for some $a^i \in \mathcal{G}$ and $Ha^i = (Ha)^i$ (prove this) it follows that every element of $\mathcal{G}/H$ is a power of $b = Ha$. Hence, $\mathcal{G}/H$ is cyclic.

18. Prove: Every finite group $\mathcal{G}$ has at least one composition series.

(i) Suppose $\mathcal{G}$ is simple; then $\mathcal{G}, U$ is a composition series.

(ii) Suppose $\mathcal{G}$ is not simple; then there exists an invariant subgroup $H \neq \mathcal{G}, U$ of $\mathcal{G}$. If H is maximal in $\mathcal{G}$ and U is maximal in H, then $\mathcal{G}, H, U$ is a composition series. Suppose H is not maximal in $\mathcal{G}$ but U is maximal in H; then there exists an invariant subgroup K of $\mathcal{G}$ such that H is an invariant subgroup of K. If K is maximal in $\mathcal{G}$ and H is maximal in K, there $\mathcal{G}, K, H, U$ is a composition series. Now suppose H is maximal in $\mathcal{G}$ but U is not maximal in H; then there exists an invariant subgroup J of H. If J is maximal in H and U is maximal in J, then $\mathcal{G}, H, J, U$ is a composition series. Next, suppose that H is not maximal in $\mathcal{G}$ and U is not maximal in H; then .... Since $\mathcal{G}$ is finite, there are only a finite number of subgroups and ultimately we must reach a composition series.

19. Consider two composition series of the cyclic group of order 60: $\mathcal{G} = \{u, a, a^2, \ldots, a^{59}\}$:

$$\mathcal{G}, H = \{u, a^2, a^4, \ldots, a^{58}\}, \quad J = \{u, a^4, a^8, \ldots, a^{56}\}, \quad K = \{u, a^{12}, a^{24}, a^{36}, a^{48}\}, \quad U = \{u\}$$

and

$$\mathcal{G}, M = \{u, a^3, a^6, \ldots, a^{57}\}, \quad N = \{u, a^{15}, a^{30}, a^{45}\}, \quad P = \{u, a^{30}\}, \quad U$$

The quotient groups are:

$$\mathcal{G}/H = \{H, Ha\}, \quad H/J = \{J, Ja^2\}, \quad J/K = \{K, Ka^4, Ka^8\}, \quad K/U = \{U, Ua^{12}, Ua^{24}, Ua^{36}, Ua^{48}\}$$

and

$$\mathcal{G}/M = \{M, Ma, Ma^2\}, \quad M/N = \{N, Na^3, Na^6, Na^9, Na^{12}\}, \quad N/P = \{P, Pa^{15}\}, \quad P/U = \{U, Ua^{30}\}$$

Then in the one-to-one correspondence: $\mathcal{G}/H \leftrightarrow N/P$, $H/J \leftrightarrow P/U$, $J/K \leftrightarrow \mathcal{G}/M$, $K/U \leftrightarrow M/N$, corresponding quotient groups are isomorphic under the mappings:

$H \leftrightarrow P$	$J \leftrightarrow U$	$K \leftrightarrow M$	$U \leftrightarrow N$
$Ha \leftrightarrow Pa^{15}$	$Ja^2 \leftrightarrow Ua^{30}$	$Ka^4 \leftrightarrow Ma$	$Ua^{12} \leftrightarrow Na^3$
		$Ka^8 \leftrightarrow Ma^2$	$Ua^{24} \leftrightarrow Na^6$
			$Ua^{36} \leftrightarrow Na^9$
			$Ua^{48} \leftrightarrow Na^{12}$

20. Prove: Let $\mathcal{G}$ be a group of order $n = rpt$, K be a subgroup of order rp of $\mathcal{G}$, and H be an invariant subgroup of order r of both K and $\mathcal{G}$. Then K is an invariant subgroup of $\mathcal{G}$ if and only if $P = K/H$ is an invariant subgroup of $S = \mathcal{G}/H$.

Let g be an arbitrary element of $\mathcal{G}$ and let $K = \{b_1, b_2, \ldots, b_{rp}\}$.

Suppose P is an invariant subgroup of S. For $Hg \in S$, we have

(i) $$(Hg)P = P(Hg)$$

Thus for any $Hb_i \in P$, there exists $Hb_j \in P$ such that

(ii) $$(Hg)(Hb_i) = (Hb_j)(Hg)$$

Moreover, $(Hg)(Hb_i) = (Hb_j)(Hg) = (Hg)(Hb_k)$ implies $Hb_i = Hb_k$. Then

(iii) $$Hb_i = (Hg^{-1})(Hb_j)(Hg) = g^{-1}(Hb_j)g$$

(iv) $$K = Hb_1 \cup Hb_2 \cup \cdots \cup Hb_p = g^{-1}Kg$$

and

(v) $$gK = Kg$$

Thus K is an invariant subgroup of $\mathcal{G}$.

Conversely, suppose K is an invariant subgroup of $\mathcal{G}$. Then, by simply reversing the steps above, we conclude that P is an invariant subgroup of S.

21. Prove: Let H and K be invariant subgroups of $\mathcal{G}$ with H an invariant subgroup of K, and let $P = K/H$ and $S = \mathcal{G}/H$. Then the quotient groups S/P and $\mathcal{G}/K$ are isomorphic.

Let $\mathcal{G}, K, H$ have the respective orders $n = rpt, rp, r$. Then K is an invariant subgroup of index t in $\mathcal{G}$ and we define

$$\mathcal{G}/K = \{Kc_1, Kc_2, \ldots, Kc_t\}, \qquad c_i \in \mathcal{G}$$

By Theorem XXX, P is an invariant subgroup of S; then P partitions S into t cosets so that we may write

$$S/P = \{P(Ha_{i_1}), P(Ha_{i_2}), \ldots, P(Ha_{i_t})\}, \qquad Ha_{i_j} \in S$$

Now the elements of $\mathcal{G}$ which make up the subgroup K, when partitioned into cosets with respect to H, constitute P. Thus each c_k is found in one and only one of the Ha_{i_j}. Then, after rearranging the cosets of S/P when necessary, we may write

$$S/P = \{P(Hc_1), P(Hc_2), \ldots, P(Hc_t)\}$$

The required mapping is $\mathcal{G}/K \leftrightarrow S/P: \quad Kc_i \leftrightarrow P(Hc_i)$

22. Prove: Let H and K be distinct maximal invariant subgroups of a group $\mathcal{G}$. Then (a) $D = H \cap K$ is an invariant subgroup of $\mathcal{G}$ and (b) H/D is isomorphic to $\mathcal{G}/K$ and K/D is isomorphic to $\mathcal{G}/H$.

(a) By Theorem X, D is a subgroup of $\mathcal{G}$. Since H and K are invariant subgroups of $\mathcal{G}$, we have for each $d \in D$ and every $g \in \mathcal{G}$,

$$g^{-1} \circ d \circ g \in H, \qquad g^{-1} \circ d \circ g \in K \qquad \text{and so} \qquad g^{-1} \circ d \circ g \in D$$

Then, for every $g \in \mathcal{G}$, $g^{-1}Dg = D$ and D is an invariant subgroup of $\mathcal{G}$.

(b) By Theorem XXV, D is an invariant subgroup of both H and K. Suppose

(i) $$H = Dh_1 \cup Dh_2 \cup \ldots \cup Dh_n, \qquad h_i \in H$$

then, since $K(Dh_i) = (KD)h_i = Kh_i$ (why?),

(ii) $$KH = Kh_1 \cup Kh_2 \cup \ldots \cup Kh_n$$

By Theorem XXVI, $HK = KH$ is a subgroup of $\mathcal{G}$. Then, since H is a proper subgroup of HK and, by hypothesis, is a maximal invariant subgroup of $\mathcal{G}$, it follows that $HK = \mathcal{G}$.

From (i) and (ii), we have

$$H/D = \{Dh_1, Dh_2, \ldots, Dh_n\} \qquad \text{and} \qquad \mathcal{G}/K = \{Kh_1, Kh_2, \ldots, Kh_n\}$$

Under the one-to-one mapping

$$Dh_i \leftrightarrow Kh_i, \qquad (i = 1, 2, 3, \ldots, n)$$

$$(Dh_i)(Dh_j) = D(h_i \circ h_j) \leftrightarrow K(h_i \circ h_j) = (Kh_i)(Kh_j)$$

and H/D is isomorphic to $\mathcal{G}/K$. It will be left for the reader to show that K/D and $\mathcal{G}/H$ are isomorphic.

23. Prove: For a finite group with distinct composition series, all series are of the same length, i.e., have the same number of elements. Moreover the quotient groups for any pair of composition series may be put into one-to-one correspondence so that corresponding quotient groups are isomorphic.

Let

(a) $$\mathcal{G}, H_1, H_2, H_3, \ldots, H_r = U$$

(b) $$\mathcal{G}, K_1, K_2, K_3, \ldots, K_s = U$$

be two distinct composition series of $\mathcal{G}$. Now the theorem is true for any group of order one. Let us assume it true for all groups of order less than that of $\mathcal{G}$. We consider two cases:

(i) $H_1 = K_1$. After removing $\mathcal{G}$ from (a) and (b), we have remaining two composition series of H_1 for which, by assumption, the theorem holds. Clearly, it will also hold when $\mathcal{G}$ is replaced in each.

(ii) $H_1 \neq K_1$. Write $D = H_1 \cap K_1$. Since $\mathcal{G}/H_1$ (also $\mathcal{G}/K_1$) is simple and, by Theorem XXXIII, is isomorphic to K_1/D (also $\mathcal{G}/K_1$ is isomorphic to H_1/D), then K_1/D (also H_1/D) is simple. Then D is the maximal invariant subgroup of both H_1 and K_1 and so $\mathcal{G}$ has the composition series

$$(a') \qquad \mathcal{G},\ H_1,\ D,\ D_1,\ D_2,\ D_3,\ \ldots,\ D_t \;=\; U$$

and

$$(b') \qquad \mathcal{G},\ K_1,\ D,\ D_1,\ D_2,\ D_3,\ \ldots,\ D_t \;=\; U$$

When the quotient groups are written in the order

$$\mathcal{G}/H_1,\ H_1/D,\ D/D_1,\ D_1/D_2,\ D_2/D_3,\ \ldots,\ D_{t-1}/D_t$$

and

$$K_1/D,\ \mathcal{G}/K_1,\ D/D_1,\ D_1/D_2,\ D_2/D_3,\ \ldots,\ D_{t-1}/D_t,$$

corresponding quotient groups are isomorphic, that is, $\mathcal{G}/H_1$ and K_1/D, H_1/D and $\mathcal{G}/K_1$, D/D_1 and $D/D_1, \ldots$ are isomorphic.

Now by (i) the quotient groups defined by (a) and (a') [also by (b) and (b')] may be put into one-to-one correspondence so that corresponding quotient groups are isomorphic. Thus the quotient groups defined by (a) and (b) are isomorphic in some order, as required.

Supplementary Problems

24. Which of the following sets form a group with respect to the indicated operation:

(a) $S = \{x : x \in I,\ x < 0\}$; addition

(b) $S = \{5x : x \in I\}$; addition

(c) $S = \{x : x \in I,\ x \text{ is odd}\}$; multiplication

(d) The n nth roots of 1; multiplication

(e) $S = \{-2, -1, 1, 2\}$; multiplication

(f) $S = \{1, -1, i, -i\}$; multiplication

(g) The set of residue classes modulo m; addition

(h) $S = \{[a] : [a] \in I/(m),\ (a, m) = 1\}$; multiplication

(i) $S = \{z : z \in C,\ |z| = 1\}$; multiplication

Ans. (a), (c), (e), do not.

25. Show that the non-zero residue classes modulo p form a group with respect to multiplication if and only if p is a prime.

26. Which of the following subsets of $I/(13)$ is a group with respect to multiplication: (a) $\{[1], [12]\}$; (b) $\{[1], [2], [4], [6], [8], [10], [12]\}$; (c) $\{[1], [5], [8], [12]\}$? *Ans.* (a), (c).

27. Consider the rectangular coordinate system in space. Denote by a, b, c respectively clockwise rotations through 180° about the X, Y, Z-axis and by u its original position. Complete the adjacent table to show that $\{u, a, b, c\}$ is a group, the Klein 4-group.

$\circ$	u	a	b	c
u	u	a	b	c
a	a			
b	b	c		
c	c	b	a	

Table 9-8

28. Prove Theorem III, Page 83.
Hint. $a^{-1} \circ x = u$ has $x = a$ and $x = (a^{-1})^{-1}$ as solutions.

29. Prove Theorem IV, Page 83.
Hint. Consider $(a \circ b) \circ (b^{-1} \circ a^{-1}) = a \circ (b \circ b^{-1}) \circ a^{-1}$

30. Prove: Theorem V, Page 83.

31. Prove: $a^{-m} = (a^m)^{-1},\ m \in I$.

32. Complete the proof of Theorem VI, Page 83.

33. Prove: Theorems IX, XI on Page 84 and Theorem XIV on Page 85.

34. Prove: Every subgroup $\mathcal{G}'$ of a group $\mathcal{G}$ has u, the identity element of $\mathcal{G}$, as identity element.

35. List all of the proper subgroups of the additive group $I/(18)$.

36. Let $\mathcal{G}$ be a group with respect to $\circ$ and a be an arbitrary element of $\mathcal{G}$. Show that

$$H = \{x : x \in \mathcal{G},\ x \circ a = a \circ x\}$$

is a subgroup of $\mathcal{G}$.

37. Prove: Every proper subgroup of an abelian group is abelian. State the converse and show by an example that it is false.

38. Prove: The order of $a \in \mathcal{G}$ is the order of the cyclic subgroup generated by a.

39. Find the order of each of the elements (*a*) (123), (*b*) (1432), (*c*) (12)(34) of S_4.
Ans. (*a*) 3, (*b*) 4, (*c*) 2

40. Verify that the subset A_n of all even permutations in S_n forms a subgroup of S_n. Show that each element of A_n leaves the polynomial of Problem 12, Chapter 2, Page 27, unchanged.

41. Show that the set $\{x : x \in I,\ 5 \mid x\}$ is a subgroup of the additive group I.

42. Form an operation table to discover whether $\{(1), (12)(34), (13)(24), (14)(23)\}$ is a regular permutation group on four symbols.

43. Determine the subset of S_4 which leaves (*a*) the element 2 invariant, (*b*) the elements 2 and 4 invariant, (*c*) $x_1x_2 + x_3x_4$ invariant, (*d*) $x_1x_2 + x_3 + x_4$ invariant.
Ans. (*a*) $\{(1), (13), (14), (34), (134), (143)\}$ (*c*) $\{(1), (12), (34), (12)(34), (13)(24), (14)(23), (1423), (1324)\}$
(*b*) $\{(1), (13)\}$ (*d*) $\{(1), (12), (34), (12)(34)\}$

44. Prove the second part of Theorem XV, Page 86. *Hint.* Use $[m] \leftrightarrow a^m$.

45. Show that the Klein 4-group is isomorphic to the subgroup $P = \{(1), (12)(34), (13)(24), (14)(23)\}$ of S_4.

46. Show that the group of Example 7 is isomorphic to the permutation group

$$P = \{(1), (12)(35)(46), (14)(25)(36), (13)(26)(45), (156)(243), (165)(234)\}$$

on six symbols.

47. Show that the non-zero elements $I/(13)$ under multiplication form a cyclic group isomorphic to the additive group $I/(12)$. Find all isomorphisms between the two groups.

48. Prove: The only groups of order 4 are the cyclic group of order 4 and the Klein 4-group.
Hint. $\mathcal{G} = \{u, a, b, c\}$ either has an element of order 4 or all of its elements except u have order 2. In the latter case, $a \circ b \neq a, b, u$ by the Cancellation Laws.

49. Let S be a subgroup of a group $\mathcal{G}$ and define $T = \{x : x \in \mathcal{G},\ Sx = xS\}$. Prove that T is a subgroup of $\mathcal{G}$.

50. Prove: Two right cosets Ha and Hb of a subgroup H of a group $\mathcal{G}$ are identical if and only if $ab^{-1} \in H$.

51. Prove: $a \in Hb$ implies $Ha = Hb$ where H is a subgroup of $\mathcal{G}$ and $a, b \in \mathcal{G}$.

52. List all cosets of the subgroup $\{(1), (12)(34)\}$ in the octic group.

53. Form the operation table for the symmetric group S_3 on three symbols. List its proper subgroups and obtain right and left cosets for each. Is S_3 simple?

54. Obtain the symmetric group of Problem 53 using the properties of symmetry of an equilateral triangle.

55. Obtain the subgroup $\{u, \rho^2, \sigma^2, \tau^2\}$ of S_4 using the symmetry properties of a non-square rectangle.

56. Obtain the alternating group A_4 of S_4 using the symmetry properties of a regular tetrahedron.

57. Prove: Theorem XXV, Page 89.

58. Show that $K = \{u, \rho^2, \sigma^2, \tau^2\}$ is an invariant subgroup of S_4. Obtain S_4/K and write out in full the homomorphism $S_4 \to S_4/K: x \to Kx$.

Partial answer. $U \to K$, $(12) \to K(12)$, $(13) \to K(13)$, $\ldots$, $(24) \to K(13)$, $(34) \to K(12)$, $\ldots$.

59. Use $K = \{u, \rho^2, \sigma^2, \tau^2\}$, an invariant subgroup of S_4 and $H = \{u, \sigma^2\}$, an invariant subgroup of K, to show that a proper invariant subgroup of a proper invariant subgroup of a group $\mathcal{G}$ is not necessarily an invariant subgroup of $\mathcal{G}$.

60. Prove: The additive group $I/(m)$ is a quotient group of the additive group I.

61. Prove: If H is an invariant subgroup of a group $\mathcal{G}$, the quotient group $\mathcal{G}/H$ is cyclic if the index of H in $\mathcal{G}$ is a prime.

62. Show that the mapping $\begin{cases} (1), \rho^2, \sigma^2, \tau^2 \to u \\ \alpha, \beta^2, \gamma, \delta^2 \to a \\ \alpha^2, \beta, \gamma^2, \delta \to a^2 \end{cases}$ defines a homomorphism of A_4 onto $\mathcal{G} = \{u, a, a^2\}$. Note that the subset of A_4 which maps onto the identity element of $\mathcal{G}$ is an invariant subgroup of A_4.

63. Prove: In a homomorphism of a group $\mathcal{G}$ onto a group $\mathcal{G}'$, let H be the set of all elements of $\mathcal{G}$ which map into $u' \in \mathcal{G}'$. Then the quotient group of $\mathcal{G}/H$ is isomorphic to $\mathcal{G}'$.

64. Set up a homomorphism of the octic group onto $\{u, a\}$.

65. When $H = \{u, \alpha, \alpha^2\}$ and $K = \{u, \beta, \beta^2\}$ are subgroups of S_4, show that $HK \neq KH$. Use HK and KH to verify: In general, the product of two subgroups of a group $\mathcal{G}$ is not a subgroup of $\mathcal{G}$.

66. Prove: If $H = \{h_1, h_2, \ldots, h_r\}$ and $K = \{b_1, b_2, \ldots, b_p\}$ are subgroups of a group $\mathcal{G}$ and one is invariant, then (*a*) $HK = KH$, (*b*) HK is a subgroup of $\mathcal{G}$.

67. Prove: If H and K are invariant subgroups of $\mathcal{G}$, so also is HK.

68. Let $\mathcal{G}$, with group operation $\circ$ and identity element u, and $\mathcal{G}'$, with group operation $\square$ and unity element u', be given groups and form

$$J \;=\; \mathcal{G} \times \mathcal{G}' \;=\; \{(g, g'): g \in \mathcal{G}, g' \in \mathcal{G}'\}$$

Define the "product" of pairs of elements $(g, g'), (h, h') \in J$ by

(i) $$(g, g')(h, h') \;=\; (g \circ h, g' \square h')$$

(*a*) Show that J is a group under the operation defined in (i).

(*b*) Show that $S = \{(g, u'): g \in \mathcal{G}\}$ and $T = \{(u, g'): g' \in \mathcal{G}'\}$ are subgroups of J.

(*c*) Show that the mappings

$$S \to \mathcal{G}: (g, u') \to g \qquad \text{and} \qquad T \to \mathcal{G}': (u, g') \to g'$$

are isomorphisms.

69. For $\mathcal{G}$ and $\mathcal{G}'$ of Problem 68, define $U = \{u\}$ and $U' = \{u'\}$; also $\bar{\mathcal{G}} = \mathcal{G} \times U'$ and $\bar{\mathcal{G}}' = U \times \mathcal{G}'$. Prove:

(*a*) $\bar{\mathcal{G}}$ and $\bar{\mathcal{G}}'$ are invariant subgroups of J.

(*b*) $J/\bar{\mathcal{G}}$ is isomorphic to $U \times \mathcal{G}'$, and $J/\bar{\mathcal{G}}'$ is isomorphic to $\mathcal{G} \times U'$.

(*c*) $\bar{\mathcal{G}}$ and $\bar{\mathcal{G}}'$ have only (u, u') in common.

(*d*) Every element of $\bar{\mathcal{G}}$ commutes with every element of $\bar{\mathcal{G}}'$.

(*e*) Every element of J can be expressed uniquely as the product of an element of $\bar{\mathcal{G}}$ by an element of $\bar{\mathcal{G}}'$.

70. Show that S_4, A_4, $\{u, \rho^2, \sigma^2, \tau^2\}$, $\{u, \sigma^2\}$, U is a composition series of S_4. Find another in addition to that of Example 13(*b*), Page 90.

71. For the cyclic group $\mathcal{G}$ of order 36 generated by a:

(i) Show that $a^2, a^3, a^4, a^6, a^9, a^{12}, a^{18}$ generate invariant subgroups $\mathcal{G}_{18}, \mathcal{G}_{12}, \mathcal{G}_9, \mathcal{G}_6, \mathcal{G}_4, \mathcal{G}_3, \mathcal{G}_2$ respectively of $\mathcal{G}$.

(ii) $\mathcal{G}, \mathcal{G}_{18}, \mathcal{G}_9, \mathcal{G}_3, U$ is a composition series of $\mathcal{G}$. There are six composition series of $\mathcal{G}$ in all; list them.

72. Prove: Theorem XXXII, Page 91.

73. Write the operation table to show that $\bar{Q} = \{1, -1, i, -i, j, -j, k, -k\}$ satisfying $i^2 = j^2 = k^2 = -1$, $ij = k = -ji$, $jk = i = -kj$, $ki = j = -ik$ forms a group.

74. Prove: A non-commutative group $\mathcal{G}$, with group operation $\circ$ has at least six elements.

Hint. (1) $\mathcal{G}$ has at least 3 elements: u, the identity, and two non-commuting elements a and b.

(2) $\mathcal{G}$ has at least 5 elements: $u, a, b, a \circ b, b \circ a$. Suppose it had only 4. Then $a \circ b \neq b \circ a$ implies $a \circ b$ or $b \circ a$ must equal some one of u, a, b.

(3) $\mathcal{G}$ has at least 6 elements: $u, a, b, a \circ b, b \circ a$, and either a^2 or $a \circ b \circ a$.

75. Construct the operation tables for each of the non-commutative groups with 6 elements.

76. Consider $S = \{u, a, a^2, a^3, b, ab, a^2b, a^3b\}$ with $a^4 = u$. Verify:

(*a*) If $b^2 = u$, then either $ba = ab$ or $ba = a^3b$. Write the operation tables A_8, when $ba = ab$, and D_8, when $ba = a^3b$, of the resulting groups.

(*b*) If $b^2 = a$ or $b^2 = a^3$, the resulting groups are isomorphic to C_8, the cyclic group of order 8.

(*c*) If $b^2 = a^2$, then either $ba = ab$ or $ba = a^3b$. Write the operation tables A'_8, when $ba = ab$, and Q_8, when $ba = a^3b$.

(*d*) A_8 and A'_8 are isomorphic.

(*e*) D_8 is isomorphic to the octic group.

(*f*) Q_8 is isomorphic to the (quaternion) group $\bar{Q}$ of Problem 73.

(*g*) Q_8 has only one composition series.

77. Obtain another pair of composition series of the group of Problem 19; set up a one-to-one correspondence between the quotient groups and write the mappings under which corresponding quotient groups are isomorphic.

Chapter 10

Rings

RINGS

A non-empty set $\mathcal{R}$ is said to form a *ring* with respect to the binary operations addition ($+$) and multiplication ($\cdot$) provided, for arbitrary $a, b, c \in \mathcal{R}$, the following properties hold:

P$_1$: $(a+b)+c = a+(b+c)$ (associative law of addition)

P$_2$: $a+b = b+a$ (commutative law of addition)

P$_3$: There exists $z \in \mathcal{R}$ such that $a+z = a$. (existence of an additive identity (zero))

P$_4$: For each $a \in \mathcal{R}$ there exists $-a \in \mathcal{R}$ such that $a+(-a) = z$. (existence of additive inverses)

P$_5$: $(a \cdot b) \cdot c = a \cdot (b \cdot c)$ (associative law of multiplication)

P$_6$: $a(b+c) = a \cdot b + a \cdot c$

P$_7$: $(b+c)a = b \cdot a + c \cdot a$ (distributive laws)

Example 1: Since the properties enumerated above are only a partial list of the properties common to I, Q, R, and C under ordinary addition and multiplication, it follows that these systems are examples of rings.

Example 2: The set $S = \{x + y\sqrt[3]{3} + z\sqrt[3]{9} : x,y,z \in Q\}$ is a ring with respect to addition and multiplication on R. To prove this, we first show that S is closed with respect to these operations. We have, for $a + b\sqrt[3]{3} + c\sqrt[3]{9},\ d + e\sqrt[3]{3} + f\sqrt[3]{9} \in S$,

$$(a + b\sqrt[3]{3} + c\sqrt[3]{9}) + (d + e\sqrt[3]{3} + f\sqrt[3]{9}) = (a+d) + (b+e)\sqrt[3]{3} + (c+f)\sqrt[3]{9} \in S$$

and

$$(a + b\sqrt[3]{3} + c\sqrt[3]{9})(d + e\sqrt[3]{3} + f\sqrt[3]{9}) = (ad + 3bf + 3ce) + (ae + bd + 3cf)\sqrt[3]{3} + (af + be + cd)\sqrt[3]{9} \in S$$

Next, we note that P_1, P_2, P_5-P_7 hold since S is a subset of the ring R. Finally, $0 = 0 + 0\sqrt[3]{3} + 0\sqrt[3]{9}$ satisfies P_3, and for each $x + y\sqrt[3]{3} + z\sqrt[3]{9} \in S$ there exists $-x - y\sqrt[3]{3} - z\sqrt[3]{9} \in S$ which satisfies P_4. Thus S has all of the required properties of a ring.

Example 3: (*a*) The set $S = \{a, b\}$ with addition and multiplication defined by the tables

+	a	b
a	a	b
b	b	a

and

·	a	b
a	a	a
b	a	b

is a ring.

(*b*) The set $T = \{a, b, c, d\}$ with addition and multiplication defined by

+	a	b	c	d
a	a	b	c	d
b	b	a	d	c
c	c	d	a	b
d	d	c	b	a

and

·	a	b	c	d
a	a	a	a	a
b	a	b	a	b
c	a	c	a	c
d	a	d	a	d

is a ring.

For these rings the zero element is a and each element is its own additive inverse.

See also Problems 1-3.

In Examples 1 and 2 the binary operations on the rings (the ring operations) coincide with ordinary addition and multiplication on the various number systems involved; in Example 3 the ring operations have no meaning beyond the given tables. In this example there can be no confusion in using familiar symbols to denote the ring operations. However, when there is a possibility of confusion, we shall use $\oplus$ and $\odot$ to indicate the ring operations.

Example 4: Consider the set of rational numbers Q. Clearly addition ($\oplus$) and multiplication ($\odot$) defined by

$$a \oplus b = a \cdot b \quad \text{and} \quad a \odot b = a + b \qquad \text{for all } a, b \in Q$$

where $+$ and $\cdot$ are ordinary addition and multiplication on rational numbers, are binary operations on Q. Now the fact that $\mathbf{P}_1$, $\mathbf{P}_2$, and $\mathbf{P}_5$ hold is immediate; also, $\mathbf{P}_3$ holds with $z = 1$. We leave it for the reader to show that $\mathbf{P}_4$, $\mathbf{P}_6$, and $\mathbf{P}_7$ do not hold and so Q is not a ring with respect to $\oplus$ and $\odot$.

PROPERTIES OF RINGS

The elementary properties of rings are analogous to those properties of I which do not depend upon either the commutative law of multiplication or the existence of a multiplicative identity element. We call attention here to some of these properties:

(i) Every ring is an abelian additive group.

(ii) There exists a *unique* additive identity element z, (the *zero* of the ring).

See Theorem III, Chapter 2, Page 20.

(iii) Each element has a *unique* additive inverse, (the *negative* of that element).

See Theorem IV, Chapter 2, Page 20.

(iv) The Cancellation Law for addition holds.

(v) $-(-a) = a, \quad -(a+b) = (-a) + (-b)$ for all a, b of the ring.

(vi) $a \cdot z = z \cdot a = z$ (For a proof, see Problem 4.)

(vii) $a(-b) = -(ab) = (-a)b$

SUBRINGS

Let $\mathcal{R}$ be a ring. A non-empty subset S of the set $\mathcal{R}$, which is itself a ring with respect to the binary operations on $\mathcal{R}$ is called a *subring* of $\mathcal{R}$. When S is a subring of a ring $\mathcal{R}$, it is evident that S is a subgroup of the additive group $\mathcal{R}$.

Example 5: (*a*) From Example 1 it follows that I is a subring of the rings Q, R, C; that Q is a subring of R, C; and R is a subring of C.

(*b*) In Example 2, S is a subring of R.

(*c*) In Example 3(*b*), $T_1 = \{a\}$, $T_2 = \{a, b\}$ are subrings of T. Why is $T_3 = \{a, b, c\}$ not a subring of T?

The subrings $\{z\}$ and $\mathcal{R}$ itself of a ring $\mathcal{R}$ are called *improper*; other subrings, if any, of $\mathcal{R}$ are called *proper*.

We leave for the reader the proof of

Theorem I. Let $\mathcal{R}$ be a ring and S be a proper subset of the set $\mathcal{R}$. Then S is a subring of $\mathcal{R}$ if and only if

(*a*) S is closed with respect to the ring operations.

(*b*) for each $a \in S$, we have $-a \in S$.

TYPES OF RINGS

A ring for which multiplication is commutative is called a *commutative* ring.

Example 6: The rings of Examples 1, 2, 3(*a*) are commutative; the ring of Example 3(*b*) is non-commutative, i.e., $b \cdot c = a$ but $c \cdot b = c$.

A ring having a multiplicative identity element (*unit element* or *unity*) is called a *ring with identity element* or *ring with unity.*

Example 7: For each of the rings of Examples 1 and 2, the unity is 1. The unity of the ring of Example 3(*a*) is b; the ring of Example 3(*b*) has no unity.

Let $\mathcal{R}$ be a ring with unity u. Then u is its own multiplicative inverse ($u^{-1} = u$), but other non-zero elements of $\mathcal{R}$ may or may not have multiplicative inverses. Multiplicative inverses, when they exist, are always unique.

Example 8: (*a*) The ring of Problem 1 is a non-commutative ring without unity.

(*b*) The ring of Problem 2 is a commutative ring with unity $u = h$. Here the non-zero elements b, e, f have no multiplicative inverses; the inverses of c, d, g, h are g, d, c, h respectively.

(*c*) The ring of Problem 3 has as unity $u = (1,0,0,1)$. (Show this.) Since $(1,0,1,0)(0,0,0,1) = (0,0,0,0)$ while $(0,0,0,1)(1,0,1,0) = (0,0,1,0)$, the ring is non-commutative. The existence of multiplicative inverses is discussed in Problem 5.

CHARACTERISTIC

Let $\mathcal{R}$ be a ring with zero element z and suppose that there exists a positive integer n such that $na = a + a + a + \cdots + a = z$ for every $a \in \mathcal{R}$. The smallest such positive integer n is called the *characteristic* of $\mathcal{R}$. If no such integer exists, $\mathcal{R}$ is said to have *characteristic zero.*

Example 9: (*a*) The rings I, Q, R, C have characteristic zero since for these rings $na = n \cdot a$.

(*b*) In Problem 1 we have $a + a = b + b = \cdots = h + h = a$, the zero of the ring, and the characteristic of the ring is two.

(*c*) The ring of Problem 2 has characteristic four.

DIVISORS OF ZERO

Let $\mathcal{R}$ be a ring with zero element z. An element $a \neq z$ of $\mathcal{R}$ is called a *divisor of zero* if there exists an element $b \neq z$ of $\mathcal{R}$ such that $a \cdot b = z$ or $b \cdot a = z$.

Example 10: (*a*) The rings I, Q, R, C have no divisors of zero, that is, in each system $ab = 0$ always implies $a = 0$ or $b = 0$.

(*b*) For the ring of Problem 3, we have seen in Example 8(*c*) that $(1,0,1,0)$ and $(0,0,0,1)$ are divisors of zero.

(*c*) The ring of Problem 2 has divisors of zero since $b \cdot e = a$. Find all divisors of zero for this ring.

HOMOMORPHISMS AND ISOMORPHISMS

A homomorphism (isomorphism) of the additive group of a ring $\mathcal{R}$ into (onto) the additive group of a ring $\mathcal{R}'$ which also preserves the second operation, multiplication, is called a homomorphism (isomorphism) of $\mathcal{R}$ into (onto) $\mathcal{R}'$.

Example 11: Consider the ring $\mathcal{R} = \{a, b, c, d\}$ with addition and multiplication tables

+	a	b	c	d
a	a	b	c	d
b	b	a	d	c
c	c	d	a	b
d	d	c	b	a

$\cdot$	a	b	c	d
a	a	a	a	a
b	a	b	c	d
c	a	c	d	b
d	a	d	b	c

and the ring $\mathcal{R}' = \{p, q, r, s\}$ with addition and multiplication tables

+	p	q	r	s
p	r	s	p	q
q	s	r	q	p
r	p	q	r	s
s	q	p	s	r

$\cdot$	p	q	r	s
p	s	p	r	q
q	p	q	r	s
r	r	r	r	r
s	q	s	r	p

The one-to-one mapping

$$a \leftrightarrow r, \quad b \leftrightarrow q, \quad c \leftrightarrow s, \quad d \leftrightarrow p$$

carries $\mathcal{R}$ onto $\mathcal{R}'$ (also, $\mathcal{R}'$ onto $\mathcal{R}$) and at the same time preserves all binary operations; for example,

$$d = b + c \leftrightarrow q + s = p$$

$$b = c \cdot d \leftrightarrow s \cdot p = q, \quad \text{etc.}$$

Thus, $\mathcal{R}$ and $\mathcal{R}'$ are isomorphic rings.

Using the isomorphic rings $\mathcal{R}$ and $\mathcal{R}'$ of Example 11, it is easy to verify

Theorem II. In any isomorphism of a ring $\mathcal{R}$ onto a ring $\mathcal{R}'$:

(*a*) if z is the zero of $\mathcal{R}$ and z' is the zero of $\mathcal{R}'$, we have $z \leftrightarrow z'$.

(*b*) if $\mathcal{R} \leftrightarrow \mathcal{R}': a \leftrightarrow a'$, then $-a \leftrightarrow -a'$.

(*c*) if u is the unity of $\mathcal{R}$ and u' is the unity of $\mathcal{R}'$, we have $u \leftrightarrow u'$.

(*d*) if $\mathcal{R}$ is a commutative ring, so also is $\mathcal{R}'$.

IDEALS

Let $\mathcal{R}$ be a ring with zero element z. A subgroup S of $\mathcal{R}$, having the property $r \cdot x \in S$ $(x \cdot r \in S)$ for all $x \in S$ and $r \in \mathcal{R}$, is called a *left* (*right*) *ideal* in $\mathcal{R}$. Clearly, $\{z\}$ and $\mathcal{R}$ itself are both left and right ideals in $\mathcal{R}$; they are called *improper* left (right) ideals in $\mathcal{R}$. All other left (right) ideals in $\mathcal{R}$, if any, are called *proper*.

A subgroup $\mathcal{J}$ of $\mathcal{R}$, which is both a left *and* right ideal in $\mathcal{R}$, that is, for all $x \in \mathcal{J}$ and $r \in \mathcal{R}$ both $r \cdot x \in \mathcal{J}$ and $x \cdot r \in \mathcal{J}$, is called an *ideal* (*invariant subring*) in $\mathcal{R}$. Clearly, every left (right) ideal in a commutative ring $\mathcal{R}$ is an ideal in $\mathcal{R}$.

For every ring $\mathcal{R}$, the ideals $\{z\}$ and $\mathcal{R}$ itself are called *improper* ideals in $\mathcal{R}$; any other ideals in $\mathcal{R}$ are called *proper*. A ring having no proper ideals is called a *simple* ring.

Example 12: (*a*) For the ring S of Problem 1, $\{a, b, c, d\}$ is a proper right ideal in S (examine the first four rows of the multiplication table) but not a left ideal (examine the first four columns of the same table). The proper ideals in S are $\{a, c\}$, $\{a, e\}$, $\{a, g\}$, and $\{a, c, e, g\}$.

(*b*) In the commutative ring I, the subgroup P of all integral multiples of any integer p is an ideal in I.

(*c*) For every fixed $a, b \in Q$, the subgroup $J = \{(ar, br, as, bs) : r, s \in Q\}$ is a left ideal in the ring M of Problem 3 and $K = \{(ar, as, br, bs) : r, s \in Q\}$ is a right ideal in M since, for every $(m, n, p, q) \in M$,

$$(m, n, p, q) \cdot (ar, br, as, bs) = \big(a(mr + ns),\ b(mr + ns),\ a(pr + qs),\ b(pr + qs)\big) \in J$$

and

$$(ar, as, br, bs) \cdot (m, n, p, q) = \big(a(mr + ps),\ a(nr + qs),\ b(mr + ps),\ b(nr + qs)\big) \in K$$

Example 12(b) illustrates

Theorem III. If p is an arbitrary element of a commutative ring $\mathcal{R}$, then $P = \{p \cdot r : r \in \mathcal{R}\}$ is an ideal in $\mathcal{R}$.

For a proof, see Problem 9.

In Example 12(a), each element x of the left ideal $\{a, c, e, g\}$ has the property that it is an element of S for which $r \cdot x = a$, the zero element of S, for every $r \in S$. This illustrates

Theorem IV. Let $\mathcal{R}$ be a ring with zero element z; then

$$T = \{x : x \in \mathcal{R},\ r \cdot x = z\ (x \cdot r = z) \text{ for all } r \in \mathcal{R}\}$$

is a left (right) ideal in $\mathcal{R}$.

Let $P, Q, S, T, \ldots$ be any collection of ideals in a ring $\mathcal{R}$ and define $\mathcal{J} = P \cap Q \cap S \cap T \cap \cdots$. Since each ideal of the collection is an abelian additive group, so also, by Theorem X, Chapter 9, Page 84, is $\mathcal{J}$. Moreover, for any $x \in \mathcal{J}$ and $r \in \mathcal{R}$, the products $x \cdot r$ and $r \cdot x$ belong to each ideal of the collection and, hence, to $\mathcal{J}$. We have proved

Theorem V. The intersection of any collection of ideals in a ring is an ideal in the ring.

In Problem 10, we prove

Theorem VI. In any homomorphism of a ring $\mathcal{R}$ onto another ring $\mathcal{R}'$ the set S of elements of $\mathcal{R}$ which are mapped on z', the zero element of $\mathcal{R}'$, is an ideal in $\mathcal{R}$.

Example 13: Consider the ring $G = \{a + bi : a, b \in I\}$ of Problem 8.

(a) The set of residue classes modulo 2 of G is $H = \{[0], [1], [i], [1+i]\}$. (Note that $1 - i \equiv 1 + i \pmod 2$.) From the operation tables for addition and multiplication modulo 2, it will be found that H is a commutative ring with unity; also, H has divisors of zero although G does not.

The mapping $G \to H : g \to [g]$ is a homomorphism in which $S = \{2g : g \in G\}$, an ideal in G, is mapped on $[0]$, the zero element of H.

(b) The set of residue classes modulo 3 of G is

$$K = \{[0], [1], [i], [2], [2i], [1+i], [2+i], [1+2i], [2+2i]\}$$

It can be shown as in (a) that K is a commutative ring with unity but is without divisors of zero.

PRINCIPAL IDEALS

Let $\mathcal{R}$ be a ring and K be a right ideal in $\mathcal{R}$ with the further property

$$K = \{a \cdot r : r \in \mathcal{R},\ a \text{ is some fixed element of } K\}$$

We shall then call K a *principal right ideal* in $\mathcal{R}$ and say that it is generated by the element a of K. Principal left ideals and principal ideals are defined analogously.

Example 14: (a) In the ring S of Problem 1, the subring $\{a, g\}$ is a principal right ideal in S generated by the element g (see the row of the multiplication table opposite g). Since $r \cdot g = a$ for every $r \in S$ (see the column of the multiplication table headed g), $\{a, g\}$ is not a principal left ideal and, hence, not a principal ideal in S.

(b) In the commutative ring S of Problem 2, the ideal $\{a, b, e, f\}$ in S is a principal ideal and may be thought of as generated by either b or f.

(c) In the ring S of Problem 1, the right ideal $\{a, b, c, d\}$ in S is not a principal right ideal since it cannot be generated by any one of its elements.

(d) For any $m \in I$, $J = \{mx : x \in I\}$ is a principal ideal in I.

In the ring I, consider the principal ideal K generated by the element 12. It is clear that K is generated also by the element -12. Since K can be generated by no other of its elements, let it be defined as the principal ideal generated by 12. The generator 12 of K, besides being an element of K, is also an element of each of the principal ideals: A generated by 6, B generated by 4, C generated by 3, D generated by 2, and I itself. Now $K \subset A$, $K \subset B$, $K \subset C$, $K \subset D$, $K \subset I$; moreover, 12 is not contained in any other principal ideal of I. Thus K is the intersection of all principal ideals in I in which 12 is an element.

It follows readily that any principal ideal in I generated by the integer m is contained in every principal ideal in I generated by a factor of m. In particular, if m is a prime the only principal ideal in I which properly contains the principal ideal generated by m is I.

Every ring $\mathcal{R}$ has at least one principal ideal, namely, the *null* ideal $\{z\}$ where z is the zero element of $\mathcal{R}$. Every ring with unity has at least two principal ideals, namely, $\{z\}$ and the ideal $\mathcal{R}$ generated by the unity.

Let $\mathcal{R}$ be a commutative ring. If every ideal in $\mathcal{R}$ is a principal ideal, we shall call $\mathcal{R}$ a *principal ideal ring*. For example, consider any ideal $\mathcal{J} \neq \{0\}$ in the ring of integers I. If $a \neq 0 \in \mathcal{J}$ so also is $-a$. Then $\mathcal{J}$ contains positive integers and, since I^+ is well-ordered, contains a least positive integer, say e. For any $b \in \mathcal{J}$, we have by the Division Algorithm of Chapter 5, Page 50,

$$b = e \cdot q + r, \qquad q, r \in I, \quad 0 \leqq r < e$$

Now $e \cdot q \in \mathcal{J}$; hence, $r = 0$ and $b = e \cdot q$. Thus, $\mathcal{J}$ is a principal ideal in I and we have proved

The ring I is a principal ideal ring.

PRIME AND MAXIMAL IDEALS

An ideal $\mathcal{J}$ in a commutative ring $\mathcal{R}$ is said to be a *prime ideal* if, for arbitrary elements r, s of $\mathcal{R}$, the fact that $r \cdot s \in \mathcal{J}$ implies either $r \in \mathcal{J}$ or $s \in \mathcal{J}$.

Example 15: In the ring I,

(a) the ideal $J = \{7r : r \in I\}$, also written as $J = (7)$, is a prime ideal since if $a \cdot b \in J$ either $7 \mid a$ or $7 \mid b$; hence, either $a \in J$ or $b \in J$.

(b) the ideal $K = \{14r : r \in I\}$ or $K = (14)$ is not a prime ideal since, for example, $28 = 4 \cdot 7 \in K$ but neither 4 nor 7 is in K.

Example 15 illustrates

Theorem VII. In the ring I a proper ideal $\mathcal{J} = \{mr : r \in I, m \neq 0\}$ is a prime ideal if and only if m is a prime integer.

A proper ideal $\mathcal{J}$ in a commutative ring $\mathcal{R}$ is called *maximal* if there exists no proper ideal in $\mathcal{R}$ which properly contains $\mathcal{J}$.

Example 16: (a) The ideal J of Example 15 is a maximal ideal in I since the only ideal in I which properly contains J is I itself.

(b) The ideal K of Example 15 is not a maximal ideal since K is properly contained in J which, in turn, is properly contained in I.

QUOTIENT RINGS

Since the additive group of a ring $\mathcal{R}$ is abelian, all of its subgroups are invariant subgroups. Thus, any ideal $\mathcal{J}$ in the ring is an invariant subgroup of the additive group $\mathcal{R}$ and the quotient group $\mathcal{R}/\mathcal{J} = \{r + \mathcal{J} : r \in \mathcal{R}\}$ is the set of all distinct cosets of $\mathcal{J}$ in $\mathcal{R}$.

(*Note.* The use of $r+\mathcal{J}$ instead of the familiar $r\mathcal{J}$ for a coset is in a sense unnecessary since, by definition, $r\mathcal{J} = \{r \circ a:\ a \in \mathcal{J}\}$ and the operation here *is* addition. Nevertheless, we shall use it.) In the section titled Quotient Groups of Chapter 9, Page 88, addition (+) on the cosets (of an additive group) was well-defined by

$$(x + \mathcal{J}) + (y + \mathcal{J}) \quad = \quad (x + y) + \mathcal{J}$$

We now define multiplication ($\cdot$) on the cosets by

$$(x + \mathcal{J}) \cdot (y + \mathcal{J}) \quad = \quad (x \cdot y) + \mathcal{J}$$

and establish that it too is well-defined. For this purpose suppose $x' = x+s$ and $y' = y+t$ are elements of the additive group $\mathcal{R}$ such that $x'+\mathcal{J}$ and $y'+\mathcal{J}$ are other representations of $x+\mathcal{J}$ and $y+\mathcal{J}$ respectively. From

$$x' + \mathcal{J} \quad = \quad (x+s) + \mathcal{J} \quad = \quad (x + \mathcal{J}) + (s + \mathcal{J}) \quad = \quad x + \mathcal{J}$$

it follows that s (and similarly t) $\in \mathcal{J}$. Then

$$(x' + \mathcal{J}) \cdot (y' + \mathcal{J}) \quad = \quad (x' \cdot y') + \mathcal{J} \quad = \quad [(x \cdot y) + (x \cdot t) + (s \cdot y) + (s \cdot t)] + \mathcal{J} \quad = \quad (x \cdot y) + \mathcal{J}$$

since $x \cdot t,\ s \cdot y,\ s \cdot t \in \mathcal{J}$, and multiplication is well-defined. (We have continued to call $x + \mathcal{J}$ a coset; in ring theory, it is called a *residue class* of $\mathcal{J}$ in the ring $\mathcal{R}$.)

Example 17: Consider the ideal $\mathcal{J} = \{3r:\ r \in I\}$ of the ring I and the quotient group $I/\mathcal{J} = \{\mathcal{J}, 1+\mathcal{J}, 2+\mathcal{J}\}$. It is clear that the elements of $I/\mathcal{J}$ are simply the residue classes of $I/(3)$ and, thus, constitute a ring with respect to addition and multiplication modulo 3.

Example 17 illustrates

Theorem VIII. If $\mathcal{J}$ is an ideal in a ring $\mathcal{R}$, the quotient group $\mathcal{R}/\mathcal{J}$ is a ring with respect to addition and multiplication of cosets (residue classes) as defined above.

It is customary to designate this ring by $\mathcal{R}/\mathcal{J}$ and to call it the *quotient* or *factor ring* of $\mathcal{R}$ relative to $\mathcal{J}$.

From the definition of addition and multiplication of residue classes, it follows that

(*a*) The mapping $\mathcal{R} \to \mathcal{R}/\mathcal{J}:\ a \to a + \mathcal{J}$ is a homomorphism of $\mathcal{R}$ onto $\mathcal{R}/\mathcal{J}$.

(*b*) $\mathcal{J}$ is the zero element of the ring $\mathcal{R}/\mathcal{J}$.

(*c*) If $\mathcal{R}$ is a commutative ring, so also is $\mathcal{R}/\mathcal{J}$.

(*d*) If $\mathcal{R}$ has a unity element u so also has $\mathcal{R}/\mathcal{J}$, namely, $u + \mathcal{J}$.

(*e*) If $\mathcal{R}$ is without divisors of zero, $\mathcal{R}/\mathcal{J}$ may or may not have divisors of zero. For, while

$$(a + \mathcal{J}) \cdot (b + \mathcal{J}) \quad = \quad a \cdot b + \mathcal{J} \quad = \quad \mathcal{J}$$

indicates $a \cdot b \in \mathcal{J}$, it does not necessarily imply either $a \in \mathcal{J}$ or $b \in \mathcal{J}$.

EUCLIDEAN RINGS

In the next chapter we shall be concerned with various types of rings, for example, commutative rings, rings with unity, rings without divisors of zero, commutative rings with unity, ... obtained by adding to the basic properties of a ring one or more other properties (see Page 71) of R. There are other types of rings and we end this chapter with a brief study of one of them. By a *Euclidean ring* is meant:

Any commutative ring $\mathcal{R}$ having the property that to each $x \in \mathcal{R}$ a non-negative integer $\theta(x)$ can be assigned such that

(i) $\theta(x) = 0$ if and only if $x = z$, the zero element of $\mathcal{R}$.

(ii) $\theta(x \cdot y) \geqq \theta(x)$ when $x \cdot y \neq z$.

(iii) For every $x \in \mathcal{R}$ and $y \neq z \in \mathcal{R}$,

$$x = y \cdot q + r \qquad q, r \in \mathcal{R}, \quad 0 \leqq \theta(r) < \theta(y)$$

Example 18: I is a Euclidean ring. This follows easily by using $\theta(x) = |x|$ for every $x \in I$.

See also Problem 12.

There follow

Theorem IX. Every Euclidean ring $\mathcal{R}$ is a principal ideal ring.

Theorem X. Every Euclidean ring has a unity.

Solved Problems

1. The set $S = \{a, b, c, d, e, f, g, h\}$ with addition and multiplication defined by

+	a	b	c	d	e	f	g	h
a	a	b	c	d	e	f	g	h
b	b	a	d	c	f	e	h	g
c	c	d	a	b	g	h	e	f
d	d	c	b	a	h	g	f	e
e	e	f	g	h	a	b	c	d
f	f	e	h	g	b	a	d	c
g	g	h	e	f	c	d	a	b
h	h	g	f	e	d	c	b	a

·	a	b	c	d	e	f	g	h
a	a	a	a	a	a	a	a	a
b	a	b	a	b	a	b	a	b
c	a	c	a	c	a	c	a	c
d	a	d	a	d	a	d	a	d
e	a	e	a	e	a	e	a	e
f	a	f	a	f	a	f	a	f
g	a	g	a	g	a	g	a	g
h	a	h	a	h	a	h	a	h

is a ring. The complete verification that P_1 and P_5-P_7, Page 101, are satisfied is a considerable chore but the reader is urged to do a bit of "spot checking". The zero element is a and each element is its own additive inverse.

2. The set S of Problem 1 with addition and multiplication defined by

+	a	b	c	d	e	f	g	h
a	a	b	c	d	e	f	g	h
b	b	a	d	c	f	e	h	g
c	c	d	e	f	g	h	a	b
d	d	c	f	e	h	g	b	a
e	e	f	g	h	a	b	c	d
f	f	e	h	g	b	a	d	c
g	g	h	a	b	c	d	e	f
h	h	g	b	a	d	c	f	e

·	a	b	c	d	e	f	g	h
a	a	a	a	a	a	a	a	a
b	a	e	f	b	a	e	f	b
c	a	f	d	g	e	b	h	c
d	a	b	g	h	e	f	c	d
e	a	a	e	e	a	a	e	e
f	a	e	b	f	a	e	b	f
g	a	f	h	c	e	b	d	g
h	a	b	c	d	e	f	g	h

is a ring. What is the zero element? Find the additive inverse of each element.

3. Prove: The set $M = \{(a, b, c, d): a, b, c, d \in Q\}$ with addition and multiplication defined by

$$(a, b, c, d) + (e, f, g, h) = (a+e, b+f, c+g, d+h)$$

$$(a, b, c, d)(e, f, g, h) = (ae+bg, af+bh, ce+dg, cf+dh)$$

for all $(a, b, c, d), (e, f, g, h) \in M$ is a ring.

The associative and commutative laws for ring addition are immediate consequences of the associative and commutative laws of addition on Q. The zero element of M is $(0,0,0,0)$ and the additive inverse of (a,b,c,d) is $(-a,-b,-c,-d) \in M$. The associative law for ring multiplication is verified as follows:

$$\begin{aligned}
&[(a,b,c,d)(e,f,g,h)](i,j,k,l) \\
&= \big((ae+bg)i+(af+bh)k,\ (ae+bg)j+(af+bh)l,\ (ce+dg)i+(cf+dh)k,\ (ce+dg)j+(cf+dh)l\big) \\
&= \big(a(ei+fk)+b(gi+hk),\ a(ej+fl)+b(gj+hl),\ c(ei+fk)+d(gi+hk),\ c(ej+fl)+d(gj+hl)\big) \\
&= (a,b,c,d)(ei+fk,\ ej+fl,\ gi+hk,\ gj+hl) \\
&= (a,b,c,d)[(e,f,g,h)(i,j,k,l)]
\end{aligned}$$

for all $(a,b,c,d), (e,f,g,h), (i,j,k,l) \in M$.

The computations required to verify the distributive laws will be left for the reader.

4. Prove: If $\mathcal{R}$ is a ring with zero element z, then for all $a \in \mathcal{R}$, $a \cdot z = z \cdot a = z$.

Since $a+z = a$, it follows that

$$a \cdot a = (a+z)a = (a \cdot a) + z \cdot a$$

Now $a \cdot a = (a \cdot a) + z$; hence, $(a \cdot a) + z \cdot a = (a \cdot a) + z$. Then, using the Cancellation Law, we have $z \cdot a = z$. Similarly, $a \cdot a = a(a+z) = a \cdot a + a \cdot z$ and $a \cdot z = z$.

5. Investigate the possibility of multiplicative inverses of elements of the ring M of Problem 3.

For any element $(a,b,c,d) \neq (0,0,0,0)$ of M, set

$$(a,b,c,d)(p,q,r,s) = (ap+br,\ aq+bs,\ cp+dr,\ cq+ds) = (1,0,0,1)$$

the unity of M, and examine the equations

$$\text{(i)} \begin{cases} ap + br = 1 \\ cp + dr = 0 \end{cases} \qquad \text{(ii)} \begin{cases} aq + bs = 0 \\ cq + ds = 1 \end{cases}$$

for solutions p, q, r, s.

From (i), we have $(ad-bc)p = d$; thus, provided $ad-bc \neq 0$, $p = \dfrac{d}{ad-bc}$ and $r = \dfrac{-c}{ad-bc}$. Similarly, from (ii), we find $q = \dfrac{-b}{ad-bc}$ and $s = \dfrac{a}{ad-bc}$. We conclude that only those elements $(a,b,c,d) \in M$ for which $ad-bc \neq 0$ have multiplicative inverses.

6. Show that $P = \{(a,b,-b,a):\ a,b \in I\}$ with addition and multiplication defined by

$$(a,b,-b,a) + (c,d,-d,c) = (a+c,\ b+d,\ -b-d,\ a+c)$$

and

$$(a,b,-b,a)(c,d,-d,c) = (ac-bd,\ ad+bc,\ -ad-bc,\ ac-bd)$$

is a commutative subring of the non-commutative ring M of Problem 3.

First, we note that P is a subset of M and that the operations defined on P are precisely those defined on M. Now P is closed with respect to these operations; moreover, $(-a,-b,b,-a) \in P$ whenever $(a,b,-b,a) \in P$. Thus, by Theorem I, P is a subring of M. Finally, for arbitrary $(a,b,-b,a), (c,d,-d,c) \in P$ we have

$$(a,b,-b,a)(c,d,-d,c) = (c,d,-d,c)(a,b,-b,a)$$

and P is a commutative ring.

7. Consider the mapping $(a,b,-b,a) \to a$ of the ring P of Problem 6 into the ring I of integers.

The reader will show that the mapping carries

$$(a,b,-b,a) + (c,d,-d,c) \ \to\ a + c$$

and

$$(a,b,-b,a) \cdot (c,d,-d,c) \ \to\ ac - bd$$

Now the additive groups of P and I are homomorphic. (Why not isomorphic?) However, since $ac-bd \neq ac$ generally, the rings P and I are not homomorphic under this mapping.

8. A complex number $a+bi$, where $a, b \in I$, is called a *Gaussian integer*. (In Problem 26, the reader is asked to show that the set $G = \{a+bi: a, b \in I\}$ of all Gaussian integers is a ring with respect to ordinary addition and multiplication on C.) Show that the ring P of Problem 6 and G are isomorphic.

Consider the mapping $(a, b, -b, a) \rightarrow a+bi$ of P into G. The mapping is clearly one-to-one; moreover since

$$(a, b, -b, a) + (c, d, -d, c) = (a+c, b+d, -b-d, a+c) \rightarrow (a+c) + (b+d)i = (a+bi) + (c+di)$$

and

$$(a, b, -b, a)(c, d, -d, c) = (ac-bd, ad+bc, -ad-bc, ac-bd) \rightarrow (ac-bd) + (ad+bc)i = (a+bi)(c+di)$$

all binary operations are preserved. Thus P and G are isomorphic.

9. Prove: If p is an arbitrary element of a commutative ring $\mathcal{R}$, then $P = \{p \cdot r: r \in \mathcal{R}\}$ is an ideal in $\mathcal{R}$.

We are to prove that P is a subgroup of the additive group $\mathcal{R}$ such that $(p \cdot r)s \in P$ for all $s \in \mathcal{R}$.

For all $r,s \in \mathcal{R}$, we have

(i) $p \cdot r + p \cdot s = p(r+s) \in P$, since $r+s \in \mathcal{R}$; thus P is closed with respect to addition.

(ii) $-(p \cdot r) = p(-r) \in P$ whenever $p \cdot r \in P$, since $-r \in \mathcal{R}$ whenever $r \in \mathcal{R}$; by Theorem VII, Chapter 9, Page 84, P is a subgroup of the additive group.

(iii) $(p \cdot r)s = p(r \cdot s) \in P$ since $(r \cdot s) \in \mathcal{R}$.

The proof is complete.

10. Prove: In any homomorphism of a ring $\mathcal{R}$ with multiplication denoted by $\cdot$, into another ring $\mathcal{R}'$ with multiplication denoted by $\square$, the set S of elements of $\mathcal{R}$ which are mapped on z', the zero element of $\mathcal{R}'$, is an ideal in $\mathcal{R}$.

By Theorem XXI, Chapter 9, Page 88, S is a subgroup of $\mathcal{R}'$; hence, for arbitrary $a,b,c \in S$, Properties $\mathbf{P}_1$-$\mathbf{P}_4$, Page 101, hold and ring addition is a binary operation on S.

Since all elements of S are elements of $\mathcal{R}$, Properties $\mathbf{P}_5$-$\mathbf{P}_7$ hold. Now for all $a,b \in S$, $a \cdot b \rightarrow z'$; hence, $a \cdot b \in S$ and ring multiplication is a binary operation on S.

Finally, for every $a \in S$ and $g \in \mathcal{R}$, we have

$$a \cdot g \rightarrow z' \square g' = z' \quad \text{and} \quad g \cdot a \rightarrow g' \square z' = z'$$

Thus S is an ideal in $\mathcal{R}$.

11. Prove: The set $\mathcal{R}/\mathcal{J} = \{r+\mathcal{J}: r \in \mathcal{R}\}$ of the cosets of an ideal $\mathcal{J}$ in a ring $\mathcal{R}$ is itself a ring with respect to addition and multiplication defined by

$$(x+\mathcal{J}) + (y+\mathcal{J}) = (x+y) + \mathcal{J}$$

and for all $x+\mathcal{J}, y+\mathcal{J} \in R/\mathcal{J}$

$$(x+\mathcal{J}) \cdot (y+\mathcal{J}) = (x \cdot y) + \mathcal{J}$$

Since $\mathcal{J}$ is an invariant subgroup of the group $\mathcal{R}$, it follows that $\mathcal{R}/\mathcal{J}$ is a group with respect to addition. It is clear from the definition of multiplication that closure is assured. There remains then to show that the Associative Law and the Distributive Laws hold. We find for all $w+\mathcal{J}$, $x+\mathcal{J}$, $y+\mathcal{J} \in \mathcal{R}/\mathcal{J}$,

$$\begin{aligned}
[(w+\mathcal{J}) \cdot (x+\mathcal{J})] \cdot (y+\mathcal{J}) &= (w \cdot x + \mathcal{J}) \cdot (y+\mathcal{J}) = (w \cdot x) \cdot y + \mathcal{J} = w \cdot (x \cdot y) + \mathcal{J} \\
&= (w+\mathcal{J}) \cdot (x \cdot y + \mathcal{J}) = (w+\mathcal{J}) \cdot [(x+\mathcal{J}) \cdot (y+\mathcal{J})],
\end{aligned}$$

$$\begin{aligned}
(w+\mathcal{J}) \cdot [(x+\mathcal{J}) + (y+\mathcal{J})] &= (w+\mathcal{J}) \cdot [(x+y) + \mathcal{J}] = [w \cdot (x+y)] + \mathcal{J} \\
&= (w \cdot x + w \cdot y) + \mathcal{J} = (w \cdot x + \mathcal{J}) + (w \cdot y + \mathcal{J}) \\
&= (w+\mathcal{J}) \cdot (x+\mathcal{J}) + (w+\mathcal{J}) \cdot (y+\mathcal{J})
\end{aligned}$$

and, in a similar manner,

$$[(x+\mathcal{J}) + (y+\mathcal{J})] \cdot (w+\mathcal{J}) = (x+\mathcal{J}) \cdot (w+\mathcal{J}) + (y+\mathcal{J}) \cdot (w+\mathcal{J})$$

12. Prove: The ring $G = \{a + bi : a, b \in I\}$ is a Euclidean ring.

Define $\theta(\alpha + \beta i) = \alpha^2 + \beta^2$ for every $\alpha + \beta i \in G$. It is easily verified that the properties (i) and (ii), Page 108, for a Euclidean ring hold. (Note also that $\theta(a + bi)$ is simply the square of the amplitude of $a + bi$ and, hence, is defined for all elements of C.)

For every $x \in G$ and $y \neq z \in G$, compute $x \cdot y^{-1} = s + ti$. Now if every $s + ti \in G$, the theorem would follow readily; however, this is not the case as the reader will show by taking $x = 1 + i$ and $y = 2 + 3i$.

Suppose then for a given x and y that $s + ti \notin G$. Let $c + di \in G$ be such that $|c - s| \leq \frac{1}{2}$ and $|d - t| \leq \frac{1}{2}$, and write $x = y(c + di) + r$. Then

$$\begin{aligned}\theta(r) \;=\; \theta[x - y(c + di)] &= \theta[x - y(s + ti) + y(s + ti) - y(c + di)] \\ &= \theta[y\{(s - c) + (t - d)i\}] \leq \tfrac{1}{2}\theta(y) < \theta(y)\end{aligned}$$

Thus (iii) holds and G is a Euclidean ring.

Supplementary Problems

13. Show that $S = \{2x : x \in I\}$ with addition and multiplication as defined on I is a ring while $T = \{2x + 1 : x \in I\}$ is not.

14. Verify that S of Problem 2 is a commutative ring with unity $= h$.

15. When $a, b \in I$ define $a \oplus b = a + b + 1$ and $a \odot b = a + b + ab$. Show that I is a commutative ring with respect to $\oplus$ and $\odot$. What is the zero of the ring? Does it have a unit element?

16. Verify that $S = \{a, b, c, d, e, f, g\}$ with addition and multiplication defined by

+	a	b	c	d	e	f	g
a	a	b	c	d	e	f	g
b	b	c	d	e	f	g	a
c	c	d	e	f	g	a	b
d	d	e	f	g	a	b	c
e	e	f	g	a	b	c	d
f	f	g	a	b	c	d	e
g	g	a	b	c	d	e	f

•	a	b	c	d	e	f	g
a	a	a	a	a	a	a	a
b	a	b	c	d	e	f	g
c	a	c	e	g	b	d	f
d	a	d	g	c	f	b	e
e	a	e	b	f	c	g	d
f	a	f	d	b	g	e	c
g	a	g	f	e	d	c	b

is a ring. What is its unity? its characteristic? Does it have divisors of zero? Is it a simple ring? Show that it is isomorphic to the ring $I/(7)$.

17. Show that $\bar{Q} = \{(z_1, z_2, -\bar{z}_2, \bar{z}_1) : z_1, z_2 \in C\}$ with addition and multiplication defined as in Problem 3 is a non-commutative ring with unity $(1, 0, 0, 1)$. Verify that every element of $\bar{Q}$ with the exception of the zero element $(z_1 = z_2 = 0 + 0i)$ has an inverse in the form $\{\bar{z}_1/\Delta, -z_2/\Delta, \bar{z}_2/\Delta, z_1/\Delta\}$, where $\Delta = |z_1|^2 + |z_2|^2$, and thus the non-zero elements of $\bar{Q}$ form a multiplicative group.

18. Prove: In any ring $\mathcal{R}$,

(a) $-(-a) = a$ for every $a \in \mathcal{R}$

(b) $a(-b) = -(ab) = (-a)b$ for all $a, b \in \mathcal{R}$.

Hint. (a) $a + [(-a) - (-a)] = a + z = a$

19. Consider $\mathcal{R}$, the set of all subsets of a given set S and define, for all $A, B \in \mathcal{R}$,

$$A \oplus B = A \cup B - A \cap B \qquad A \odot B = A \cap B$$

Show that $\mathcal{R}$ is a commutative ring with unity.

20. Show that $S = \{(a, b, -b, a) : a, b \in Q\}$ with addition and multiplication defined as in Problem 6 is a ring. What is its zero? its unity? Is it a commutative ring? Follow through as in Problem 5 to show that every element except $(0, 0, 0, 0)$ has a multiplicative inverse.

21. Complete the operation tables for the ring $\mathcal{R} = \{a, b, c, d\}$:

$+$	a	b	c	d
a	a	b	c	d
b	b	a	d	c
c	c	d	a	b
d	d	c	b	a

$\cdot$	a	b	c	d
a	a	a	a	a
b	a	b		
c	a			a
d	a	b	c	

Is $\mathcal{R}$ a commutative ring? Does it have a unity? What is its characteristic?
Hint. $c \cdot b = (b + d) \cdot b$; $c \cdot c = c \cdot (b + d)$; etc.

22. Complete the operation tables for the ring $\mathcal{B} = \{a, b, c, d\}$:

$+$	a	b	c	d
a	a	b	c	d
b	b	a	d	c
c	c	d	a	b
d	d	c	b	a

$\cdot$	a	b	c	d
a	a	a	a	a
b	a	b		
c	a			c
d	a	b	c	

Is $\mathcal{B}$ a commutative ring? Does it have a unity? What is its characteristic? Verify that $x^2 = x$ for every $x \in \mathcal{B}$. A ring having this property is called a *Boolean ring.*

23. Prove: If $\mathcal{B}$ is a Boolean ring, then (*a*) it has characteristic two, (*b*) it is a commutative ring.
Hint. Consider $(x + y)^2 = x + y$ when $y = x$ and when $y \neq x$.

24. Let $\mathcal{R}$ be a ring with unity and let a and b be elements of $\mathcal{R}$, with multiplicative inverses a^{-1} and b^{-1} respectively. Show that $(a \cdot b)^{-1} = b^{-1} a^{-1}$.

25. Show that $\{a\}$, $\{a, b\}$, $\{a, b, c, d\}$ are subrings of the ring S of Problem 1.

26. Show that $G = \{a + bi : a, b \in I\}$ with respect to addition and multiplication defined on C is a subring of the ring C.

27. Prove Theorem I, Page 102.

28. (*a*) Verify that $\mathcal{R} = \{(z_1, z_2, z_3, z_4) : z_1, z_2, z_3, z_4 \in C\}$ with addition and multiplication defined as in Problem 3 is a ring with unity $(1, 0, 0, 1)$. Is it a commutative ring?

(*b*) Show that the subset $S = \{(z_1, z_2, -z_2, z_1) : z_1, z_2 \in C\}$ of $\mathcal{R}$ with addition and multiplication defined as on $\mathcal{R}$ is a subring of $\mathcal{R}$.

29. List all 15 subrings of S of Problem 1.

30. Prove: Every subring of a ring $\mathcal{R}$ is a subgroup of the additive group $\mathcal{R}$.

31. Prove: A subset S of a ring $\mathcal{R}$ is a subring of $\mathcal{R}$ provided $a - b$ and $a \cdot b \in S$ whenever $a, b \in S$.

32. Verify that the set $I/(n)$ of integers modulo n is a commutative ring with unity. When is the ring without divisors of zero? What is the characteristic of the ring $I/(5)$? of the ring $I/(6)$?

33. Show that the ring $I/(2)$ is isomorphic to the ring of Example 3(*a*).

34. Prove Theorem II, Page 104.

35. (*a*) Show that $M_1 = \{(a, 0, c, d) : a, c, d \in Q\}$ and $M_2 = \{(a, 0, 0, d) : a, d \in Q\}$ with addition and multiplication defined as in Problem 3 are subrings of M of Problem 3.

(*b*) Show that the mapping $M_1 \to M_2 : (x, 0, y, w) \to (x, 0, 0, w)$ is a homomorphism.

(*c*) Show that the subset $\{(0, 0, y, 0) : y \in Q\}$ of elements of M_1 which in (*b*) are mapped into $(0, 0, 0, 0) \in M_2$ is a proper ideal in M_1.

(*d*) Find a homomorphism of M_1 into another of its subrings and, as in (*c*), obtain another proper ideal in M_1.

36. Prove: In any homomorphism of a ring $\mathcal{R}$ onto a ring $\mathcal{R}'$, having z' as identity element, let

$$\mathcal{J} = \{x : x \in \mathcal{R},\ x \to z'\}$$

Then the ring $\mathcal{R}/\mathcal{J}$ is isomorphic to $\mathcal{R}'$.
Hint. Consider the mapping $a + \mathcal{J} \to a'$ where a' is the image of $a \in \mathcal{R}$ in the homomorphism.

37. Let a, b be commutative elements of a ring $\mathcal{R}$ of characteristic two. Show that $(a+b)^2 = a^2 + b^2 = (a-b)^2$.

38. Let $\mathcal{R}$ be a ring with ring operations $+$ and $\cdot$, let $(a, r), (b, s) \in \mathcal{R} \times I$. Show that

(i) $\mathcal{R} \times I$ is closed with respect to addition ($\oplus$) and multiplication ($\odot$) defined by

$$(a, r) \oplus (b, s) = (a + b,\ r + s)$$
$$(a, r) \odot (b, s) = (a \cdot b + rb + sa,\ rs)$$

(ii) $\mathcal{R} \times I$ has $(z, 0)$ as zero element and $(z, 1)$ as unity.

(iii) $\mathcal{R} \times I$ is a ring with respect to $\oplus$ and $\odot$.

(iv) $\mathcal{R} \times \{0\}$ is an ideal in $\mathcal{R} \times I$.

(v) The mapping $\mathcal{R} \leftrightarrow \mathcal{R} \times \{0\} : x \leftrightarrow (x, 0)$ is an isomorphism.

39. Prove Theorem IX, Page 108. *Hint.* For any ideal $\mathcal{J} \neq \{z\}$ in $\mathcal{R}$, select the least $\theta(y)$, say $\theta(b)$, for all non-zero elements $y \in \mathcal{J}$. For every $x \in \mathcal{J}$ write $x = b \cdot q + r$ with $q, r \in \mathcal{J}$ and either $r = z$ or $\theta(r) < \theta(b)$.

40. Prove Theorem X, Page 108. *Hint.* Suppose $\mathcal{R}$ is generated by a; then $a = a \cdot s = s \cdot a$ for some $s \in \mathcal{R}$. For any $b \in \mathcal{R}$, $b = q \cdot a = q(a \cdot s) = b \cdot s$, etc.

Chapter 11

Integral Domains, Division Rings, Fields

INTEGRAL DOMAINS

A commutative ring $\mathcal{D}$, with unity and having no divisors of zero, is called an *integral domain.*

Example 1: (*a*) The rings I, Q, R, and C are integral domains.

(*b*) The rings of Problems 1 and 2, Chapter 10, Page 108, are not integral domains; in each, for example, $f \cdot e = a$, the zero element of the ring.

(*c*) The set $S = \{r + s\sqrt{17} : r, s \in I\}$ with addition and multiplication defined as on R is an integral domain. That S is closed with respect to addition and multiplication is shown by

$$(a + b\sqrt{17}) + (c + d\sqrt{17}) = (a + c) + (b + d)\sqrt{17} \in S$$
$$(a + b\sqrt{17})(c + d\sqrt{17}) = (ac + 17bd) + (ad + bc)\sqrt{17} \in S$$

for all $(a + b\sqrt{17}), (c + d\sqrt{17}) \in S$. Since S is a subset of R, S is without divisors of zero; also, the associative laws, commutative laws and distributive laws hold. The zero element of S is $0 \in R$ and every $a + b\sqrt{17} \in S$ has an additive inverse $-a - b\sqrt{17} \in S$. Thus, S is an integral domain.

(*d*) The ring $S = \{a, b, c, d, e, f, g, h\}$ with addition and multiplication defined by

+	a	b	c	d	e	f	g	h
a	a	b	c	d	e	f	g	h
b	b	a	d	c	f	e	h	g
c	c	d	a	b	g	h	e	f
d	d	c	b	a	h	g	f	e
e	e	f	g	h	a	b	c	d
f	f	e	h	g	b	a	d	c
g	g	h	e	f	c	d	a	b
h	h	g	f	e	d	c	b	a

·	a	b	c	d	e	f	g	h
a	a	a	a	a	a	a	a	a
b	a	b	c	d	e	f	g	h
c	a	c	h	f	g	e	b	d
d	a	d	f	g	c	b	h	e
e	a	e	g	c	d	h	f	b
f	a	f	e	b	h	c	d	g
g	a	g	b	h	f	d	e	c
h	a	h	d	e	b	g	c	f

Table 11-1

is an integral domain. Note that the non-zero elements of S form an abelian multiplicative group. We shall see later that this is a common property of all finite integral domains.

A word of caution is necessary here. The term integral domain is used by some to denote any ring without divisors of zero and by others to denote any commutative ring without divisors of zero.

See Problem 1.

The Cancellation Law for Addition holds in every integral domain $\mathcal{D}$ since every element of $\mathcal{D}$ has an additive inverse. In Problem 2 we show that the Cancellation Law for Multiplication also holds in $\mathcal{D}$ in spite of the fact that the non-zero elements of $\mathcal{D}$ do not necessarily have multiplicative inverses. As a result "having no divisors of zero" in the definition of an integral domain may be replaced by "for which the Cancellation Law for Multiplication holds".

In Problem 3, we prove

Theorem I. Let $\mathcal{D}$ be an integral domain and $\mathcal{J}$ be an ideal in $\mathcal{D}$. Then $\mathcal{D}/\mathcal{J}$ is an integral domain if and only if $\mathcal{J}$ is a prime ideal in $\mathcal{D}$.

UNIT, ASSOCIATE, DIVISOR

Let $\mathcal{D}$ be an integral domain. An element v of $\mathcal{D}$ having a multiplicative inverse in $\mathcal{D}$ is called a *unit* (*regular element*) of $\mathcal{D}$. An element b of $\mathcal{D}$ is called an *associate* of $a \in \mathcal{D}$ if $b = v \cdot a$ where v is some unit of $\mathcal{D}$.

Example 2: (a) The only units of I are ± 1; the only associates of $a \in I$ are $\pm a$.

(b) Consider the integral domain $\mathcal{D} = \{r + s\sqrt{17}: r,s \in I\}$. Now $\alpha = a + b\sqrt{17} \in \mathcal{D}$ is a unit if and only if there exists $x + y\sqrt{17} \in \mathcal{D}$ such that

$$(a + b\sqrt{17})(x + y\sqrt{17}) \quad = \quad (ax + 17by) + (bx + ay)\sqrt{17} \quad = \quad 1 \quad = \quad 1 + 0\sqrt{17}$$

From $\begin{cases} ax + 17by = 1 \\ bx + ay \quad = 0 \end{cases}$ we obtain $x = \dfrac{a}{a^2 - 17b^2}$ and $y = \dfrac{-b}{a^2 - 17b^2}$. Now $x + y\sqrt{17} \in \mathcal{D}$, i.e., $x,y \in I$, if and only if $a^2 - 17b^2 = \pm 1$; hence, α is a unit if and only if $a^2 - 17b^2 = \pm 1$. Thus, ± 1, $4 \pm \sqrt{17}$, $-4 \pm \sqrt{17}$ are units in $\mathcal{D}$ while $2 - \sqrt{17}$ and $-9 - 2\sqrt{17} = (2 - \sqrt{17})(4 + \sqrt{17})$ are associates in $\mathcal{D}$.

(c) Every non-zero element of $I/(7) = \{0, 1, 2, 3, 4, 5, 6\}$ is a unit of $I/(7)$ since $1 \cdot 1 \equiv 1 (\text{mod } 7)$, $2 \cdot 4 \equiv 1 (\text{mod } 7)$, etc.

See Problem 4.

An element a of $\mathcal{D}$ is a *divisor* of $b \in \mathcal{D}$ provided there exists an element c of $\mathcal{D}$ such that $b = a \cdot c$. Every non-zero element b of $\mathcal{D}$ has as divisors its associates in $\mathcal{D}$ and the units of $\mathcal{D}$. These divisors are called *trivial* (*improper*); all other divisors, if any, are called *non-trivial* (*proper*). A non-zero, non-unit element b of $\mathcal{D}$, having only trivial divisors, is called a *prime* (*irreducible element*) of $\mathcal{D}$. An element b of $\mathcal{D}$, having non-trivial divisors, is called a *reducible element* of $\mathcal{D}$. For example, 15 has non-trivial divisors over I but not over Q; 7 is a prime in I but not in Q.

See Problem 5.

There follows

Theorem II. If $\mathcal{D}$ is an integral domain which is also a Euclidean ring then, for $a \neq z$, $b \neq z$ of $\mathcal{D}$,

$$\theta(a \cdot b) = \theta(a) \quad \text{if and only if } b \text{ is a unit of } \mathcal{D}$$

SUBDOMAINS

A subset $\mathcal{D}'$ of an integral domain $\mathcal{D}$, which is itself an integral domain with respect to the ring operations of $\mathcal{D}$, is called a *subdomain* of $\mathcal{D}$. It will be left for the reader to show that z and u, the zero and unity elements of $\mathcal{D}$, are also the zero and unity elements of any subdomain of $\mathcal{D}$.

One of the more interesting subdomains of an integral domain $\mathcal{D}$ (see Problem 6) is

$$\mathcal{D}' \quad = \quad \{nu: n \in I\}$$

where nu has the same meaning as in Chapter 10. For, if $\mathcal{D}''$ be any other subdomain of $\mathcal{D}$, then $\mathcal{D}'$ is a subdomain of $\mathcal{D}''$ and hence, in the sense of inclusion, $\mathcal{D}'$ is the *least* subdomain in $\mathcal{D}$. Thus,

Theorem III. If $\mathcal{D}$ is an integral domain, the subset $\mathcal{D}' = \{nu: n \in I\}$ is its least subdomain.

By the *characteristic* of an integral domain $\mathcal{D}$ we shall mean the characteristic, as defined in Chapter 10, of the ring $\mathcal{D}$. The integral domains of Example 1(a) are then of characteristic zero while that of Example 1(d) has characteristic two. In Problem 7, we prove

Theorem IV. The characteristic of an integral domain is either zero or a prime.

Let $\mathscr{D}$ be an integral domain having $\mathscr{D}'$ as its least subdomain and consider the mapping

$$I \to \mathscr{D}' : n \to nu$$

If $\mathscr{D}$ is of characteristic zero, the mapping is an isomorphism of I onto $\mathscr{D}'$; hence, in $\mathscr{D}$ we may always replace $\mathscr{D}'$ by I. If $\mathscr{D}$ is of characteristic p (a prime), the mapping

$$I/(p) \to \mathscr{D}' : [n] \to nu$$

is an isomorphism of $I/(p)$ onto $\mathscr{D}'$.

ORDERED INTEGRAL DOMAINS

An integral domain $\mathscr{D}$ which contains a subset $\mathscr{D}^+$ having the properties:

(i) $\mathscr{D}^+$ is closed with respect to addition and multiplication as defined on $\mathscr{D}$,

(ii) for every $a \in \mathscr{D}$, one and only one of

$$a = z \qquad a \in \mathscr{D}^+ \qquad -a \in \mathscr{D}^+$$

holds,

is called an *ordered integral domain.* The elements of $\mathscr{D}^+$ are called the *positive* elements of D; all other non-zero elements of $\mathscr{D}$ are called *negative* elements of $\mathscr{D}$.

Example 3: The integral domains of Example 1(*a*) are ordered integral domains. In each, the set $\mathscr{D}^+$ consists of the positive elements as defined in the chapter in which the domain was first considered.

Let $\mathscr{D}$ be an ordered integral domain and, for all $a, b \in \mathscr{D}$, define

$$a > b \quad \text{when} \quad a - b \in \mathscr{D}^+$$

and

$$a < b \quad \text{if and only if} \quad b > a$$

Since $a > z$ means $a \in \mathscr{D}^+$ and $a < z$ means $-a \in \mathscr{D}^+$, it follows that, if $a \neq z$, then $a^2 \in \mathscr{D}^+$. In particular, $u \in \mathscr{D}^+$.

Suppose now that $\mathscr{D}$ is an ordered integral domain with $\mathscr{D}^+$ well-ordered; then u is the least element of $\mathscr{D}^+$. For, should there exist $a \in \mathscr{D}^+$ with $z < a < u$, then $z < a^2 < au = a$. Now $a^2 \in \mathscr{D}^+$ so that $\mathscr{D}^+$ has no least element, a contradiction.

In Problem 8, we prove

Theorem V. If $\mathscr{D}$ is an ordered integral domain with $\mathscr{D}^+$ well-ordered, then

(i) $\mathscr{D}^+ = \{pu : p \in I^+\}$

(ii) $\mathscr{D} = \{mu : m \in I\}$

Moreover, the representation of any $a \in \mathscr{D}$ as $a = mu$ is unique.

There follow

Theorem VI. Two ordered integral domains $\mathscr{D}_1$ and $\mathscr{D}_2$, whose respective sets of positive elements $\mathscr{D}_1^+$ and $\mathscr{D}_2^+$ are well-ordered, are isomorphic.

and

Theorem VII. Apart from notation, the ring of integers I is the only ordered integral domain whose set of positive elements is well-ordered.

DIVISION ALGORITHM

Let $\mathscr{D}$ be an integral domain and suppose $d \in \mathscr{D}$ is a common divisor of the non-zero elements $a, b \in \mathscr{D}$. We call d a *greatest common divisor* of a and b provided for any other common divisor $d' \in \mathcal{D}$, we have $d' \mid d$. When $\mathcal{D}$ is also a Euclidean ring, $d' \mid d$ is equivalent to $\theta(d) > \theta(d')$.

(To show that this definition conforms with that of the greatest common divisor of two integers as given in Chapter 5, suppose $\pm d$ are the greatest common divisors of $a, b \in I$ and let d' be any other common divisor. Since for $n \in I$, $\theta(n) = |n|$, it follows that $\theta(d) = \theta(-d)$ while $\theta(d) > \theta(d')$.)

We state for an integral domain which is also a Euclidean ring

The Division Algorithm. Let $a \neq z$ and b be in $\mathcal{D}$, an integral domain which is also a Euclidean ring. There exist unique $q, r \in \mathcal{D}$ such that

$$b = q \cdot a + r, \qquad 0 \leq \theta(r) < \theta(q)$$

See Problem 5, Chapter 5.

UNIQUE FACTORIZATION

In Chapter 5 it was shown that every integer $a > 1$ can be expressed uniquely (except for order of the factors) as a product of positive primes. Suppose $a = p_1 \cdot p_2 \cdot p_3$ is such a factorization. Then

$$-a = -p_1 \cdot p_2 \cdot p_3 = p_1(-p_2)p_3 = p_1 \cdot p_2(-p_3) = (-1)p_1 \cdot p_2 \cdot p_3 = (-1)p_1 \cdot (-1)p_2 \cdot (-1)p_3$$

and this factorization in *primes* can be considered unique up to the use of unit elements as factors. We may then restate the unique factorization theorem for integers as follows:

> Any non-zero, non-unit element of I can be expressed uniquely (up to the order of factors and the use of unit elements as factors) as the product of prime elements of I. In this form we shall show later that the unique factorization theorem holds in any integral domain which is also a Euclidean ring.

In Problem 9, we prove

Theorem VIII. Let J and K, each distinct from $\{z\}$, be principal ideals in an integral domain $\mathcal{D}$. Then $J = K$ if and only if their generators are associate elements in $\mathcal{D}$.

In Problem 10, we prove

Theorem IX. Let $a, b, p \in \mathcal{D}$, an integral domain which is also a principal ideal ring, such that $p \mid a \cdot b$. Then if p is a prime element in $\mathcal{D}$, $p \mid a$ or $p \mid b$.

A proof that the unique factorization theorem holds in an integral domain which is also a Euclidean ring (sometimes called a Euclidean domain) is given in Problem 11.

As a consequence of Theorem IX, we have

Theorem X. In an integral domain $\mathcal{D}$ in which the unique factorization theorem holds, every prime element in $\mathcal{D}$ generates a prime ideal.

DIVISION RINGS

A ring $\mathcal{S}$, whose non-zero elements form a multiplicative group, is called a *division ring* (*skew field* or *sfield*). Every division ring has a unity and each of its non-zero elements has a multiplicative inverse. Multiplication, however, is not necessarily commutative.

Example 4: (*a*) The rings Q, R, and C are division rings. Since multiplication is commutative, they are examples of commutative division rings.

(*b*) The ring $\bar{Q}$ of Problem 17, Chapter 10, is a non-commutative division ring.

(*c*) The ring I is not a division ring. (Why?)

Let $\mathcal{D}$ be an integral domain having a finite number of elements. For any $b \neq z \in \mathcal{D}$, we have

$$\{b \cdot x : x \in \mathcal{D}\} \quad = \quad \mathcal{D}$$

since otherwise b would be a divisor of zero. Thus, $b \cdot x = u$ for some $x \in \mathcal{D}$ and b has a multiplicative inverse in $\mathcal{D}$. We have proved

Theorem XI. Every integral domain, having a finite number of elements, is a commutative division ring.

We now prove

Theorem XII. Every division ring is a simple ring.

For, suppose $\mathcal{J} \neq \{z\}$ is an ideal of a division ring $\mathcal{S}$. If $a \neq z \in \mathcal{J}$, we have $a^{-1} \in \mathcal{S}$ and $a \cdot a^{-1} = u \in \mathcal{J}$. Then for every $b \in \mathcal{S}$, $b \cdot u = b \in \mathcal{J}$; hence, $\mathcal{J} = \mathcal{S}$.

FIELDS

A ring $\mathcal{F}$ whose non-zero elements form an abelian multiplicative group is called a *field*.

Example 5: (*a*) The rings, Q, R, and C are fields.

(*b*) The ring S of Example 1(*d*) is a field.

(*c*) The ring M of Problem 3, Chapter 10, Page 108, is not a field.

Every field is an integral domain; hence, from Theorem IV, Page 115, there follows

Theorem XIII. The characteristic of a field is either zero or is a prime.

Since every commutative division ring is a field, we have (see Theorem XI)

Theorem XIV. Every integral domain having a finite number of elements is a field.

Any subset $\mathcal{F}'$ of a field $\mathcal{F}$, which is itself a field, is called a *subfield* of $\mathcal{F}$.

Example 6: Q is a subfield of the fields R and C; also, R is a subfield of C.

See also Problem 12.

Let $\mathcal{F}$ be a field of characteristic zero. Its least subdomain, I, is not a subfield. However, for each $b \neq 0$, $b \in I$, we have $b^{-1} \in \mathcal{F}$; hence, for all $a, b \in I$ with $b \neq 0$, it follows that $a \cdot b^{-1} = a/b \in \mathcal{F}$. Thus, Q is the least subfield of $\mathcal{F}$. Let $\mathcal{F}$ be a field of characteristic p, a prime. Then $I/(p)$, the least subdomain of $\mathcal{F}$ is the least subfield of $\mathcal{F}$.

A field $\mathcal{F}$ which has no proper subfield $\mathcal{F}'$ is called a *prime field*. Thus Q is the prime field of characteristic zero and $I/(p)$ is the prime field of characteristic p, where p is a prime.

We state without proof

Theorem XV. Let $\mathcal{F}$ be a prime field. If $\mathcal{F}$ has characteristic zero, it is isomorphic to Q; If $\mathcal{F}$ has characteristic p, a prime, it is isomorphic to $I/(p)$.

In Problem 13, we prove

Theorem XVI. Let $\mathcal{D}$ be an integral domain and $\mathcal{J}$ an ideal in $\mathcal{D}$. Then $\mathcal{D}/\mathcal{J}$ is a field if and only if $\mathcal{J}$ is a maximal ideal in $\mathcal{D}$.

Solved Problems

1. Prove: The ring $I/(m)$ is an integral domain if and only if m is a prime.

Suppose m is a prime p. If $[r]$ and $[s]$ are elements of $I/(p)$ such that $[r]\cdot[s] = [0]$, then $r\cdot s \equiv 0(\text{mod } p)$ and $r \equiv 0(\text{mod } p)$ or $s \equiv 0(\text{mod } p)$. Hence, $[r] = [0]$ or $[s] = [0]$; and $I/(p)$, having no divisors of zero, is an integral domain.

Suppose m is not a prime, that is, suppose $m = m_1 \cdot m_2$ with $1 < m_1, m_2 < m$. Since $[m] = [m_1]\cdot[m_2] = [0]$ while neither $[m_1] = 0$ nor $[m_2] = [0]$, it is evident that $I/(m)$ has divisors of zero and, hence, is not an integral domain.

2. Prove: For every integral domain the Cancellation Law of Multiplication

$$\text{If } a\cdot c = b\cdot c \text{ and } c \neq z, \text{ then } a = b$$

holds.

From $a\cdot c = b\cdot c$ we have $a\cdot c - b\cdot c = (a-b)\cdot c = z$. Now D has no divisors of zero; hence, $a - b = z$ and $a = b$ as required.

3. Prove: Let $\mathcal{D}$ be an integral domain and $\mathcal{J}$ be an ideal in $\mathcal{D}$. Then $\mathcal{D}/\mathcal{J}$ is an integral domain if and only if $\mathcal{J}$ is a prime ideal in $\mathcal{D}$.

The case $\mathcal{J} = \mathcal{D}$ is trivial; we consider $\mathcal{J} \subset \mathcal{D}$.

Suppose $\mathcal{J}$ is a prime ideal in $\mathcal{D}$. Since $\mathcal{D}$ is a commutative ring with unity, so also is $\mathcal{D}/\mathcal{J}$. To show that $\mathcal{D}/\mathcal{J}$ is without divisors of zero, assume $a+\mathcal{J}, b+\mathcal{J} \in \mathcal{D}/\mathcal{J}$ such that

$$(a+\mathcal{J})(b+\mathcal{J}) \quad = \quad a\cdot b + \mathcal{J} \quad = \quad \mathcal{J}$$

Now $a\cdot b \in \mathcal{J}$ and, by definition of a prime ideal, either $a \in \mathcal{J}$ or $b \in \mathcal{J}$. Thus either $a+\mathcal{J}$ or $b+\mathcal{J}$ is the zero element in $\mathcal{D}/\mathcal{J}$; and $\mathcal{D}/\mathcal{J}$, being without divisors of zero, is an integral domain.

Conversely, suppose $\mathcal{D}/\mathcal{J}$ is an integral domain. Let $a \neq z$ and $b \neq z$ of $\mathcal{D}$ be such that $a\cdot b \in \mathcal{J}$. From

$$\mathcal{J} \quad = \quad a\cdot b + \mathcal{J} \quad = \quad (a+\mathcal{J})(b+\mathcal{J})$$

it follows that $a+\mathcal{J} = \mathcal{J}$ or $b+\mathcal{J} = \mathcal{J}$. Thus $a\cdot b \in \mathcal{J}$ implies either $a \in \mathcal{J}$ or $b \in \mathcal{J}$, and $\mathcal{J}$ is a prime ideal in $\mathcal{D}$.

Note. Although $\mathcal{D}$ is free of divisors of zero, this property has not been used in the above proof. Thus, in the theorem, "Let $\mathcal{D}$ be an integral domain" may be replaced by "Let $\mathcal{R}$ be a commutative ring with unity".

4. Let t be some positive integer which is not a perfect square and consider the integral domain $\mathcal{D} = \{r + s\sqrt{t}\colon r,s \in I\}$. For each $\rho = r + s\sqrt{t} \in \mathcal{D}$, define $\bar{\rho} = r - s\sqrt{t}$ and the *norm* of ρ as $N(\rho) = \rho\cdot\bar{\rho}$. From Example 2(*b*), Page 115, we infer that $\rho = a + b\sqrt{t}$ is a unit of $\mathcal{D}$ if and only if $N(\rho) = \pm 1$. Show that for $\alpha = a + b\sqrt{t} \in \mathcal{D}$ and $\beta = c + d\sqrt{t} \in \mathcal{D}$, $N(\alpha\cdot\beta) = N(\alpha)\cdot N(\beta)$.

We have $\alpha\cdot\beta = (ac+bdt) + (ad+bc)\sqrt{t}$ and $\overline{\alpha\cdot\beta} = (ac+bdt) - (ad+bc)\sqrt{t}$. Then $N(\alpha\cdot\beta) = (\alpha\cdot\beta)(\overline{\alpha\cdot\beta}) = (ac+bdt)^2 - (ad+bc)^2 t = (a^2 - b^2 t)(c^2 - d^2 t) = N(\alpha)\cdot N(\beta)$, as required.

5. In the integral domain $\mathcal{D} = \{r + s\sqrt{17}\colon r,s \in I\}$, verify: (*a*) $9 - 2\sqrt{17}$ is a prime, (*b*) $\gamma = 15 + 7\sqrt{17}$ is reducible.

(*a*) Suppose $\alpha,\beta \in \mathcal{D}$ such that $\alpha\cdot\beta = 9 - 2\sqrt{17}$. By Problem 4,

$$N(\alpha\cdot\beta) \quad = \quad N(\alpha)\cdot N(\beta) \quad = \quad N(9 - 2\sqrt{17}) \quad = \quad 13$$

Since 13 is a prime integer, it divides either $N(\alpha)$ or $N(\beta)$; hence either β or α is a unit of $\mathcal{D}$, and $9 - 2\sqrt{17}$ is a prime.

(*b*) Suppose $\alpha = a + b\sqrt{17}$, $\beta = c + d\sqrt{17} \in \mathcal{D}$ such that $\alpha\cdot\beta = \gamma = 15 + 7\sqrt{17}$; then $N(\alpha)\cdot N(\beta) = -608$. From $N(\alpha) = a^2 - 17b^2 = 19$ and $N(\beta) = c^2 - 17d^2 = -32$, we obtain $\alpha = 6 - \sqrt{17}$ and $\beta = 11 + 3\sqrt{17}$. Since α and β are neither units of $\mathcal{D}$ nor associates of γ, $15 + 7\sqrt{17}$ is reducible.

6. Show that $\mathcal{D}' = \{nu: n \in I\}$, where u is the unity of an integral domain $\mathcal{D}$, is a subdomain of $\mathcal{D}$.

For every $ru, su \in \mathcal{D}'$, we have

$$ru + su = (r+s)u \in \mathcal{D}' \quad \text{and} \quad (ru)(su) = rsu \in \mathcal{D}'$$

Hence $\mathcal{D}'$ is closed with respect to the ring operations on $\mathcal{D}$. Also,

$$0u = z \in \mathcal{D}' \quad \text{and} \quad 1u = u \in \mathcal{D}'$$

and for each $ru \in \mathcal{D}'$ there exists an additive inverse $-ru \in \mathcal{D}'$. Finally, $(ru)(su) = z$ implies $ru = z$ or $su = z$. Thus $\mathcal{D}'$ is an integral domain, a subdomain of $\mathcal{D}$.

7. Prove: The characteristic of an integral domain $\mathcal{D}$ is either zero or a prime.

From Examples 1(*a*) and 1(*d*) it is evident that there exist integral domains of characteristic zero and integral domains of characteristic $m > 0$.

Suppose $\mathcal{D}$ has characteristic $m = m_1 \cdot m_2$ with $1 < m_1, m_2 < m$. Then $mu = (m_1u)(m_2u) = z$ and either $m_1u = z$ or $m_2u = z$, a contradiction. Thus m is a prime.

8. Prove: If $\mathcal{D}$ is an ordered integral domain such that $\mathcal{D}^+$ is well-ordered, then

$$\text{(i)} \quad \mathcal{D}^+ = \{pu: p \in I^+\} \qquad \text{(ii)} \quad \mathcal{D} = \{mu: m \in I\}$$

Moreover, the representation of any $a \in \mathcal{D}$ as $a = mu$ is unique.

Since $u \in \mathcal{D}^+$ it follows by the closure property that $2u = u + u \in \mathcal{D}^+$ and, by induction, that $pu \in \mathcal{D}^+$ for all $p \in I^+$. Denote by E the set of all elements of $\mathcal{D}^+$ not included in the set $\{pu: p \in I^+\}$ and by e the least element of E. Now $u \notin E$ so that $e > u$ and, hence, $e - u \in \mathcal{D}^+$ but $e - u \notin E$. (Why?) Then $e - u = p_1u$, for some $p_1 \in I^+$, and $e = u + p_1u = (1+p_1)u = p_2u$, where $p_2 \in I^+$. But this is a contradiction; hence $E = \emptyset$, and (i) is established.

Suppose $a \in \mathcal{D}$ but $a \notin \mathcal{D}^+$; then either $a = z$ or $-a \in \mathcal{D}^+$. If $a = z$, then $a = 0u$. If $-a \in \mathcal{D}^+$ then, by (i), $-a = mu$ for some $m \in I^+$ so that $a = (-m)u$, and (ii) is established.

Clearly, if for any $a \in \mathcal{D}$ we have both $a = ru$ and $a = su$, where $r, s \in I$, then $z = a - a = ru - su = (r-s)u$ and $r = s$. Thus the representation of each $a \in \mathcal{D}$ as $a = mu$ is unique.

9. Prove: Let J and K, each distinct from $\{z\}$, be principal ideals in an integral domain $\mathcal{D}$. Then $J = K$ if and only if their generators are associate elements in $\mathcal{D}$.

Let the generators of J and K be a and b respectively.

First, suppose a and b are associates and $b = a \cdot v$, where v is a unit in $\mathcal{D}$. For any $c \in K$ there exists some $s \in \mathcal{D}$ such that

$$c = b \cdot s = (a \cdot v)s = a(v \cdot s) = a \cdot s', \qquad \text{where } s' \in \mathcal{D}$$

Then $c \in J$ and $K \subseteq J$. Now $b = a \cdot v$ implies $a = b \cdot v^{-1}$; thus, by repeating the argument with any $d \in J$, we have $J \subseteq K$. Hence, $J = K$ as required.

Conversely, suppose $J = K$. Then for some $s, t \in \mathcal{D}$ we have $a = b \cdot s$ and $b = a \cdot t$. Now

$$a = b \cdot s = (a \cdot t)s = a(t \cdot s)$$

so that

$$a - a(t \cdot s) = a(u - t \cdot s) = z$$

where u is the unity and z is the zero element in $\mathcal{D}$. Since $a \neq z$, by hypothesis, we have $u - t \cdot s = z$ so that $t \cdot s = u$ and s is a unit in $\mathcal{D}$. Thus a and b are associate elements in $\mathcal{D}$, as required.

10. Prove: Let $a, b, p \in \mathcal{D}$, an integral domain which is also a principal ideal ring, and suppose $p \mid a \cdot b$. Then if p is a prime element in $\mathcal{D}$, $p \mid a$ or $p \mid b$.

If either a or b is a unit or if a or b (or both) is an associate of p, the theorem is trivial. Suppose the contrary and, moreover, suppose $p \nmid a$. Denote by $\mathcal{J}$ the ideal in $\mathcal{D}$ which is the intersection of all ideals in $\mathcal{D}$ which contain both p and a. Since $\mathcal{J}$ is a principal ideal, suppose it is generated by $c \in \mathcal{J}$ so that $p = c \cdot x$ for some $x \in \mathcal{D}$. Then either (i) x is a unit in $\mathcal{D}$ or (ii) c is a unit in $\mathcal{D}$.

(i) Suppose x is a unit in $\mathcal{D}$; then, by Theorem VIII, p and its associate c generate the same principal ideal $\mathcal{J}$. Since $a \in \mathcal{J}$, we must have

$$a = c \cdot g = p \cdot h \qquad \text{for some } g, h \in \mathcal{D}$$

But then $p \mid a$, a contradiction; hence, x is not a unit.

(ii) Suppose c is a unit; then $c \cdot c^{-1} = u \in \mathcal{J}$ and $\mathcal{J} = \mathcal{D}$. Now there exist $s, t \in \mathcal{D}$ such that $u = p \cdot s + t \cdot a$, where u is the unity of $\mathcal{D}$. Then

$$b = u \cdot b = (p \cdot s)b + (t \cdot a)b = p(s \cdot b) + t(a \cdot b)$$

and, since $p \mid a \cdot b$, we have $p \mid b$ as required.

11. Prove: The unique factorization theorem holds in any integral domain $\mathcal{D}$ which is also a Euclidean ring.

We are to prove that every non-zero, non-unit element of $\mathcal{D}$ can be expressed uniquely (up to the order of the factors and the appearance of unit elements as factors) as the product of prime elements of $\mathcal{D}$.

Suppose $a \neq 0 \in \mathcal{D}$ for which $\theta(a) = 1$. Write $a = b \cdot c$ with b not a unit; then c is a unit and a is a prime element in $\mathcal{D}$, since otherwise

$$\theta(a) = \theta(b \cdot c) > \theta(b) \qquad \text{by Theorem II, Page 115}$$

Next, let us assume the theorem holds for all $b \in \mathcal{D}$ for which $\theta(b) < m$ and consider $c \in \mathcal{D}$ for which $\theta(c) = m$. Now if c is a prime element in $\mathcal{D}$, the theorem holds for c. Suppose, on the contrary, that c is not a prime element and write $c = d \cdot e$ where both d and e are proper divisors of c. By Theorem II, we have $\theta(d) < m$ and $\theta(e) < m$. By hypothesis, the unique factorization theorem holds for both d and e so that we have, say,

$$c = d \cdot e = p_1 \cdot p_2 \cdot p_3 \cdots p_s$$

Since this factorization of c arises from the choice d, e of proper divisors, it may not be unique.

Suppose that for another choice of proper divisors we obtained $c = q_1 \cdot q_2 \cdot q_3 \cdots q_t$. Consider the prime factor p_1 of c. By Theorem IX, Page 117, $p_1 \mid q_1$ or $p_1 \mid (q_2 \cdot q_3 \cdots q_t)$; if $p_1 \nmid q_1$, then $p_1 \mid q_2$ or $p_1 \mid (q_3 \cdots q_t)$; if .... Suppose $p_1 \mid q_j$. Then $q_j = f \cdot p_1$ where f is a unit in $\mathcal{D}$ since, otherwise, q_j would not be a prime element in $\mathcal{D}$. Repeating the argument on

$$p_2 \cdot p_3 \cdots p_s = f^{-1} \cdot q_1 \cdot q_2 \cdots q_{j-1} \cdot q_{j+1} \cdots q_t$$

we find, say, $p_2 \mid q_k$ so that $q_k = g \cdot p_2$ with g a unit in $\mathcal{D}$. Continuing in this fashion, we ultimately find that, apart from the order of the factors and the appearance of unit elements, the factorization of c is unique. This completes the proof of the theorem by induction on m (see Problem 27, Chapter 3, Page 37).

12. Prove: $S = \{x + y\sqrt[3]{3} + z\sqrt[3]{9} : x, y, z \in Q\}$ is a subfield of R.

From Example 2, Chapter 10, Page 101, S is a subring of the ring R. Since the commutative law holds in R and $1 = 1 + 0\sqrt[3]{3} + 0\sqrt[3]{9}$ is the multiplicative identity, it is necessary only to verify that for $x + y\sqrt[3]{3} + z\sqrt[3]{9} \neq 0 \in S$, the multiplicative inverse $\dfrac{x^2 - 3yz}{D} + \dfrac{3z^2 - xy}{D}\sqrt[3]{3} + \dfrac{y^2 - xz}{D}\sqrt[3]{9}$, where $D = x^3 + 3y^3 + 9z^3 - 9xyz$, is in S.

13. Prove: Let $\mathcal{D}$ be an integral domain and $\mathcal{J}$ an ideal in $\mathcal{D}$. Then $\mathcal{D}/\mathcal{J}$ is a field if and only if $\mathcal{J}$ is a maximal ideal in $\mathcal{D}$.

First, suppose $\mathcal{J}$ is a maximal ideal in $\mathcal{D}$; then $\mathcal{J} \subset \mathcal{D}$ and (see Problem 3) $\mathcal{D}/\mathcal{J}$ is a commutative ring with unity. To prove $\mathcal{D}/\mathcal{J}$ is a field, we must show that every non-zero element has a multiplicative inverse.

For any $q \in \mathcal{D} - \mathcal{J}$, consider the subset

$$S = \{a + q \cdot x : a \in \mathcal{J}, x \in \mathcal{D}\}$$

of $\mathcal{D}$. For any $y \in \mathcal{D}$ and $a + q \cdot x \in S$, we have $(a + q \cdot x)y = a \cdot y + q(x \cdot y) \in S$ since $a \cdot y \in \mathcal{J}$; similarly, $y(a + q \cdot x) \in S$. Then S is an ideal in $\mathcal{D}$ and, since $\mathcal{J} \subset S$, we have $S = \mathcal{D}$. Thus any $r \in \mathcal{D}$ may be written as $r = a + q \cdot e$, where $e \in \mathcal{D}$. Suppose for u, the unity of $\mathcal{D}$, we find

$$u = a + q \cdot f, \qquad f \in \mathcal{D}$$

From $$u+\mathcal{J} = (a+\mathcal{J})+(q+\mathcal{J})\cdot(f+\mathcal{J}) = (q+\mathcal{J})\cdot(f+\mathcal{J})$$
it follows that $f+\mathcal{J}$ is the multiplicative inverse of $q+\mathcal{J}$. Since q is an arbitrary element of $\mathcal{D}-\mathcal{J}$, the ring of cosets $\mathcal{D}/\mathcal{J}$ is a field.

Conversely, suppose $\mathcal{D}/\mathcal{J}$ is a field. We shall assume $\mathcal{J}$ not maximal in $\mathcal{D}$ and obtain a contradiction. Let then J be an ideal in $\mathcal{D}$ such that $\mathcal{J}\subset J\subset\mathcal{D}$.

For any $a\in\mathcal{D}$ and any $p\in J-\mathcal{J}$, define $(p+\mathcal{J})^{-1}\cdot(a+\mathcal{J}) = s+\mathcal{J}$; then
$$a+\mathcal{J} = (p+\mathcal{J})\cdot(s+\mathcal{J})$$
Now $a-p\cdot s\in\mathcal{J}$ and, since $\mathcal{J}\subset J$, $a-p\cdot s\in J$. But $p\in J$; hence $a\in J$, and $J=\mathcal{D}$, a contradiction of $J\subset\mathcal{D}$. Thus $\mathcal{J}$ is maximal in $\mathcal{D}$.

The note in Problem 3, Page 119, also applies here.

Supplementary Problems

14. Enumerate the properties of a set necessary to define an integral domain.

15. Which of the following sets are integral domains, assuming addition and multiplication defined as on R:

(a) $\{2a+1:\ a\in I\}$

(b) $\{2a:\ a\in I\}$

(c) $\{a\sqrt{3}:\ a\in I\}$

(d) $\{r\sqrt{3}:\ r\in Q\}$

(e) $\{a+b\sqrt{3}:\ a,b\in I\}$

(f) $\{r+s\sqrt{3}:\ r,s\in Q\}$

(g) $\{a+b\sqrt{2}+c\sqrt{5}+d\sqrt{10}:\ a,b,c,d\in I\}$

16. For the set G of Gaussian integers (see Problem 8, Chapter 10, Page 110), verify:

(a) G is an integral domain.

(b) $\alpha=a+bi$ is a unit if and only if $N(\alpha)=a^2+b^2=1$.

(c) The only units are $\pm1,\pm i$.

17. Define $S=\{(a_1,a_2,a_3,a_4):\ a_i\in R\}$ with addition and multiplication defined respectively by
$$(a_1,a_2,a_3,a_4)+(b_1,b_2,b_3,b_4) = (a_1+b_1,a_2+b_2,a_3+b_3,a_4+b_4)$$
and
$$(a_1,a_2,a_3,a_4)(b_1,b_2,b_3,b_4) = (a_1\cdot b_1,a_2\cdot b_2,a_3\cdot b_3,a_4\cdot b_4)$$
Show that S is not an integral domain.

18. In the integral domain $\mathcal{D}$ of Example 2(b), Page 115, verify:

(a) $33\pm8\sqrt{17}$ and $-33\pm8\sqrt{17}$ are units.

(b) $48-11\sqrt{17}$ and $379-92\sqrt{17}$ are associates of $5+4\sqrt{17}$.

(c) $8 = 2\cdot2\cdot2 = 2(8+2\sqrt{17})(-8+2\sqrt{17}) = (5+\sqrt{17})(5-\sqrt{17})$, in which each factor is a prime; hence, unique factorization in primes is not a property of $\mathcal{D}$.

19. Prove: The relation of association is an equivalence relation.

20. Prove: If, for $\alpha\in\mathcal{D}$, $N(\alpha)$ is a prime integer then α is a prime element of $\mathcal{D}$.

21. Prove: A ring $\mathcal{R}$ having the property that for each $a\neq z, b\in\mathcal{R}$ there exists $r\in\mathcal{R}$ such that $a\cdot r=b$ is a division ring.

22. Let $\mathcal{D}'=\{[0],[5]\}$ and $\mathcal{D}''=\{[0],[2],[4],[6],[8]\}$ be subsets of $\mathcal{D}=I/(10)$. Show:

(a) $\mathcal{D}'$ and $\mathcal{D}''$ are subdomains of $\mathcal{D}$.

(b) $\mathcal{D}'$ and $I/(2)$ are isomorphic; also, $\mathcal{D}''$ and $I/(5)$ are isomorphic.

(c) Every $a\in\mathcal{D}$ can be written uniquely as $a=a'+a''$ where $a'\in\mathcal{D}'$ and $a''\in\mathcal{D}''$.

(d) For $a,b\in\mathcal{D}$, with $a=a'+a''$ and $b=b'+b''$, $a+b=(a'+b')+(a''+b'')$ and $a\cdot b=a'\cdot b'+a''\cdot b''$.

23. Prove: Theorem II.
Hint. If b is a unit then $\theta(a) = \theta[b^{-1}(a \cdot b)] \geq \theta(a \cdot b)$. If b is not a unit, consider $a = q(a \cdot b) + r$, where either $r = z$ or $\theta(r) < \theta(a \cdot b)$, for $a \neq z \in \mathcal{D}$.

24. Prove: The set S of all units of an integral domain is a multiplicative group.

25. Let $\mathcal{D}$ be an integral domain of characteristic p and $\mathcal{D}' = \{x^p : x \in \mathcal{D}\}$. Prove:
(*a*) $(a \pm b)^p = a^p \pm b^p$ (*b*) The mapping $\mathcal{D} \to \mathcal{D}' : x \to x^p$ is an isomorphism.

26. Show that for all $a \neq z, b$ of any division ring, the equation $ax = b$ has a solution.

27. The set $\mathcal{Q} = \{(q_1 + q_2 i + q_3 j + q_4 k) : q_1, q_2, q_3, q_4 \in R\}$ of *quaternions* with addition and multiplication defined by

$$(a_1 + a_2 i + a_3 j + a_4 k) + (b_1 + b_2 i + b_3 j + b_4 k) = (a_1 + b_1) + (a_2 + b_2)i + (a_3 + b_3)j + (a_4 + b_4)k$$

and

$$(a_1 + a_2 i + a_3 j + a_4 k) \cdot (b_1 + b_2 i + b_3 j + b_4 k) = (a_1 b_1 - a_2 b_2 - a_3 b_3 - a_4 b_4) + (a_1 b_2 + a_2 b_1 + a_3 b_4 - a_4 b_3)i$$
$$+ (a_1 b_3 - a_2 b_4 + a_3 b_1 + a_4 b_2)j + (a_1 b_4 + a_2 b_3 - a_3 b_2 + a_4 b_1)k$$

is to be proved a non-commutative division ring. Verify:

(*a*) The subsets $Q_2 = \{(q_1 + q_2 i + 0j + 0k)\}$, $Q_3 = \{(q_1 + 0i + q_3 j + 0k)\}$, and $Q_4 = \{(q_1 + 0i + 0j + q_4 k)\}$ of $\mathcal{Q}$ combine as does the set C of complex numbers; thus, $i^2 = j^2 = k^2 = -1$.

(*b*) q_1, q_2, q_3, q_4 commute with i, j, k.

(*c*) $ij = k,\ jk = i,\ ki = j$

(*d*) $ji = -k,\ kj = -i,\ ik = -j$

(*e*) With $\bar{Q}$ as defined in Problem 17, Chapter 10, Page 111, the mapping

$$\bar{Q} \to \mathcal{Q} : (q_1 + q_2 i,\ q_3 + q_4 i,\ -q_3 + q_4 i,\ q_1 - q_2 i) \to (q_1 + q_2 i + q_3 j + q_4 k)$$

is an isomorphism.

(*f*) $\mathcal{Q}$ is a non-commutative division ring (see Example 4, Page 117).

28. Prove: A field is a commutative ring whose non-zero elements have multiplicative inverses.

29. Show that $P = \{(a, b, -b, a) : a, b \in R\}$ with addition and multiplication defined by

$$(a, b, -b, a) + (c, d, -d, c) = (a + c,\ b + d,\ -b - d,\ a + c)$$

and

$$(a, b, -b, a)(c, d, -d, c) = (ac - bd,\ ad + bc,\ -ad - bc,\ ac - bd)$$

is a field. Show that P is isomorphic to C, the field of complex numbers.

30. (*a*) Show that $\{a + b\sqrt{3} : a, b \in Q\}$ and $\{a + b\sqrt{2} + c\sqrt{5} + d\sqrt{10} : a, b, c, d \in Q\}$ are subfields of R.
(*b*) Show that $\{a + b\sqrt[3]{2} : a, b \in Q\}$ is not a subfield of R.

31. Prove: $S = \{a + br : a, b \in R,\ r = -\frac{1}{2}(1 + \sqrt{3}\,i)\}$ is a subfield of C.
Hint. The multiplicative inverse of $a + br \neq 0 \in S$ is $\dfrac{a - b}{a^2 - ab + b^2} - \dfrac{b}{a^2 - ab + b^2} r \in S$.

32. (*a*) Show that the subsets $S = \{[0], [5], [10]\}$ and $T = \{[0], [3], [6], [9], [12]\}$ of the ring $I/(15)$ are integral domains with respect to the binary operations on $I/(15)$.
(*b*) Show that S is isomorphic to $I/(3)$ and, hence, is a field of characteristic 3.
(*c*) Show that T is a field of characteristic 5.

33. Consider the ideals $A = \{2g : g \in G\}$, $B = \{5g : g \in G\}$, $E = \{7g : g \in G\}$, and $F = \{(1 + i)g : g \in G\}$ of G, the ring of Gaussian integers. (*a*) Show that $G/A = G/(2)$ and $G/B = G/(5)$ are not integral domains. (*b*) Show that G/E is a field of characteristic 7 and G/F is a field of characteristic 2.

34. Prove: A field contains no proper ideals.

35. Show that Problems 3 and 13 imply: If $\mathcal{J}$ is a maximal ideal in a commutative ring $\mathcal{R}$ with unity, then $\mathcal{J}$ is a prime ideal in $\mathcal{R}$.

Chapter 12

Polynomials

INTRODUCTION

A considerable part of elementary algebra is concerned with certain types of functions, for example,

$$1 + 2x + 3x^2 \qquad x + x^5 \qquad 5 - 4x^2 + 3x^{10}$$

called polynomials in x. The coefficients in these examples are integers although it is not necessary that they always be. In elementary calculus, the range of values of x (domain of definition of the function) is R. In algebra, the range is C; for instance, the values of x for which $1+2x+3x^2$ is 0 are $-\frac{1}{3} \pm \frac{\sqrt{2}}{3}i$.

In the light of Chapter 2, any polynomial in x can be thought of as a mapping of a set S (range of x) onto a set T (range of values of the polynomial). Consider, for example, the polynomial $1 + \sqrt{2}\,x - 3x^2$. If $S = I$, then $T \subset R$ and the same is true if $S = Q$ or $S = R$; if $S = C$, then $T \subset C$.

As in previous chapters, equality implies "identical with"; thus two polynomials in x are equal if they have identical form. For example, $a + bx = c + dx$ if and only if $a = c$ and $b = d$. (Note that $a + bx = c + dx$ is never to be considered here as an equation in x.)

It has been our experience that the images of each value of $x \in S$ are the same elements of T when $\alpha(x) = \beta(x)$ and, *in general*, are distinct elements of T when $\alpha(x) \neq \beta(x)$. However, as will be seen from Example 1 below, this familiar state of affairs is somehow dependent upon the range of x.

Example 1: Consider the polynomials $\alpha(x) = [1]x$ and $\beta(x) = [1]x^5$, where $[1] \in I/(5)$, and suppose the range of x to be the field $I/(5) = \{[0], [1], [2], [3], [4]\}$. Clearly, $\alpha(x)$ and $\beta(x)$ differ in form (are not equal polynomials); yet, as is easily verified, their images for each $x \in I/(5)$ are identical.

Example 1 suggests that in our study of polynomials we begin by considering them as forms.

POLYNOMIAL FORMS

Let $\mathcal{R}$ be a ring and let x, called an *indeterminate,* be any symbol not found in $\mathcal{R}$. By a *polynomial in x over $\mathcal{R}$* will be meant any expression of the form

$$\alpha(x) \;=\; a_0x^0 + a_1x^1 + a_2x^2 + \cdots \;=\; \sum a_kx^k, \qquad a_i \in \mathcal{R}$$

in which only a finite number of the a's are different from z, the zero element of $\mathcal{R}$. Two polynomials in x over $\mathcal{R}$, $\alpha(x)$ defined above, and

$$\beta(x) \;=\; b_0x^0 + b_1x^1 + b_2x^2 + \cdots \;=\; \sum b_kx^k, \qquad b_i \in \mathcal{R}$$

will be called *equal,* $\alpha(x) = \beta(x)$, provided $a_k = b_k$ for all values of k.

In any polynomial, as $\alpha(x)$, each of the components $a_0x^0, a_1x^1, a_2x^2, \ldots$ will be called a *term*; in any term, as a_ix^i, a_i will be called the *coefficient* of the term. The terms of $\alpha(x)$ and $\beta(x)$ have been written in a prescribed (but natural) order and we shall continue this practice. Then i, the superscript of x, is merely an indicator of the position of the term a_ix^i in the polynomial. Likewise, juxtaposition of a_i and x^i in the term a_ix^i is not to be construed as indicating multiplication and the plus signs between terms are to be thought of as helpful connectives rather than operators. In fact, we might very well have written the polynomial $\alpha(x)$ above as $\alpha = (a_0, a_1, a_2, \ldots)$.

If in a polynomial, as $\alpha(x)$, the coefficient $a_n \neq z$ while all coefficients of terms which follow are z, we say that $\alpha(x)$ is of *degree n* and call a_n its *leading* coefficient. In particular, the polynomial $a_0x^0 + zx^1 + zx^2 + \cdots$ is of degree zero with leading coefficient a_0 when $a_0 \neq z$ and has *no degree* (and *no leading coefficient*) when $a_0 = z$.

Denote by $\mathcal{R}[x]$ the set of all polynomials in x over $\mathcal{R}$ and, for arbitrary $\alpha(x), \beta(x) \in \mathcal{R}[x]$, define addition (+) and multiplication (·) on $\mathcal{R}[x]$ by

$$\alpha(x) + \beta(x) = (a_0 + b_0)x^0 + (a_1 + b_1)x^1 + (a_2 + b_2)x^2 + \cdots = \sum (a_k + b_k)x^k$$

and

$$\alpha(x) \cdot \beta(x) = a_0b_0x^0 + (a_0b_1 + a_1b_0)x^1 + (a_0b_2 + a_1b_1 + a_2b_0)x^2 + \cdots = \sum c_kx^k, \quad \text{where } c_k = \sum_0^k a_i b_{k-i}$$

(Note that multiplication of elements of $\mathcal{R}$ is indicated here by juxtaposition.)

The reader may find it helpful to see these definitions written out in full as

$$\alpha(x) \boxplus \beta(x) = (a_0 \oplus b_0)x^0 + (a_1 \oplus b_1)x^1 + (a_2 \oplus b_2)x^2 + \cdots$$

and

$$\alpha(x) \boxdot \beta(x) = (a_0 \odot b_0)x^0 + (a_0 \odot b_1 \oplus a_1 \odot b_0)x^1 + (a_0 \odot b_2 \oplus a_1 \odot b_1 \oplus a_2 \odot b_0)x^2 + \cdots$$

in which $\boxplus$ and $\boxdot$ are the newly defined operations on $\mathcal{R}[x]$, $\oplus$ and $\odot$ are the binary operations on $\mathcal{R}$ and, again, + is a connective.

It is clear that both the sum and product of elements of $\mathcal{R}[x]$ are elements of $\mathcal{R}[x]$, i.e., have only a finite number of terms with non-zero coefficients $\in \mathcal{R}$. It is easy to verify that addition on $\mathcal{R}[x]$ is both associative and commutative and that multiplication is associative and distributive with respect to addition. Moreover, the *zero polynomial*

$$zx^0 + zx^1 + zx^2 + \cdots = \sum zx^k \in \mathcal{R}[x]$$

is the *additive identity* or *zero element* of $\mathcal{R}[x]$ while

$$-\alpha(x) = -a_0x^0 + (-a_1)x^1 + (-a_2)x^2 + \cdots = \sum (-a_k)x^k \in \mathcal{R}[x]$$

is the *additive inverse* of $\alpha(x)$. Thus,

Theorem I. The set of all polynomials $\mathcal{R}[x]$ in x over $\mathcal{R}$ is a ring with respect to addition and multiplication as defined above.

Let $\alpha(x)$ and $\beta(x)$ have respective degrees m and n. If $m \neq n$, the degree of $\alpha(x) + \beta(x)$ is the larger of m,n; if $m = n$, the degree of $\alpha(x) + \beta(x)$ is *at most m* (why?). The degree of $\alpha(x) \cdot \beta(x)$ is at most $m + n$ since a_mb_n may be z. However, if $\mathcal{R}$ is free of divisors of zero, the degree of the product is $m + n$. (Whenever convenient we shall follow practice and write a polynomial of degree m as consisting of no more than $m + 1$ terms.)

Consider now the subset $S = \{rx^0 : r \in \mathcal{R}\}$ of $\mathcal{R}[x]$ consisting of the zero polynomial and all polynomials of degree zero. It is easily verified that the mapping

$$\mathcal{R} \to S: r \to rx^0$$

is an isomorphism. As a consequence, we may hereafter write a_0 for a_0x^0 in any polynomial $\alpha(x) \in \mathcal{R}[x]$.

MONIC POLYNOMIALS

Let $\mathcal{R}$ be a ring with unity u. Then $u = ux^0$ is the unity of $\mathcal{R}[x]$ since $ux^0 \cdot \alpha(x) = \alpha(x)$ for every $\alpha(x) \in \mathcal{R}[x]$. Also, writing $x = ux^1 = zx^0 + ux^1$, we have $x \in \mathcal{R}[x]$. Now $a_k(x \cdot x \cdot x \cdots \text{to } k \text{ factors}) = a_kx^k \in \mathcal{R}[x]$ so that in $\alpha(x) = a_0 + a_1x + a_2x^2 + \cdots$ we may consider the superscript i in a_ix^i as truly an exponent, juxtaposition in any term a_ix^i as (polynomial) ring multiplication, and the connective $+$ as (polynomial) ring addition.

Any polynomial $\alpha(x)$ of degree m over $\mathcal{R}$ with leading coefficient u, the unity of $\mathcal{R}$, will be called *monic*.

Example 2: (*a*) The polynomials 1, $x+3$, and $x^2 - 5x + 4$ are monic while $2x^2 - x + 5$ is not a monic polynomial over I (or any ring having I as a subring).

(*b*) The polynomials b, $bx + f$, and $bx^2 + dx + e$ are monic polynomials in $S[x]$ over the ring S of Example 1(*d*), Chapter 11, Page 114.

DIVISION

In Problem 1 we prove the first part of

Theorem II. Let $\mathcal{R}$ be a ring with unity u, $\alpha(x) = a_0 + a_1x + a_2x^2 + \cdots + a_mx^m \in \mathcal{R}[x]$ be either the zero polynomial or a polynomial of degree m, and $\beta(x) = b_0 + b_1x + b_2x^2 + \cdots + ux^n \in \mathcal{R}[x]$ be a monic polynomial of degree n. Then there exist unique polynomials $q_R(x), r_R(x); q_L(x), r_L(x) \in \mathcal{R}[x]$ with $r_R(x), r_L(x)$ either the zero polynomial or of degree $< n$ such that

$$\text{(i)} \qquad \alpha(x) = q_R(x) \cdot \beta(x) + r_R(x)$$

and

$$\text{(ii)} \qquad \alpha(x) = \beta(x) \cdot q_L(x) + r_L(x)$$

In (i) of Theorem II we say that $\alpha(x)$ has been divided *on the right* by $\beta(x)$ to obtain the *right quotient* $q_R(x)$ and *right remainder* $r_R(x)$. Similarly, in (ii) we say that $\alpha(x)$ has been divided *on the left* by $\beta(x)$ to obtain the *left quotient* $q_L(x)$ and *left remainder* $r_L(x)$. When $r_R(x) = z$ $(r_L(x) = z)$, we call $\beta(x)$ a right (left) *divisor* of $\alpha(x)$.

For the special case $\beta(x) = ux - b = x - b$, Theorem II yields (see Problem 2),

Theorem III. The right and left remainders when $\alpha(x)$ is divided by $x - b$, $b \in \mathcal{R}$, are respectively

$$r_R = a_0 + a_1b + a_2b^2 + \cdots + a_nb^n$$

and

$$r_L = a_0 + ba_1 + b^2a_2 + \cdots + b^na_n$$

There follows

Theorem IV. A polynomial $\alpha(x)$ has $x - b$ as right (left) divisor if and only if $r_R = z$ $(r_L = z)$.

Examples illustrating Theorems II-IV when $\mathcal{R}$ is non-commutative will be deferred until Chapter 15. The remainder of this chapter will be devoted to the study of certain polynomial rings $\mathcal{R}[x]$ obtained by further specializing the coefficient ring $\mathcal{R}$.

COMMUTATIVE POLYNOMIAL RINGS WITH UNITY

Let $\mathcal{R}$ be a commutative ring with unity. Then $\mathcal{R}[x]$ is a commutative ring with unity (what is its unity?) and Theorems II-IV may be restated without distinction between right and left quotients (we replace $q_R(x) = q_L(x)$ by $q(x)$), remainders (we replace $r_R(x) = r_L(x)$ by $r(x)$), and divisors. Thus (i) and (ii) of Theorem II may be replaced by

$$\text{(iii)} \qquad \alpha(x) \;=\; q(x)\cdot\beta(x) + r(x)$$

and, in particular, we have

Theorem IV′. In a commutative polynomial ring with unity, a polynomial $\alpha(x)$ of degree m has $x - b$ as divisor if and only if the remainder

$$(a) \qquad r \;=\; a_0 + a_1b + a_2b^2 + \cdots + a_mb^m \;=\; z$$

When, as in Theorem IV′, $r = z$ then b is called a *zero* (*root*) of the polynomial $\alpha(x)$.

Example 3: (*a*) The polynomial $x^2 - 4$ over I has 2 and -2 as zeros since $(2)^2 - 4 = 0$ and $(-2)^2 - 4 = 0$.

(*b*) The polynomial $[3]x^2 - [4]$ over the ring $I/(8)$ has $[2]$ and $[6]$ as zeros while the polynomial $[1]x^2 - [1]$ over $I/(8)$ has $[1], [3], [5], [7]$ as zeros.

When $\mathcal{R}$ is without divisors of zero so also is $\mathcal{R}[x]$. For, suppose $\alpha(x)$ and $\beta(x)$ are elements of $\mathcal{R}[x]$, of respective degrees m and n, and that

$$\alpha(x)\cdot\beta(x) \;=\; a_0b_0 + (a_0b_1 + a_1b_0)x + \cdots + a_mb_nx^{m+n} \;=\; z$$

Then each coefficient in the product and, in particular a_mb_n, is z. But $\mathcal{R}$ is without divisors of zero; hence $a_mb_n = z$ if and only if $a_m = z$ or $b_n = z$. Since this contradicts the assumption that $\alpha(x)$ and $\beta(x)$ have degrees m and n, $\mathcal{R}[x]$ is without divisors of zero.

There follows

Theorem V. A polynomial ring $\mathcal{R}[x]$ is an integral domain if and only if the coefficient ring $\mathcal{R}$ is an integral domain.

SUBSTITUTION PROCESS

An examination of the remainder

$$(a) \qquad r \;=\; a_0 + a_1b + a_2b^2 + \cdots + a_mb^m$$

in Theorem IV′ shows that it may be obtained mechanically by replacing x by b throughout $\alpha(x)$ and, of course, interpreting juxtaposition of elements as indicating multiplication in $\mathcal{R}$. Thus, by defining $f(b)$ to mean the expression obtained by substituting b for x throughout $f(x)$, we may (and will hereafter) replace r in (*a*) by $\alpha(b)$. This is, to be sure, the familiar substitution process in elementary algebra where (let it be noted) x is considered a variable rather than an indeterminate.

It will be left for the reader to show that the substitution process will not lead to future difficulties, that is, to show that for a given $b \in \mathcal{R}$, the mapping

$$f(x) \to f(b) \qquad \text{for all } f(x) \in \mathcal{R}[x]$$

is a homomorphism of $\mathcal{R}[x]$ onto $\mathcal{R}$.

THE POLYNOMIAL DOMAIN $\mathcal{F}[x]$

The most important polynomial domains arise when the coefficient ring is a field $\mathcal{F}$. We recall that every non-zero element of a field $\mathcal{F}$ is a unit of $\mathcal{F}$ and restate for the integral domain $\mathcal{F}[x]$ the principal results of the sections above as follows:

The Division Algorithm. If $\alpha(x), \beta(x) \in \mathcal{F}[x]$ where $\beta(x) \neq z$, there exist unique polynomials $q(x), r(x)$ with $r(x)$ either the zero polynomial or of degree less than that of $\beta(x)$, such that

$$\alpha(x) = q(x) \cdot \beta(x) + r(x)$$

For a proof, see Problem 4.

When $r(x)$ is the zero polynomial, $\beta(x)$ is called a *divisor* of $\alpha(x)$ and we write $\beta(x) \mid \alpha(x)$.

The Remainder Theorem. If $\alpha(x)$, $x - b \in \mathcal{F}[x]$, the remainder when $\alpha(x)$ is divided by $x - b$ is $\alpha(b)$.

The Factor Theorem. If $\alpha(x) \in \mathcal{F}[x]$ and $b \in \mathcal{F}$, then $x - b$ is a *factor* of $\alpha(x)$ if and only if $\alpha(b) = z$, that is, $x - b$ is a factor of $\alpha(x)$ if and only if b is a zero of $\alpha(x)$.

There follow

Theorem VI. Let $\alpha(x) \in \mathcal{F}[x]$ have degree $m > 0$ and leading coefficient a. If the distinct elements $b_1, b_2, \ldots, b_m$ of $\mathcal{F}$ are zeros of $\alpha(x)$, then

$$\alpha(x) = a(x - b_1)(x - b_2) \cdots (x - b_m)$$

For a proof, see Problem 5.

Theorem VII. Every polynomial $\alpha(x) \in \mathcal{F}[x]$ of degree $m > 0$ has at most m distinct zeros in $\mathcal{F}$.

Example 4: (*a*) The polynomial $2x^2 + 7x - 15 \in Q[x]$ has the zeros $3/2, -5, \in Q$.

(*b*) The polynomial $x^2 + 2x + 3 \in C[x]$ has the zeros $-1 + \sqrt{2}\,i$ and $-1 - \sqrt{2}\,i$ over C. However, $x^2 + 2x + 3 \in Q[x]$ has no zeros in Q.

Theorem VIII. Let $\alpha(x), \beta(x) \in \mathcal{F}[x]$ be such that $\alpha(s) = \beta(s)$ for every $s \in \mathcal{F}$. Then, if the number of elements in $\mathcal{F}$ exceeds the degrees of both $\alpha(x)$ and $\beta(x)$, we have necessarily $\alpha(x) = \beta(x)$.

For a proof, see Problem 6.

Example 5: It is now clear that the polynomials of Example 1 are *distinct*, whether considered as functions or as forms, since the number of elements of $\mathcal{F} = I/(5)$ does not exceed the degree of both polynomials. What then appeared in Example 1 to be a contradiction of the reader's past experience was due, of course, to the fact that his past experience had been limited solely to infinite fields.

PRIME POLYNOMIALS

It is not difficult to show that the only units of a polynomial domain $\mathcal{F}[x]$ are the non-zero elements (i.e., the units) of the coefficient ring $\mathcal{F}$. Thus the only associates of $\alpha(x) \in \mathcal{F}[x]$ are the elements $v \cdot \alpha(x)$ of $\mathcal{F}[x]$ in which v is any unit of $\mathcal{F}$.

Since for any $v \neq z \in \mathcal{F}$ and any $\alpha(x) \in \mathcal{F}[x]$,

$$\alpha(x) = v^{-1} \cdot \alpha(x) \cdot v$$

while, whenever $\alpha(x) = q(x) \cdot \beta(x)$,

$$\alpha(x) = [v^{-1} q(x)] \cdot [v \cdot \beta(x)]$$

it follows that (*a*) every unit of $\mathcal{F}$ and every associate of $\alpha(x)$ is a divisor of $\alpha(x)$ and (*b*) if $\beta(x) \mid \alpha(x)$ so also does every associate of $\beta(x)$. The units of $\mathcal{F}$ and the associates of $\alpha(x)$ are called *trivial divisors* of $\alpha(x)$. Other divisors of $\alpha(x)$, if any, are called *non-trivial divisors*.

A polynomial $\alpha(x) \in \mathcal{F}[x]$ of degree $m \geqq 1$ is called a *prime (irreducible) polynomial* over $\mathcal{F}$ if its only divisors are trivial.

Example 6: (a) The polynomial $3x^2+2x+1 \in R[x]$ is a prime polynomial over R.

(b) Every polynomial $ax+b \in \mathcal{F}[x]$, with $a \neq z$, is a prime polynomial over $\mathcal{F}$.

THE POLYNOMIAL DOMAIN $C[x]$

Consider an arbitrary polynomial

$$\beta(x) = b_0 + b_1x + b_2x^2 + \cdots + b_mx^m \in C[x]$$

of degree $m \geqq 1$. We shall be concerned in this section with a number of elementary theorems having to do with the zeros of such polynomials and, in particular, with the subset of all polynomials of $C[x]$ whose coefficients are rational numbers. Most of the theorems will be found in any College Algebra text stated, however, in terms of roots of equations rather than in terms of zeros of polynomials.

Suppose $r \in C$ is a zero of $\beta(x)$. Then $\beta(r)=0$ and, since $b_m^{-1} \in C$, also $b_m^{-1} \cdot \beta(r) = 0$. Thus the zeros of $\beta(x)$ are precisely those of its monic associate

$$\alpha(x) = b_m^{-1} \cdot \beta(x) = a_0 + a_1x + a_2x^2 + \cdots + a_{m-1}x^{m-1} + x^m$$

Whenever more convenient, we shall deal with monic polynomials.

It is well-known that when $m=1$, $\alpha(x)=a_0+x$ has $-a_0$ as zero and when $m=2$, $\alpha(x)=a_0+a_1x+x^2$ has $\frac{1}{2}(-a_1-\sqrt{a_1^2-4a_0})$ and $\frac{1}{2}(-a_1+\sqrt{a_1^2-4a_0})$ as zeros. In Chapter 8, it was shown how to find the n roots of any $a \in C$; thus every polynomial $x^n - a \in C[x]$ has at least n zeros over C. There exist formulas (see Problems 16-19) which yield the zeros of all polynomials of degrees 3 and 4. It is also known that no formulas can be devised for arbitrary polynomials of degree $m \geqq 5$.

By Theorem VII any polynomial $\alpha(x)$ of degree $m \geqq 1$ can have no more than m distinct zeros. In the paragraph above, $\alpha(x)=a_0+a_1x+x^2$ will have two distinct zeros if and only if its discriminant $a_1^2-4a_0 \neq 0$. We shall then call each a *simple zero* of $\alpha(x)$. However, if $a_1^2-4a_0=0$, each formula yields $-\frac{1}{2}a_1$ as a zero. We shall then call $-\frac{1}{2}a_1$ a zero of *multiplicity* two of $\alpha(x)$ and exhibit the zeros as $-\frac{1}{2}a_1, -\frac{1}{2}a_1$.

Example 7: (a) The polynomial $x^3+x^2-5x+3 = (x-1)^2(x+3)$ has -3 as simple zero and 1 as zero of multiplicity two.

(b) The polynomial $x^4-x^3-3x^2+5x-2 = (x-1)^3(x+2)$ has -2 as simple zero and 1 as zero of multiplicity three.

The so-called

Fundamental Theorem of Algebra. Every polynomial $\alpha(x) \in C[x]$ of degree $m \geqq 1$ has at least one zero in C.

will be assumed here as a postulate. There follows, by induction

Theorem IX. Every polynomial $\alpha(x) \in C[x]$ of degree $m \geqq 1$ has precisely m zeros over C, with the understanding that any zero of multiplicity n is to be counted as n of the m zeros.

and, hence,

Theorem X. Any $\alpha(x) \in C[x]$ of degree $m \geqq 1$ is either of the first degree or may be written as the product of polynomials $\in C[x]$ each of the first degree.

Except for the special cases noted above, the problem of finding the zeros of a given polynomial is a difficult one and will not be considered here. In the remainder of this section we shall limit our attention to certain subsets of $C[x]$ obtained by restricting the ring of coefficients.

First, let us suppose that

$$\alpha(x) \;=\; a_0 + a_1x + a_2x^2 + \cdots + a_mx^m \;\in\; R[x]$$

of degree $m \geqq 1$ has $r = a + bi$ as zero, i.e.,

$$\alpha(r) \;=\; a_0 + a_1r + a_2r^2 + \cdots + a_mr^m \;=\; s + ti \;=\; 0$$

By Problem 2, Chapter 8, Page 79, we have

$$\alpha(\bar{r}) \;=\; a_0 + a_1\bar{r} + a_2\bar{r}^2 + \cdots + a_m\bar{r}^m \;=\; \overline{s + ti} \;=\; 0$$

so that

Theorem XI. If $r \in C$ is a zero of any polynomial $\alpha(x)$ with real coefficients, then $\bar{r}$ is also a zero of $\alpha(x)$.

Let $r = a + bi$, with $b \neq 0$, be a zero of $\alpha(x)$. By Theorem XI $\bar{r} = a - bi$ is also a zero and we may write

$$\begin{aligned}\alpha(x) \;&=\; [x - (a+bi)][x - (a-bi)] \cdot \alpha_1(x)\\ &=\; [x^2 - 2ax + a^2 + b^2] \cdot \alpha_1(x)\end{aligned}$$

where $\alpha_1(x)$ is a polynomial of degree two less than that of $\alpha(x)$ and has real coefficients. Since a quadratic polynomial with real coefficients will have imaginary zeros if and only if its discriminant is negative, we have

Theorem XII. The polynomials of the first degree and the quadratic polynomials with negative discriminant are the only polynomials $\in R[x]$ which are primes over R.

and

Theorem XIII. A polynomial of odd degree $\in R[x]$ necessarily has a real zero.

Suppose next that

$$\beta(x) \;=\; b_0 + b_1x + b_2x^2 + \cdots + b_mx^m \;\in\; Q[x]$$

Let c be the greatest common divisor of the numerators of the b's and d be the least common multiple of the denominators of the b's; then

$$\alpha(x) \;=\; \frac{d}{c} \cdot \beta(x) \;=\; a_0 + a_1x + a_2x^2 + \cdots + a_mx^m \;\in\; Q[x]$$

has integral coefficients whose only common divisors are ± 1, the units of I. Moreover, $\beta(x)$ and $\alpha(x)$ have precisely the same zeros.

If $r \in Q$ is a zero of $\alpha(x)$, i.e. if

$$\alpha(r) \;=\; a_0 + a_1r + a_2r^2 + \cdots + a_mr^m \;=\; 0$$

there follow readily

(i) if $r \in I$, then $r \mid a_0$

(ii) if $r = s/t$, a common fraction in lowest terms, then

$$t^m \cdot \alpha(s/t) \;=\; a_0t^m + a_1st^{m-1} + a_2s^2t^{m-2} + \cdots + a_{m-1}s^{m-1}t + a_ms^m \;=\; 0$$

so that $s \mid a_0$ and $t \mid a_m$. We have proved

Theorem XIV. Let $\alpha(x) \;=\; a_0 + a_1x + a_2x^2 + \cdots + a_mx^m$ be a polynomial of degree $m \geqq 1$ having integral coefficients. If $s/t \in Q$, with $(s,t) = 1$, is a zero of $\alpha(x)$, then $s \mid a_0$ and $t \mid a_m$.

Example 8: (a) The possible rational zeros of

$$\alpha(x) = 3x^3 + 2x^2 - 7x + 2$$

are $\pm 1, \pm 2, \pm\frac{1}{3}, \pm\frac{2}{3}$. Now $\alpha(1) = 0$, $\alpha(-1) \neq 0$, $\alpha(2) \neq 0$, $\alpha(-2) = 0$, $\alpha(\frac{1}{3}) = 0$, $\alpha(-\frac{1}{3}) \neq 0$, $\alpha(\frac{2}{3}) \neq 0$, $\alpha(-\frac{2}{3}) \neq 0$ so that the rational zeros are $1, -2, \frac{1}{3}$ and $\alpha(x) = 3(x-1)(x+2)(x-\frac{1}{3})$.

Note. By Theorem VII, $\alpha(x)$ can have no more than three distinct zeros. Thus once these have been found, all other possibilities untested may be discarded. It was not necessary here to test the possibilities $-\frac{1}{3}, \frac{2}{3}, -\frac{2}{3}$.

(b) The possible rational zeros of

$$\alpha(x) = 4x^5 - 4x^4 - 5x^3 + 5x^2 + x - 1$$

are $\pm 1, \pm\frac{1}{2}, \pm\frac{1}{4}$. Now $\alpha(1) = 0$, $\alpha(-1) = 0$, $\alpha(\frac{1}{2}) = 0$, $\alpha(-\frac{1}{2}) = 0$ so that

$$\alpha(x) = 4(x-1)(x+1)(x-\tfrac{1}{2})(x+\tfrac{1}{2})\cdot(x-1)$$

and the rational zeros are $1, 1, -1, \frac{1}{2}, -\frac{1}{2}$.

(c) The possible rational zeros of

$$\alpha(x) = x^4 - 2x^3 - 5x^2 + 4x + 6$$

are $\pm 1, \pm 2, \pm 3, \pm 6$. For these, only $\alpha(-1) = 0$ and $\alpha(3) = 0$ so that

$$\alpha(x) = (x+1)(x-3)(x^2-2)$$

Since none of the possible zeros $\pm 1, \pm 2$ of $x^2 - 2$ are zeros, it follows that $x^2 - 2$ is a prime polynomial over Q and the only rational zeros of $\alpha(x)$ are $-1, 3$.

(d) Of the possible rational zeros: $\pm 1, \pm\frac{1}{2}, \pm\frac{1}{3}, \pm\frac{1}{6}$, of $\alpha(x) = 6x^4 - 5x^3 + 7x^2 - 5x + 1$ only $\frac{1}{2}$ and $\frac{1}{3}$ are zeros. Then $\alpha(x) = 6(x-\frac{1}{2})(x-\frac{1}{3})(x^2+1)$ so that $x^2 + 1$ is a prime polynomial over Q, and the rational zeros of $\alpha(x)$ are $\frac{1}{2}, \frac{1}{3}$.

(e) The possible rational zeros of

$$\alpha(x) = 3x^4 - 6x^3 + 4x^2 - 10x + 2$$

are $\pm 1, \pm 2, \pm\frac{1}{3}, \pm\frac{2}{3}$. Since none, in fact, is a zero, $\alpha(x)$ is a prime polynomial over Q.

GREATEST COMMON DIVISOR

Let $\alpha(x)$ and $\beta(x)$ be non-zero polynomials in $\mathcal{F}[x]$. The polynomial $d(x) \in \mathcal{F}[x]$ having the properties

(1) $d(x)$ is monic,

(2) $d(x) \mid \alpha(x)$ and $d(x) \mid \beta(x)$,

(3) for every $c(x) \in \mathcal{F}[x]$ such that $c(x) \mid \alpha(x)$ and $c(x) \mid \beta(x)$, we have $c(x) \mid d(x)$,

is called the greatest common divisor of $\alpha(x)$ and $\beta(x)$.

It is evident (see Problem 7) that the greatest common divisor of two polynomials in $\mathcal{F}[x]$ can be found in the same manner as the greatest common divisor of two integers in Chapter 5. For the sake of variety, we prove in Problem 8

Theorem XV. Let the non-zero polynomials $\alpha(x)$ and $\beta(x)$ be in $\mathcal{F}[x]$. The monic polynomial

$$(b) \qquad d(x) = s(x)\cdot\alpha(x) + t(x)\cdot\beta(x), \qquad s(x), t(x) \in \mathcal{F}[x]$$

of least degree is the greatest common divisor of $\alpha(x)$ and $\beta(x)$.

There follow

Theorem XVI. Let $\alpha(x)$ of degree $m \geqq 2$ and $\beta(x)$ of degree $n \geqq 2$ be in $\mathcal{F}[x]$. Then non-zero polynomials $\mu(x)$ of degree at most $n-1$ and $\nu(x)$ of degree at most $m-1$ exist in $\mathcal{F}[x]$ such that

$$(c) \qquad \mu(x)\cdot\alpha(x) + \nu(x)\cdot\beta(x) = z$$

if and only if $\alpha(x)$ and $\beta(x)$ are not relatively prime.

For a proof, see Problem 9.

and

Theorem XVII. If $\alpha(x), \beta(x), p(x) \in \mathcal{F}[x]$ with $\alpha(x)$ and $p(x)$ relatively prime, then $p(x) \mid \alpha(x) \cdot \beta(x)$ implies $p(x) \mid \beta(x)$.

In Problem 10, we prove

The Unique Factorization Theorem. Any polynomial $\alpha(x)$, of degree $m \geqq 1$ and with leading coefficient a, in $\mathcal{F}[x]$ can be written as

$$\alpha(x) = a \cdot [p_1(x)]^{m_1} \cdot [p_2(x)]^{m_2} \cdots [p_j(x)]^{m_j}$$

where the $p_i(x)$ are monic prime polynomials over $\mathcal{F}$ and the m_i are positive integers. Moreover, except for the order of the factors, the factorization is unique.

Example 9: Decompose $\alpha(x) = 4x^4 + 3x^3 + 4x^2 + 4x + 6$ over $I/(7)$ into a product of prime polynomials.

We have, with the understanding that all coefficients are residue classes modulo 7,

$$\begin{aligned} \alpha(x) &= 4x^4 + 3x^3 + 4x^2 + 4x + 6 = 4x^4 + 24x^3 + 4x^2 + 4x + 20 \\ &= 4(x^4 + 6x^3 + x^2 + x + 5) = 4(x+1)(x^3 + 5x^2 + 3x + 5) \\ &= 4(x+1)(x+3)(x^2 + 2x + 4) = 4(x+1)(x+3)(x+3)(x+6) \\ &= 4(x+1)(x+3)^2(x+6) \end{aligned}$$

PROPERTIES OF THE POLYNOMIAL DOMAIN $\mathcal{F}[x]$

The ring of polynomials $\mathcal{F}[x]$ over a field $\mathcal{F}$ has a number of properties which parallel those of the ring I of integers. For example, each has prime elements, each is a Euclidean ring, (see Problem 11), and each is a principal ideal ring (see Theorem IX, Chapter 10, Page 108). Moreover, and this will be our primary concern here, $\mathcal{F}[x]$ may be partitioned by any polynomial $\lambda(x) \in \mathcal{F}[x]$ of degree $n \geqq 1$ into a ring

$$\mathcal{F}[x]/(\lambda(x)) = \{[\alpha(x)], [\beta(x)], \ldots\}$$

of equivalence classes just as I was partitioned into the ring $I/(m)$. For any $\alpha(x), \beta(x) \in \mathcal{F}[x]$ we define

(i) $$[\alpha(x)] = \{\alpha(x) + \mu(x) \cdot \lambda(x) : \mu(x) \in \mathcal{F}[x]\}$$

Then $\alpha(x) \in [\alpha(x)]$, since the zero element of $\mathcal{F}$ is also an element of $\mathcal{F}[x]$, and $[\alpha(x)] = [\beta(x)]$ if and only if $\alpha(x) \equiv \beta(x) \pmod{\lambda(x)}$, i.e. if and only if $\lambda(x) \mid (\alpha(x) - \beta(x))$.

We now define addition and multiplication on these equivalence classes by

$$[\alpha(x)] + [\beta(x)] = [\alpha(x) + \beta(x)]$$

and

$$[\alpha(x)] \cdot [\beta(x)] = [\alpha(x) \cdot \beta(x)]$$

respectively and leave for the reader to prove

(*a*) Addition and multiplication are well-defined operations on $\mathcal{F}[x]/(\lambda(x))$.

(*b*) $\mathcal{F}[x]/(\lambda(x))$ has $[z]$ as zero element and $[u]$ as unity, where z and u respectively are the zero and unity of $\mathcal{F}$.

(*c*) $\mathcal{F}[x]/(\lambda(x))$ is a commutative ring with unity.

In Problem 12, we prove

Theorem XVIII. The ring $\mathcal{F}[x]/(\lambda(x))$ contains a subring which is isomorphic to the field $\mathcal{F}$.

If $\lambda(x)$ is of degree 1, it is clear that $\mathcal{F}[x]/(\lambda(x))$ is the field $\mathcal{F}$; if $\lambda(x)$ is of degree 2, $\mathcal{F}[x]/(\lambda(x))$ consists of $\mathcal{F}$ together with all equivalence classes $\{[a_0 + a_1x] : a_0, a_1 \in \mathcal{F}, a_1 \neq z\}$; in general, if $\lambda(x)$ is of degree n, we have

$$\mathcal{F}[x]/(\lambda(x)) = \{[a_0 + a_1x + a_2x^2 + \cdots + a_{n-1}x^{n-1}] : a_i \in \mathcal{F}\}$$

Now the definitions of addition and multiplication on equivalence classes and the isomorphism: $a_i \leftrightarrow [a_i]$ imply

$$\begin{aligned}[a_0 + a_1x + a_2x^2 + \cdots + a_{n-1}x^{n-1}] &= [a_0] + [a_1][x] + [a_2]\cdot[x]^2 + \cdots + [a_{n-1}]\cdot[x]^{n-1} \\ &= a_0 + a_1[x] + a_2[x]^2 + \cdots + a_{n-1}[x]^{n-1}\end{aligned}$$

As a final simplification, let $[x]$ be replaced by ξ so that we have

$$\mathcal{F}[x]/(\lambda(x)) = \{a_0 + a_1\xi + a_2\xi^2 + \cdots + a_{n-1}\xi^{n-1} : a_i \in \mathcal{F}\}$$

In Problem 13, we prove

Theorem XIX. The ring $\mathcal{F}[x]/(\lambda(x))$ is a field if and only if $\lambda(x)$ is a prime polynomial over $\mathcal{F}$.

Example 10: Consider $\lambda(x) = x^2 - 3 \in Q[x]$, a prime polynomial over Q. Now

$$Q[x]/(x^2 - 3) = \{a_0 + a_1\xi : a_0, a_1 \in Q\}$$

is a field with respect to addition and multiplication defined as usual except that in multiplication ξ^2 is to be replaced by 3. It is easy to show that the mapping

$$a_0 + a_1\xi \leftrightarrow a_0 + a_1\sqrt{3}$$

is an isomorphism of $Q[x]/(x^2 - 3)$ onto

$$Q[\sqrt{3}] = \{a_0 + a_1\sqrt{3} : a_0, a_1 \in Q\},$$

the set of all polynomials in $\sqrt{3}$ over Q. Clearly $Q[\sqrt{3}] \subset R$ so that $Q[\sqrt{3}]$ is the smallest field in which $x^2 - 3$ factors completely.

The polynomial $x^2 - 3$ of Example 10 is the monic polynomial over Q of least degree having $\sqrt{3}$ as a root. Being unique, it is called the *minimum polynomial of* $\sqrt{3}$ over Q. Note that the minimum polynomial of $\sqrt{3}$ over R is $x - \sqrt{3}$.

Example 11: Let $\mathcal{F} = I/(3) = \{0, 1, 2\}$ and take $\lambda(x) = x^2 + 1$, a prime polynomial over $\mathcal{F}$. Construct the addition and multiplication table for the field $\mathcal{F}[x]/(\lambda(x))$.

Here
$$\begin{aligned}\mathcal{F}[x]/(\lambda(x)) &= \{a_0 + a_1\xi : a_0, a_1 \in \mathcal{F}\} \\ &= \{0, 1, 2, \xi, 2\xi, 1+\xi, 1+2\xi, 2+\xi, 2+2\xi\}\end{aligned}$$

Since $\lambda(\xi) = \xi^2 + 1 = [0]$, we have $\xi^2 = [-1] = [2]$, or 2. The required tables are:

$+$	0	1	2	ξ	2ξ	$1+\xi$	$1+2\xi$	$2+\xi$	$2+2\xi$
0	0	1	2	ξ	2ξ	$1+\xi$	$1+2\xi$	$2+\xi$	$2+2\xi$
1	1	2	0	$1+\xi$	$1+2\xi$	$2+\xi$	$2+2\xi$	ξ	2ξ
2	2	0	1	$2+\xi$	$2+2\xi$	ξ	2ξ	$1+\xi$	$1+2\xi$
ξ	ξ	$1+\xi$	$2+\xi$	2ξ	0	$1+2\xi$	1	$2+2\xi$	2
2ξ	2ξ	$1+2\xi$	$2+2\xi$	0	ξ	1	$1+\xi$	2	$2+\xi$
$1+\xi$	$1+\xi$	$2+\xi$	ξ	$1+2\xi$	1	$2+2\xi$	2	2ξ	0
$1+2\xi$	$1+2\xi$	$2+2\xi$	2ξ	1	$1+\xi$	2	$2+\xi$	0	ξ
$2+\xi$	$2+\xi$	ξ	$1+\xi$	$2+2\xi$	2	2ξ	0	$1+2\xi$	1
$2+2\xi$	$2+2\xi$	2ξ	$1+2\xi$	2	$2+\xi$	0	ξ	1	$1+\xi$

Table 12-1

$\cdot$	0	1	2	ξ	2ξ	$1+\xi$	$1+2\xi$	$2+\xi$	$2+2\xi$
0	0	0	0	0	0	0	0	0	0
1	0	1	2	ξ	2ξ	$1+\xi$	$1+2\xi$	$2+\xi$	$2+2\xi$
2	0	2	1	2ξ	ξ	$2+2\xi$	$2+\xi$	$1+2\xi$	$1+\xi$
ξ	0	ξ	2ξ	2	1	$2+\xi$	$1+\xi$	$2+2\xi$	$1+2\xi$
2ξ	0	2ξ	ξ	1	2	$1+2\xi$	$2+2\xi$	$1+\xi$	$2+\xi$
$1+\xi$	0	$1+\xi$	$2+2\xi$	$2+\xi$	$1+2\xi$	2ξ	2	1	ξ
$1+2\xi$	0	$1+2\xi$	$2+\xi$	$1+\xi$	$2+2\xi$	2	ξ	2ξ	1
$2+\xi$	0	$2+\xi$	$1+2\xi$	$2+2\xi$	$1+\xi$	1	2ξ	ξ	2
$2+2\xi$	0	$2+2\xi$	$1+\xi$	$1+2\xi$	$2+\xi$	ξ	1	2	2ξ

Table 12-2

Example 12: Let $\mathcal{F} = Q$ and take $\lambda(x) = x^3 + x + 1$, a prime polynomial over $\mathcal{F}$. Find the multiplicative inverse of $\xi^2 + \xi + 1 \in \mathcal{F}[x]/(\lambda(x))$.

Here $\mathcal{F}[x]/(\lambda(x)) = \{a_0 + a_1\xi + a_2\xi^2 : a_0, a_1, a_2 \in Q\}$ and, since $\lambda(\xi) = \xi^3 + \xi + 1 = 0$, we have $\xi^3 = -1 - \xi$ and $\xi^4 = -\xi - \xi^2$. One procedure for finding the required inverse is:

set $(a_0 + a_1\xi + a_2\xi^2)(1 + \xi + \xi^2) = 1$,
multiply out and substitute for ξ^3 and ξ^4,
equate the corresponding coefficients of ξ^0, ξ, ξ^2
and solve for a_0, a_1, a_2.

This usually proves more tedious than to follow the proof of the existence of the inverse in Problem 13. Thus, using the division algorithm, we find

$$1 = \tfrac{1}{3}(\xi^3 + \xi + 1)(1 - \xi) + \tfrac{1}{3}(\xi^2 + \xi + 1)(\xi^2 - 2\xi + 2)$$

Then $$(\xi^3 + \xi + 1) \mid [1 - \tfrac{1}{3}(\xi^2 + \xi + 1)(\xi^2 - 2\xi + 2)]$$

so that $$\tfrac{1}{3}(\xi^2 + \xi + 1)(\xi^2 - 2\xi + 2) = 1$$

and $\frac{1}{3}(\xi^2 - 2\xi + 2)$ is the required inverse.

Example 13: Show that the field $R[x]/(x^2 + 1)$ is isomorphic to C.

We have $R[x]/(x^2 + 1) = \{a_0 + a_1\xi : a_0, a_1 \in R\}$. Since $\xi^2 = -1$, the mapping

$$a_0 + a_1\xi \rightarrow a_0 + a_1 i$$

is an isomorphism of $R[x]/(x^2 + 1)$ onto C. We have then a second method of constructing the field of complex numbers from the field of real numbers. It is, however, not possible to use such a procedure to construct the field of real numbers from the rationals.

Example 13 illustrates

Theorem XX. If $\alpha(x)$ of degree $m \geqq 2$ is an element in $\mathcal{F}[x]$, then there exists a field $\mathcal{F}'$, where $\mathcal{F} \subset \mathcal{F}'$, in which $\alpha(x)$ has a zero.

For a proof, see Problem 14.

It is to be noted that over the field $\mathcal{F}'$ of Theorem XX, $\alpha(x)$ may or may not be written as the product of m factors each of degree one. However, if $\alpha(x)$ does not factor completely over $\mathcal{F}'$, it has a prime factor of degree $n \geqq 2$ which may be used to obtain a field $\mathcal{F}''$, with $\mathcal{F} \subset \mathcal{F}' \subset \mathcal{F}''$, in which $\alpha(x)$ has another zero. Since $\alpha(x)$ has only a finite number of zeros, the procedure can always be repeated a sufficient number of times to ultimately produce a field $\mathcal{F}^{(i)}$ in which $\alpha(x)$ factors completely.

See Problem 15.

Solved Problems

1. Prove: Let $\mathcal{R}$ be a ring with unity u; let

$$\alpha(x) = a_0 + a_1x + a_2x^2 + \cdots + a_mx^m \in \mathcal{R}[x]$$

be either the zero polynomial or of degree m; and let

$$\beta(x) = b_0 + b_1x + b_2x^2 + \cdots + ux^n \in \mathcal{R}[x]$$

be a monic polynomial of degree n. Then there exist unique polynomials $q_R(x), r_R(x) \in \mathcal{R}[x]$ with $r_R(x)$ either the zero polynomial or of degree $< n$ such that

$$\text{(i)} \qquad \alpha(x) = q_R(x) \cdot \beta(x) + r_R(x)$$

If $\alpha(x)$ is the zero polynomial or if $n > m$, then (i) holds with $q_R(x) = z$ and $r_R(x) = \alpha(x)$.

Let $n \leq m$. The theorem is again trivial if $m = 0$ or if $m = 1$ and $n = 0$. For the case $m = n = 1$, take $\alpha(x) = a_0 + a_1x$ and $\beta(x) = b_0 + ux$. Then

$$\alpha(x) = a_0 + a_1x = a_1(b_0 + ux) + (a_0 - a_1b_0)$$

and the theorem is true with $q_R(x) = a_1$ and $r_R(x) = a_0 - a_1b_0$.

We shall now use the induction principle of Problem 27, Chapter 3, Page 37. For this purpose we assume the theorem true for all $\alpha(x)$ of degree $\leq m-1$ and consider $\alpha(x)$ of degree m. Now $\gamma(x) = \alpha(x) - a_mx^{n-m} \cdot \beta(x) \in \mathcal{R}[x]$ and has degree $< m$. By assumption,

$$\gamma(x) = \delta(x) \cdot \beta(x) + r(x)$$

with $r(x)$ of degree at most $n-1$. Then

$$\begin{aligned} \alpha(x) = \gamma(x) + a_mx^{n-m} \cdot \beta(x) &= \left(\delta(x) + a_mx^{n-m}\right) \cdot \beta(x) + r(x) \\ &= q_R(x) \cdot \beta(x) + r_R(x) \end{aligned}$$

where $q_R(x) = \delta(x) + a_mx^{n-m}$ and $r_R(x) = r(x)$, as required.

To prove uniqueness, suppose

$$\alpha(x) = q_R(x) \cdot \beta(x) + r_R(x) = q'_R(x) \cdot \beta(x) + r'_R(x)$$

Then

$$\left(q_R(x) - q'_R(x)\right) \cdot \beta(x) = r'_R(x) - r_R(x)$$

Now $r'_R(x) - r_R(x)$ has degree at most $n-1$ while, unless $q_R(x) - q'_R(x) = z$, $\left(q_R(x) - q'_R(x)\right) \cdot \beta(x)$ has degree at least n. Thus $q_R(x) - q'_R(x) = z$ and then $r'_R(x) - r_R(x) = z$, which establishes uniqueness.

2. Prove: The right remainder when $\alpha(x) = a_0 + a_1x + a_2x^2 + \cdots + a_mx^m \in \mathcal{R}[x]$ is divided by $x - b$, $b \in \mathcal{R}$, is $r_R = a_0 + a_1b + a_2b^2 + \cdots + a_mb^m$.

Consider

$$\begin{aligned} \alpha(x) - r_R &= a_1(x-b) + a_2(x^2 - b^2) + \cdots + a_m(x^m - b^m) \\ &= \{a_1 + a_2(x+b) + \cdots + a_m(x^{m-1} + bx^{m-2} + \cdots + b^{m-1})\} \cdot (x - b) \\ &= q_R(x) \cdot (x-b) \end{aligned}$$

Then

$$\alpha(x) = q_R(x) \cdot (x-b) + r_R$$

By Problem 1 the right remainder is unique; hence r_R is the required right remainder.

3. In the polynomial ring $S[x]$ over S, the ring of Example 1(*d*), Chapter 11, Page 114,

(*a*) Find the remainder when $\alpha(x) = cx^4 + dx^3 + cx^2 + hx + g$ is divided by $\beta(x) = bx^2 + fx + d$.

(*b*) Verify that f is a zero of $\gamma(x) = cx^4 + dx^3 + cx^2 + bx + d$.

(a) We proceed as in ordinary division, with one variation. Since S is of characteristic two, $s+(-t)=s+t$ for all $s,t \in S$; hence, in the step "change the signs and add" we shall simply add.

The remainder is $r(x) = gx+f$.

$$\begin{array}{r|l} & cx^2 + hx + b \\ \hline bx^2+fx+d & cx^4 + dx^3 + cx^2 + hx + g \\ & cx^4 + ex^3 + fx^2 \\ \hline & hx^3 + hx^2 + hx \\ & hx^3 + gx^2 + ex \\ \hline & bx^2 + dx + g \\ & bx^2 + fx + d \\ \hline & gx + f \end{array}$$

(b) Here $f^2 = c$, $f^3 = cf = e$, and $f^4 = h$. Then

$$\begin{aligned} \gamma(f) &= ch + de + c^2 + bf + d \\ &= d + c + h + f + d = a \end{aligned}$$

as required.

4. Prove the Division Algorithm as stated for the polynomial domain $\mathcal{F}[x]$.

Note. The requirement that $\beta(x)$ be monic in Theorem II, Page 126, and its restatement for commutative rings with unity was necessary.

Suppose now that $\alpha(x), \beta(x) \in \mathcal{F}[x]$ with $b_n \neq u$ the leading coefficient of $\beta(x)$. Then, since b_n^{-1} always exists in $\mathcal{F}$ and $\beta'(x) = b_n^{-1} \cdot \beta(x)$ is monic, we may write

$$\begin{aligned} \alpha(x) &= q'(x) \cdot \beta'(x) + r(x) \\ &= [b_n^{-1} q'(x)] \cdot [b_n \cdot \beta'(x)] + r(x) \\ &= q(x) \cdot \beta(x) + r(x) \end{aligned}$$

with $r(x)$ either the zero polynomial or of degree less than that of $\beta(x)$.

5. Prove: Let $\alpha(x) \in \mathcal{F}[x]$ have degree $m > 0$ and leading coefficient a. If the distinct elements $b_1, b_2, \ldots, b_m$ of $\mathcal{F}$ are zeros of $\alpha(x)$, then $\alpha(x) = a(x-b_1)(x-b_2)\cdots(x-b_m)$.

Suppose $m=1$ so that $\alpha(x) = ax + a_1$ has, say, b_1 as zero. Then $\alpha(b_1) = ab_1 + a_1 = z$, $a_1 = -ab_1$, and

$$\alpha(x) = ax + a_1 = ax - ab_1 = a(x - b_1)$$

The theorem is true for $m = 1$.

Now assume the theorem to be true for $m = k$ and consider $\alpha(x)$ of degree $k+1$ with zeros $b_1, b_2, \ldots, b_{k+1}$. Since b_1 is a zero of $\alpha(x)$, we have by the Factor Theorem

$$\alpha(x) = q(x) \cdot (x - b_1)$$

where $q(x)$ is of degree k with leading coefficient a. Since $\alpha(b_j) = q(b_j) \cdot (b_j - b_1) = z$ for $j = 2, 3, \ldots, k+1$ and since $b_j - b_1 \neq z$ for all $j \neq 1$, it follows that $b_2, b_3, \ldots, b_{k+1}$ are k distinct zeros of $q(x)$. By assumption,

$$q(x) = a(x-b_2)(x-b_3)\cdots(x-b_{k+1})$$

Then

$$\alpha(x) = a(x-b_1)(x-b_2)\cdots(x-b_{k+1})$$

and the proof by induction is complete.

6. Prove: Let $\alpha(x), \beta(x) \in \mathcal{F}[x]$ be such that $\alpha(s) = \beta(s)$ for every $s \in \mathcal{F}$. Then, if the number of elements of $\mathcal{F}$ exceeds the degree of both $\alpha(x)$ and $\beta(x)$, we have necessarily $\alpha(x) = \beta(x)$.

Set $\gamma(x) = \alpha(x) - \beta(x)$. Now $\gamma(x)$ is either the zero polynomial or is of degree p which certainly does not exceed the greater of the degrees of $\alpha(x)$ and $\beta(x)$. By hypothesis, $\gamma(s) = \alpha(s) - \beta(s)$ for every $s \in \mathcal{F}$. Then $\gamma(x) = z$ (otherwise $\gamma(x)$ would have more zeros than its degree, contrary to Theorem VII, Page 128) and $\alpha(x) = \beta(x)$ as required.

7. Find the greatest common divisor of $\alpha(x) = 6x^5 + 7x^4 - 5x^3 - 2x^2 - x + 1$ and $\beta(x) = 6x^4 - 5x^3 - 19x^2 - 13x - 5$ over Q and express it in the form

$$d(x) = s(x) \cdot \alpha(x) + t(x) \cdot \beta(x)$$

Proceeding as in the corresponding problem with integers, we find

$$\alpha(x) = (x+2) \cdot \beta(x) + (24x^3 + 49x^2 + 30x + 11) = q_1(x) \cdot \beta(x) + r_1(x)$$

$$\beta(x) = \frac{1}{32}(8x - 23) \cdot r_1(x) + \frac{93}{32}(3x^2 + 2x + 1) = q_2(x) \cdot r_1(x) + r_2(x)$$

$$r_1(x) = \frac{1}{93}(256x + 352) \cdot r_2(x)$$

Since $r_2(x)$ is not monic, it is an associate of the required greatest common divisor $d(x) = x^2 + \frac{2}{3}x + \frac{1}{3}$.

Now
$$\begin{aligned} r_2(x) &= \beta(x) - q_2(x) \cdot r_1(x) \\ &= \beta(x) - q_2(x) \cdot \alpha(x) + q_1(x) \cdot q_2(x) \cdot \beta(x) \\ &= -q_2(x) \cdot \alpha(x) + \big(1 + q_1(x) \cdot q_2(x)\big) \cdot \beta(x) \\ &= -\frac{1}{32}(8x - 23) \cdot \alpha(x) + \frac{1}{32}(8x^2 - 7x - 14) \cdot \beta(x) \end{aligned}$$

and
$$d(x) = \frac{32}{279} r_2(x) = -\frac{1}{279}(8x - 23) \cdot \alpha(x) + \frac{1}{279}(8x^2 - 7x - 14) \cdot \beta(x)$$

8. Prove: Let the non-zero polynomials $\alpha(x)$ and $\beta(x)$ be in $\mathcal{F}[x]$. The monic polynomial

$$d(x) = s_0(x) \cdot \alpha(x) + t_0(x) \cdot \beta(x), \qquad s_0(x), t_0(x) \in \mathcal{F}[x]$$

of least degree is the greatest common divisor of $\alpha(x)$ and $\beta(x)$.

Consider the set

$$S = \{s(x) \cdot \alpha(x) + t(x) \cdot \beta(x) : s(x), t(x) \in \mathcal{F}[x]\}$$

Clearly this is a non-empty subset of $\mathcal{F}[x]$ and, hence, contains a non-zero polynomial $\delta(x)$ of least degree. For any $b(x) \in S$, we have, by the Division Algorithm, $b(x) = q(x) \cdot \delta(x) + r(x)$ where $r(x) \in S$ (prove this) is either the zero polynomial or has degree less than that of $\delta(x)$. Then $r(x) = z$ and $b(x) = q(x) \cdot \delta(x)$ so that every element of S is a multiple of $\delta(x)$. Hence, $\delta(x) \mid \alpha(x)$ and $\delta(x) \mid \beta(x)$. Moreover, since $\delta(x) = \bar{s}_0(x) \cdot \alpha(x) + \bar{t}_0(x) \cdot \beta(x)$, any common divisor $c(x)$ of $\alpha(x)$ and $\beta(x)$ is a divisor of $\delta(x)$. Now if $\delta(x)$ is monic, it is the greatest common divisor $d(x)$ of $\alpha(x)$ and $\beta(x)$; otherwise, there exists a unit v such that $v \cdot \delta(x)$ is monic and $d(x) = v \cdot \delta(x)$ is the required greatest common divisor.

9. Prove: Let $\alpha(x)$ of degree $m \geqq 2$ and $\beta(x)$ of degree $n \geqq 2$ be in $\mathcal{F}[x]$. Then non-zero polynomials $\mu(x)$ of degree at most $n-1$ and $\nu(x)$ of degree at most $m-1$ exist in $\mathcal{F}[x]$ such that

$$(c) \qquad \mu(x) \cdot \alpha(x) + \nu(x) \cdot \beta(x) = z$$

if and only if $\alpha(x)$ and $\beta(x)$ are not relatively prime.

Suppose $\delta(x)$ of degree $p \geqq 1$ is the greatest common divisor of $\alpha(x)$ and $\beta(x)$, and write

$$\alpha(x) = \alpha_0(x) \cdot \delta(x), \qquad \beta(x) = \beta_0(x) \cdot \delta(x)$$

Clearly $\alpha_0(x)$ has degree $\leqq m-1$ and $\beta_0(x)$ has degree $\leqq n-1$. Moreover,

$$\beta_0(x) \cdot \alpha(x) = \beta_0(x) \cdot \alpha_0(x) \cdot \delta(x) = \alpha_0(x) \cdot [\beta_0(x) \cdot \delta(x)] = \alpha_0(x) \cdot \beta(x)$$

so that
$$\beta_0(x) \cdot \alpha(x) + [-\alpha_0(x) \cdot \beta(x)] = z$$

and we have (c) with $\mu(x) = \beta_0(x)$ and $\nu(x) = -\alpha_0(x)$.

Conversely, suppose (c) holds with $\alpha(x)$ and $\beta(x)$ relatively prime. By Theorem XV, Page 131, we have

$$u = s(x)\cdot\alpha(x) + t(x)\cdot\beta(x) \qquad \text{for some } s(x), t(x) \in \mathcal{F}[x]$$

Then

$$\begin{aligned}\mu(x) &= \mu(x)\cdot s(x)\cdot\alpha(x) + \mu(x)\cdot t(x)\cdot\beta(x)\\ &= s(x)\,[-\nu(x)\cdot\beta(x)] + \mu(x)\cdot t(x)\cdot\beta(x)\\ &= \beta(x)\,[\mu(x)\cdot t(x) - \nu(x)\cdot s(x)]\end{aligned}$$

and $\beta(x) \mid \mu(x)$. But this is impossible; hence (c) does not hold if $\alpha(x)$ and $\beta(x)$ are relatively prime.

10. Prove: The unique factorization theorem holds in $\mathcal{F}[x]$.

Consider an arbitrary $\alpha(x) \in \mathcal{F}[x]$. If $\alpha(x)$ is a prime polynomial, the theorem is trivial. If $\alpha(x)$ is reducible, write $\alpha(x) = a\cdot\beta(x)\cdot\gamma(x)$ where $\beta(x)$ and $\gamma(x)$ are monic polynomials of positive degree less than that of $\alpha(x)$. Now either $\beta(x)$ and $\gamma(x)$ are prime polynomials, as required in the theorem, or one or both are reducible and may be written as the product of two monic polynomials. If all factors are primes, we have the theorem; otherwise, ... This process cannot be continued indefinitely (for example, in the extreme case we would obtain $\alpha(x)$ as the product of m polynomials each of degree one). The proof of uniqueness is left for the reader, who also may wish to use the induction procedure of Problem 27, Chapter 3, Page 37, in the first part of the proof.

11. Prove: The polynomial ring $\mathcal{F}[x]$ over the field $\mathcal{F}$ is a Euclidean ring.

For each non-zero polynomial $\alpha(x) \in \mathcal{F}[x]$, define $\theta(\alpha) = m$ where m is the degree of $\alpha(x)$. If $\alpha(x), \beta(x) \in \mathcal{F}[x]$ have respective degrees m and n, it follows that $\theta(\alpha) = m$, $\theta(\beta) = n$, $\theta(\alpha\cdot\beta) = m+n$ and, hence, $\theta(\alpha\cdot\beta) \geqq \theta(\alpha)$. Now we have already established the division algorithm:

$$\alpha(x) = q(x)\cdot\beta(x) + r(x)$$

where $r(x)$ is either z or of degree less than that of $\beta(x)$. Thus, either $r(x) = z$ or $\theta(r) < \theta(\beta)$ as required.

12. Prove: The ring $\mathcal{F}[x]/(\lambda(x))$ contains a subring which is isomorphic to the field $\mathcal{F}$.

Let a, b be distinct elements of $\mathcal{F}$; then $[a], [b]$ are distinct elements of $\mathcal{F}[x]/(\lambda(x))$ since $[a] = [b]$ if and only if $\lambda(x) \mid (a - b)$.

Now the mapping $a \to [a]$ is an isomorphism of $\mathcal{F}$ onto a subset of $\mathcal{F}[x]/(\lambda(x))$ since it is one-to-one and the operations of addition and multiplication are preserved. It will be left for the reader to show that this subset is a subring of $\mathcal{F}[x]/(\lambda(x))$.

13. Prove: The ring $\mathcal{F}[x]/(\lambda(x))$ is a field if and only if $\lambda(x)$ is a prime polynomial over $\mathcal{F}$.

Suppose $\lambda(x)$ is a prime polynomial over $\mathcal{F}$. Then for any $[\alpha(x)] \neq [z]$ of $\mathcal{F}[x]/(\lambda(x))$, we have by Theorem XV, Page 131,

$$u = \alpha(x)\cdot\beta(x) + \lambda(x)\cdot\gamma(x) \qquad \text{for some } \beta(x), \gamma(x) \in \mathcal{F}[x]$$

Now $\lambda(x) \mid u - \alpha(x)\cdot\beta(x)$ so that $[\alpha(x)]\cdot[\beta(x)] = [u]$. Hence, every non-zero element $[\alpha(x)] \in \mathcal{F}[x]/(\lambda(x))$ has a multiplicative inverse and $\mathcal{F}[x]/(\lambda(x))$ is a field.

Suppose $\lambda(x)$ of degree $m \geqq 2$ is not a prime polynomial over $\mathcal{F}$, i.e. suppose $\lambda(x) = \mu(x)\cdot\nu(x)$ where $\mu(x), \nu(x) \in \mathcal{F}[x]$ have positive degrees s and t such that $s + t = m$. Then $s < m$ so that $\lambda(x) \nmid \mu(x)$ and $[\mu(x)] \neq [z]$; similarly, $[\nu(x)] \neq [z]$. But $[\mu(x)]\cdot[\nu(x)] = [\mu(x)\cdot\nu(x)] = [\lambda(x)] = [z]$. Thus, since $[\mu(x)], [\nu(x)] \in \mathcal{F}[x]/(\lambda(x))$, it follows that $\mathcal{F}[x]/(\lambda(x))$ has divisors of zero and is not a field.

14. Prove: If $\alpha(x)$ of degree $m \geqq 2$ is an element of $\mathcal{F}[x]$, there exists a field $\mathcal{F}'$, where $\mathcal{F} \subset \mathcal{F}'$, in which $\alpha(x)$ has a zero.

The theorem is trivial if $\alpha(x)$ has a zero in $\mathcal{F}$; suppose that it does not. Then there exists a monic prime polynomial $\lambda(x) \in \mathcal{F}[x]$ of degree $n \geqq 2$ such that $\lambda(x) \mid \alpha(x)$. Since $\lambda(x)$ is prime over $\mathcal{F}$, define $\mathcal{F}' = \mathcal{F}[x]/(\lambda(x))$.

Now by Theorem XVIII, Page 132, $\mathcal{F} \subset \mathcal{F}'$ so that $\alpha(x) \in \mathcal{F}'[x]$. Also, there exists $\xi \in \mathcal{F}'$ such that $\lambda(\xi) = [z]$. Thus ξ is a zero of $\alpha(x)$ and $\mathcal{F}'$ is a field which meets the requirement of the theorem.

15. Find a field in which $x^3 - 3 \in Q[x]$ (*a*) has a factor, (*b*) factors completely.

Consider the field $\quad Q[x]/(x^3-3) \quad = \quad \{a_0 + a_1\xi + a_2\xi^2 : \; a_0, a_1, a_2 \in Q\}$

(*a*) The field just defined is isomorphic to

$$\mathcal{F}' \quad = \quad \{a_0 + a_1\sqrt[3]{3} + a_2\sqrt[3]{9} : \; a_0, a_1, a_2 \in Q\}$$

in which $x^3 - 3$ has a zero.

(*b*) Since the zeros of $x^3 - 3$ are $\sqrt[3]{3}, \sqrt[3]{3}\,\omega, \sqrt[3]{3}\,\omega^2$, it is clear that $x^3 - 3$ factors completely in $\mathcal{F}'' = \mathcal{F}'[\omega]$.

16. Derive formulas for the zeros of the cubic polynomial $\alpha(x) = a_0 + a_1x + a_2x^2 + x^3$ over C when $a_0 \neq 0$.

The derivation rests basically on two changes to new variables:

(i) If $a_2 = 0$, use $x = y$ and proceed as in (ii) below; if $a_2 \neq 0$, use $x = y + v$ and choose v so that the resulting cubic lacks the term in y^2. Since the coefficient of this term is $a_2 + 3v$, the proper relation is $x = y - a_2/3$. Let the resulting polynomial be

$$\beta(y) \quad = \quad \alpha(y - a_2/3) \quad = \quad q + py + y^3$$

If $q = 0$, the zeros of $\beta(y)$ are $0, \sqrt{-p}, -\sqrt{-p}$ and the zeros of $\alpha(x)$ are obtained by decreasing each zero of $\beta(y)$ by $a_2/3$. If $q \neq 0$ but $p = 0$, the zeros of $\beta(y)$ are the three cube roots $\rho, \omega\rho, \omega^2\rho$ (see Chapter 8) of $-q$ from which the zeros of $\alpha(x)$ are obtained as before. For the case $pq \neq 0$,

(ii) Use $y = z - p/3z$ to obtain

$$\gamma(z) \quad = \quad \beta(z - p/3z) \quad = \quad z^3 + q - p^3/27z^3 \quad = \quad \frac{z^6 + qz^3 - p^3/27}{z^3}$$

Now any zero, say s, of the polynomial $\delta(z) = z^6 + qz^3 - p^3/27$ yields the zero $s - p/3s - a_2/3$ of $\alpha(x)$; the six zeros of $\delta(z)$ yield, as can be shown, each zero of $\alpha(x)$ twice. Write

$$\delta(z) \quad = \quad [z^3 + \tfrac{1}{2}(q - \sqrt{q^2 + 4p^3/27}\,)] \cdot [z^3 + \tfrac{1}{2}(q + \sqrt{q^2 + 4p^3/27}\,)]$$

and denote the zeros of $z^3 + \frac{1}{2}(q - \sqrt{q^2 + 4p^3/27}\,)$ by $A, \omega A, \omega^2 A$. The zeros of $\alpha(x)$ are then: $A - p/3A - a_2/3$, $\omega A - \omega^2 p/3A - a_2/3$, and $\omega^2 A - \omega p/3A - a_2/3$.

17. Find the zeros of $\alpha(x) = -11 - 3x + 3x^2 + x^3$.

The substitution $x = y - 1$ yields

$$\beta(y) \quad = \quad \alpha(y-1) \quad = \quad -6 - 6y + y^3$$

In turn, the substitution $y = z + 2/z$ yields

$$\gamma(z) \quad = \quad \beta(z + 2/z) \quad = \quad z^3 + 8/z^3 - 6 \quad = \quad \frac{z^6 - 6z^3 + 8}{z^3} \quad = \quad \frac{(z^3-2)(z^3-4)}{z^3}$$

Take $A = \sqrt[3]{2}$; then the zeros of $\alpha(x)$ are:

$$\sqrt[3]{2} + \sqrt[3]{4} - 1, \quad \omega\sqrt[3]{2} + \omega^2\sqrt[3]{4} - 1, \quad \omega^2\sqrt[3]{2} + \omega\sqrt[3]{4} - 1$$

The reader will now show that, apart from order, these zeros are obtained by taking $A = \sqrt[3]{4}$.

18. Derive a procedure for finding the zeros of the quartic polynomial

$$\alpha(x) = a_0 + a_1x + a_2x^2 + a_3x^3 + x^4 \in C[x], \quad \text{when } a_0 \neq 0.$$

If $a_3 \neq 0$, use $x = y - a_3/4$ to obtain

$$\beta(y) = \alpha(y - a_3/4) = b_0 + b_1y + b_2y^2 + y^4$$

Now

$$\begin{aligned}\beta(y) &= (p + 2qy + y^2)(r - 2qy + y^2)\\ &= pr + 2q(r-p)y + (r+p-4q^2)y^2 + y^4\end{aligned}$$

provided there exist $p, q, r \in C$ satisfying

$$pr = b_0, \quad 2q(r-p) = b_1, \quad r + p - 4q^2 = b_2$$

If $b_1 = 0$, take $q = 0$; otherwise, with $q \neq 0$, we find

$$2p = b_2 + 4q^2 - b_1/2q \quad \text{and} \quad 2r = b_2 + 4q^2 + b_1/2q$$

Since $2p \cdot 2r = 4b_0$, we have

$$\text{(i)} \qquad 64q^6 + 32b_2q^4 + 4(b_2^2 - 4b_0)q^2 - b_1^2 = 0$$

Thus, considering the left member of (i) as a cubic polynomial in q^2, any of its zeros different from 0 will yield the required factorization. Then, having the four zeros of $\beta(y)$, the zeros of $\alpha(x)$ are obtained by decreasing each by $a_3/4$.

19. Find the zeros of $\alpha(x) = 35 - 16x - 4x^3 + x^4$.

Using $x = y + 1$, we obtain

$$\beta(y) = \alpha(y+1) = 16 - 24y - 6y^2 + y^4$$

Here, (i) of Problem 18 becomes

$$64q^6 - 192q^4 - 112q^2 - 576 = 16(q^2-4)(4q^4+4q^2+9) = 0$$

Take $q = 2$; then $p = 8$ and $r = 2$ so that

$$16 - 24y - 6y^2 + y^4 = (8 + 4y + y^2)(2 - 4y + y^2)$$

with zeros $-2 \pm 2i$ and $2 \pm \sqrt{2}$. The zeros of $\alpha(x)$ are: $-1 \pm 2i$ and $3 \pm \sqrt{2}$.

Supplementary Problems

20. Give an example of two polynomials in x of degree 3 with integral coefficients whose sum is of degree 2.

21. Find the sum and product of each pair of polynomials over the indicated coefficient ring. (For convenience, $[a], [b], \ldots$ have been replaced by $a, b, \ldots$.)

(a) $4 + x + 2x^2$, $1 + 2x + 3x^2$; $I/(5)$

(b) $1 + 5x + 2x^2$, $7 + 2x + 3x^2 + 4x^3$; $I/(8)$

(c) $2 + 2x + x^3$, $1 + x + x^2 + x^4$; $I/(3)$

Ans. (a) $3x$; $4 + 4x + x^2 + 2x^3 + x^4$

(c) $x^2 + x^3 + x^4$; $2 + x + x^2 + x^7$

22. In the polynomial ring $S[x]$ over S, the ring of Problem 2, Chapter 10, Page 108, verify:

(a) $(b + gx + fx^2) + (d + gx) = c + ex + fx^2$

(b) $(b + gx + fx^2)(d + cx) = b + hx + cx^2 + bx^3$

(c) $(b + gx + fx^2)(d + ex) = b + cx + bx^2$

(d) $f + bx$ and $e + ex$ are divisors of zero.

(e) c is a zero of $f + cx + fx^2 + ex^3 + dx^4$.

23. Given $\alpha(x), \beta(x), \gamma(x) \in \mathcal{F}[x]$ with respective leading coefficients a, b, c and suppose $\alpha(x) = \beta(x) \cdot \gamma(x)$. Show that $\alpha(x) = a \cdot \beta'(x) \cdot \gamma'(x)$ where $\beta'(x)$ and $\gamma'(x)$ are monic polynomials.

24. Show that $\mathcal{D}[x]$ is not a field for any integral domain $\mathcal{D}$.
Hint. Let $\alpha(x) \in \mathcal{D}[x]$ have degree > 0 and assume $\beta(x) \in \mathcal{D}[x]$ a multiplicative inverse of $\alpha(x)$. Then $\alpha(x) \cdot \beta(x)$ has degree > 0, a contradiction.

25. Factor each of the following into products of prime polynomials over (i) Q, (ii) R, (iii) C.

(a) $x^4 - 1$ (c) $6x^4 + 5x^3 + 4x^2 - 2x - 1$

(b) $x^4 - 4x^2 - x + 2$ (d) $4x^5 + 4x^4 - 13x^3 - 11x^2 + 10x + 6$

Ans. (a) $(x-1)(x+1)(x^2+1)$ over Q, R; $(x-1)(x+1)(x-i)(x+i)$ over C
(d) $(x-1)(2x+1)(2x+3)(x^2-2)$ over Q; $(x-1)(2x+1)(2x+3)(x-\sqrt{2})(x+\sqrt{2})$ over R, C

26. Factor each of the following into products of prime polynomials over the indicated field. (See note in Problem 21.)

(a) $x^2 + 1$; $I/(5)$ (c) $2x^2 + 2x + 1$; $I/(5)$

(b) $x^2 + x + 1$; $I/(3)$ (d) $3x^3 + 4x^2 + 3$; $I/(5)$

Ans. (a) $(x+2)(x+3)$, (d) $(x+2)^2(3x+2)$

27. Factor $x^4 - 1$ over (a) $I/(11)$, (b) $I/(13)$.

28. In (d) of Problem 26 obtain also $3x^3 + 4x^2 + 3 = (x+2)(x+4)(3x+1)$. Explain why this does not contradict the Unique Factorization Theorem.

29. In the polynomial ring $S[x]$ over S, the ring of Example 1(d), Chapter 11, Page 114,

(a) Show that $bx^2 + ex + g$ and $gx^2 + dx + b$ are prime polynomials.

(b) Factor $hx^4 + ex^3 + cx^2 + b$.

Ans. (b) $(bx + b)(cx + g)(gx + d)(hx + e)$

30. Find all zeros over C of the polynomials of Problem 25.

31. Find all zeros of the polynomials of Problem 26. *Ans.* (a) 2, 3; (d) 1, 3, 3

32. Find all zeros of the polynomial of Problem 29(b). *Ans.* b, e, f, d

33. List all polynomials of the form $3x^2 + cx + 4$ which are prime over $I/(5)$.
Ans. $3x^2 + 4$, $3x^2 + x + 4$, $3x^2 + 4x + 4$

34. List all polynomials of degree 4 which are prime over $I/(2)$.

35. Prove: Theorems VII, IX, and XIII.

36. Prove: If $a + b\sqrt{c}$, with $a, b, c \in Q$ and c not a perfect square, is a zero of $\alpha(x) \in I[x]$, so also is $a - b\sqrt{c}$.

37. Let $\mathcal{R}$ be a commutative ring with unity. Show that $\mathcal{R}[x]$ is a principal ideal ring. What are its prime ideals?

38. Form polynomials $\alpha(x) \in I[x]$ of least degree having among its zeros:

(a) $\sqrt{3}$ and 2 (c) 1 and $2+3\sqrt{5}$ (e) $1+i$ of multiplicity 2

(b) i and 3 (d) $-1+i$ and $2-3i$

Ans. (a) $x^3 - 2x^2 - 3x + 6$, (d) $x^4 - 2x^3 + 7x^2 + 18x + 26$

39. Verify that the minimum polynomial of $\sqrt{3} + 2i$ over R is of degree 2 and over Q is of degree 4.

40. Find the greatest common divisor of each pair $\alpha(x), \beta(x)$ over the indicated coefficient ring and express it in the form $s(x) \cdot \alpha(x) + t(x) \cdot \beta(x)$.

(a) $x^5 + x^4 - x^3 - 3x + 2, \; x^3 + 2x^2 - x - 2$; Q

(b) $3x^4 - 6x^3 + 12x^2 + 8x - 6, \; x^3 - 3x^2 + 6x - 3$; Q

(c) $x^5 - 3ix^3 - 2ix^2 - 6, \; x^2 - 2i$; C

(d) $x^5 + 3x^3 + x^2 + 2x + 2, \; x^4 + 3x^3 + 3x^2 + x + 2$; $I/(5)$

(e) $x^5 + x^3 + x, \; x^4 + 2x^3 + 2x$; $I/(3)$

(f) $cx^4 + hx^3 + ax^2 + gx + e, \; gx^3 + hx^2 + dx + g$; S of Example 1(d), Chapter 11, Page 114.

Ans. (b) $\frac{1}{447}(37x^2 - 108x + 177) \cdot \alpha(x) + \frac{1}{447}(-111x^3 + 213x^2 - 318x - 503) \cdot \beta(x)$

(d) $(x+2) \cdot \alpha(x) + (4x^2 + x + 2) \cdot \beta(x)$

(f) $(gx+h) \cdot \alpha(x) + (cx^2 + hx + e) \cdot \beta(x)$

41. Prove Theorem XVII, Page 132.

42. Show that every polynomial of degree 2 in $\mathcal{F}[x]$ of Example 11, Page 133, factors completely in $\mathcal{F}[x]/(x^2+1)$ by exhibiting the factors.

Partial Ans. $x^2 + 1 = (x+\xi)(x+2\xi)$; $x^2 + x + 2 = (x+\xi+2)(x+2\xi+2)$;

$2x^2 + x + 1 = (x+\xi+1)(2x+\xi+2)$

43. Discuss the field $Q[x]/(x^3-3)$. (See Example 10, Page 133.)

44. Find the multiplicative inverse of $\xi^2 + 2$ in (a) $Q[x]/(x^3+x+2)$, (b) $\mathcal{F}[x]/(x^2+x+1)$ when $\mathcal{F} = I/(5)$.

Ans. (a) $\frac{1}{6}(1 - 2\xi - \xi^2)$, (b) $\frac{1}{3}(\xi + 2)$

Chapter 13

Vector Spaces

INTRODUCTION

In this chapter we shall define and study a type of algebraic system called a vector space. Before making a formal definition, we recall that in elementary physics one deals with two types of quantities: (*a*) scalars (time, temperature, speed) which have magnitude only and (*b*) vectors (force, velocity, acceleration) which have both magnitude and direction. Such vectors are frequently represented by arrows. For example, consider in Fig. 13-1 a given plane in which a rectangular coordinate system has been established and a vector $\xi_1 = OP = (a, b)$ joining the origin to the point $P(a, b)$. The magnitude of ξ_1 (length of OP) is given by $r = \sqrt{a^2 + b^2}$ and the direction (the angle θ, always measured from the positive x-axis) is determined by any two of the relations $\sin\theta = b/r$, $\cos\theta = a/r$, $\tan\theta = b/a$.

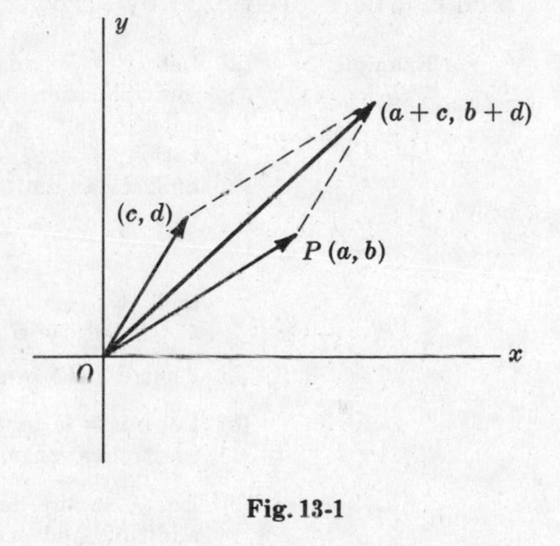

Fig. 13-1

Two operations are defined on these vectors:

Scalar Multiplication. Let the vector $\xi_1 = (a, b)$ represent a force at O. The product of the scalar 3 and the vector ξ_1 defined by $3\xi_1 = (3a, 3b)$ represents a force at O having the direction of ξ_1 and three times its magnitude. Similarly, $-2\xi_1$ represents a force at O having twice the magnitude of ξ_1 but with direction opposite that of ξ_1.

Vector Addition. If $\xi_1 = (a, b)$ and $\xi_2 = (c, d)$ represent two forces at O, their resultant ξ (the single force at O having the same effect as the two forces ξ_1 and ξ_2) is given by $\xi = \xi_1 + \xi_2 = (a + c, b + d)$ obtained by means of the parallelogram law.

In the example above it is evident that every scalar $s \in R$ and every vector $\xi \in R \times R$. There can be no confusion then in using $(+)$ to denote addition of vectors as well as addition of scalars.

Denote by V the set of all vectors in the plane, (i.e., $V = R \times R$). Now V has a zero element $\zeta = (0, 0)$ and every $\xi = (a, b) \in V$ has an additive inverse $-\xi = (-a, -b) \in V$ such that $\xi + (-\xi) = \zeta$; in fact, V is an abelian additive group. Moreover, for all $s, t \in R$ and $\xi, \eta \in V$, the following properties hold:

$$s(\xi + \eta) = s\xi + s\eta \qquad (s + t)\xi = s\xi + t\xi$$

$$s(t\xi) = (st)\xi \qquad 1\xi = \xi$$

Example 1: Consider the vectors $\xi = (1, 2)$, $\eta = (\frac{1}{2}, 0)$, $\sigma = (0, -3/2)$. Then

(*a*) $3\xi = 3(1, 2) = (3, 6)$, $2\eta = (1, 0)$, and $3\xi + 2\eta = (3, 6) + (1, 0) = (4, 6)$.

(*b*) $\xi + 2\eta = (2, 2)$, $\eta + \sigma = (\frac{1}{2}, -3/2)$, and $5(\xi + 2\eta) - 4(\eta + \sigma) = (8, 16)$.

VECTOR SPACES

Let $\mathcal{F}$ be a field and V be an abelian additive group such that there is a scalar multiplication of V by $\mathcal{F}$ which associates with each $s \in \mathcal{F}$ and $\xi \in V$ the element $s\xi \in V$. Then V is called a *vector space over* $\mathcal{F}$ provided, with u the unity of $\mathcal{F}$,

$$\text{(i)}\quad s(\xi+\eta) = s\xi + s\eta \qquad\qquad \text{(iii)}\quad s(t\xi) = (st)\xi$$

$$\text{(ii)}\quad (s+t)\xi = s\xi + t\xi \qquad\qquad \text{(iv)}\quad u\xi = \xi$$

hold for all $s,t \in \mathcal{F}$ and $\xi,\eta \in V$.

It is evident that, in fashioning the definition of a vector space, the set of all plane vectors of the section above was used as a guide. However, as will be seen from the examples below, the elements of a vector space, i.e. the vectors, are not necessarily quantities which can be represented by arrows.

Example 2: (*a*) Let $\mathcal{F} = R$ and $V = V_2(R) = \{(a_1, a_2) : a_1, a_2 \in R\}$ with addition and scalar multiplication defined as in the first section. Then, of course, V is a vector space over $\mathcal{F}$; in fact, we list the example in order to point out a simple generalization: Let $\mathcal{F} = R$ and $V = V_n(R) = \{(a_1, a_2, \ldots, a_n) : a_i \in R\}$ with addition and scalar multiplication defined by

$$\begin{aligned} \xi + \eta &= (a_1, a_2, \ldots, a_n) + (b_1, b_2, \ldots, b_n) \\ &= (a_1 + b_1, a_2 + b_2, \ldots, a_n + b_n), \qquad \xi, \eta \in V_n(R) \end{aligned}$$

and

$$s\xi = s(a_1, a_2, \ldots, a_n) = (sa_1, sa_2, \ldots, sa_n), \qquad s \in R,\ \xi \in V_n(R)$$

Then $V_n(R)$ is a vector space over R.

(*b*) Let $\mathcal{F} = R$ and $V_n(C) = \{(a_1, a_2, \ldots, a_n) : a_i \in C\}$ with addition and scalar multiplication as in (*a*). Then $V_n(C)$ is a vector space over R.

(*c*) Let $\mathcal{F}$ be any field, $V = \mathcal{F}[x]$ be the polynomial domain in x over $\mathcal{F}$, and define addition and scalar multiplication as ordinary addition and multiplication in $\mathcal{F}[x]$. Then V is a vector space over $\mathcal{F}$.

Let $\mathcal{F}$ be a field with zero element z and V be a vector space over $\mathcal{F}$. Since V is an abelian additive group, it has a unique zero element ζ and, for each element $\xi \in V$, there exists a unique additive inverse $-\xi$ such that $\xi + (-\xi) = \zeta$. By means of the distributive laws (i) and (ii), we find for all $s \in \mathcal{F}$ and $\xi \in V$,

$$s\xi + z\xi = (s+z)\xi = s\xi = s\xi + \zeta$$

and

$$s\xi + s\zeta = s(\xi + \zeta) = s\xi = s\xi + \zeta$$

Hence, $z\xi = \zeta$ and $s\zeta = \zeta$.

We state these properties, together with others which will be left for the reader to establish, as

Theorem I. In a vector space V over $\mathcal{F}$ with z the zero element of $\mathcal{F}$ and ζ the zero element of V, we have

(1) $s\zeta = \zeta$ for all $s \in \mathcal{F}$

(2) $z\xi = \zeta$ for all $\xi \in V$

(3) $(-s)\xi = s(-\xi) = -(s\xi)$ for all $s \in \mathcal{F}$ and $\xi \in V$

(4) If $s\xi = \zeta$, then $s = z$ or $\xi = \zeta$

SUBSPACE OF A VECTOR SPACE

A non-empty subset U of a vector space V over $\mathcal{F}$ is a *subspace* of V provided U is itself a vector space over $\mathcal{F}$ with respect to the operations defined on V. There follows

Theorem II. A non-empty subset U of a vector space V over $\mathcal{F}$ is a subspace of V if and only if U is closed with respect to scalar multiplication and vector addition as defined on V.

For a proof, see Problem 1.

Example 3: Consider the vector space $V = V_3(R) = \{(a, b, c) : a, b, c \in R\}$ over R. By Theorem II, the subset $U = \{(a, b, 0) : a, b \in R\}$ is a subspace of V since for all $s \in R$ and $(a, b, 0), (c, d, 0) \in U$, we have

$$(a, b, 0) + (c, d, 0) = (a + c, b + d, 0) \in U$$

and

$$s(a, b, 0) = (sa, sb, 0) \in U$$

In Example 3, V is the set of all vectors in ordinary space while U is the set of all such vectors in the XOY-plane. Similarly, $W = \{(a, 0, 0) : a \in R\}$ is the set of all vectors along the X-axis. Clearly, W is a subspace of both U and V.

Let $\xi_1, \xi_2, \ldots, \xi_m \in V$, a vector space over $\mathcal{F}$. By a *linear combination* of these m vectors is meant the vector $\xi \in V$ given by

$$\xi = \sum c_i\xi_i = c_1\xi_1 + c_2\xi_2 + \cdots + c_m\xi_m, \qquad c_i \in \mathcal{F}$$

Consider now two such linear combinations $\sum c_i\xi_i$ and $\sum d_i\xi_i$. Since

$$\sum c_i\xi_i + \sum d_i\xi_i = \sum (c_i + d_i)\xi_i$$

and, for any $s \in \mathcal{F}$,

$$s\sum c_i\xi_i = \sum (sc_i)\xi_i$$

we have, by Theorem II,

Theorem III. The set U of all linear combinations of an arbitrary set S of vectors of a (vector) space V is a subspace of V.

The subspace U of V defined in Theorem III is said to be *spanned* by S. In turn, the vectors of S are called *generators* of the space U.

Example 4: Consider the space $V_3(R)$ of Example 3 and the subspaces

$$U = \{s(1, 2, 1) + t(3, 1, 5) : s, t \in R\}$$

spanned by $\xi_1 = (1, 2, 1)$ and $\xi_2 = (3, 1, 5)$

and

$$W = \{a(1, 2, 1) + b(3, 1, 5) + c(3, -4, 7) : a, b, c \in R\}$$

spanned by ξ_1, ξ_2, and $\xi_3 = (3, -4, 7)$.

We now assert that U and W are identical subspaces of V. For, since $(3, -4, 7) = -3(1, 2, 1) + 2(3, 1, 5)$, we may write

$$\begin{aligned} W &= \{(a - 3c)(1, 2, 1) + (b + 2c)(3, 1, 5) : a, b, c \in R\} \\ &= \{s'(1, 2, 1) + t'(3, 1, 5) : s', t' \in R\} \\ &= U \end{aligned}$$

Let

$$U = \{k_1\xi_1 + k_2\xi_2 + \cdots + k_m\xi_m : k_i \in \mathcal{F}\}$$

be the space spanned by $S = \{\xi_1, \xi_2, \ldots, \xi_m\}$, a subset of vectors of V over $\mathcal{F}$. Now U contains the zero vector $\zeta \in V$ (why?); hence, if $\zeta \in S$, it may be excluded from S leaving a proper subset which also spans U. Moreover, as Example 4 indicates, if some one of the vectors, say ξ_j, of S can be written as a linear combination of other vectors of S, then ξ_j may also be excluded from S and the remaining vectors will again span U. This raises questions concerning the minimum number of vectors necessary to span a given space U and the characteristic property of such a set.

See also Problem 2.

LINEAR DEPENDENCE

A non-empty set $S = \{\xi_1, \xi_2, \ldots, \xi_m\}$ of vectors of a vector space V over $\mathcal{F}$ is called *linearly dependent* over $\mathcal{F}$ if and only if there exist elements $k_1, k_2, \ldots, k_m$ of $\mathcal{F}$, not all equal to z, such that

$$\sum k_i\xi_i = k_1\xi_1 + k_2\xi_2 + \cdots + k_m\xi_m = \zeta$$

A non-empty set $S = \{\xi_1, \xi_2, \ldots, \xi_m\}$ of vectors of V over $\mathcal{F}$ is called *linearly independent* over $\mathcal{F}$ if and only if

$$\sum k_i\xi_i = k_1\xi_1 + k_2\xi_2 + \cdots + k_m\xi_m = \zeta$$

implies *every* $k_i = z$.

Note. Once the field $\mathcal{F}$ is fixed, we shall omit thereafter the phrase "over $\mathcal{F}$"; moreover, by "the vector space $V_n(Q)$" we shall mean the vector space $V_n(Q)$ over Q, and similarly for $V_n(R)$. Also, when the field is Q or R, we shall denote the zero vector by 0. Although this overworks 0, it will always be clear from the context whether an element of the field or a vector of the space is intended.

Example 5: (*a*) The vectors $\xi_1 = (1,2,1)$ and $\xi_2 = (3,1,5)$ of Example 4 are linearly independent, since if

$$k_1\xi_1 + k_2\xi_2 = (k_1 + 3k_2, 2k_1 + k_2, k_1 + 5k_2) = 0 = (0,0,0)$$

then $k_1 + 3k_2 = 0$, $2k_1 + k_2 = 0$, $k_1 + 5k_2 = 0$. Solving the first relation for $k_1 = -3k_2$ and substituting in the second, we find $-5k_2 = 0$; then $k_2 = 0$ and $k_1 = -3k_2 = 0$.

(*b*) The vectors $\xi_1 = (1,2,1)$, $\xi_2 = (3,1,5)$ and $\xi_3 = (3,-4,7)$ are linearly dependent, since $3\xi_1 - 2\xi_2 + \xi_3 = 0$.

There follow

Theorem IV. If some one of the set $S = \{\xi_1, \xi_2, \ldots, \xi_m\}$ of vectors in V over $\mathcal{F}$ is the zero vector ζ, then necessarily S is a linearly dependent set.

Theorem V. The set of non-zero vectors $S = \{\xi_1, \xi_2, \ldots, \xi_m\}$ of V over $\mathcal{F}$ is linearly dependent if and only if some one of them, say ξ_j, can be expressed as a linear combination of the vectors $\xi_1, \xi_2, \ldots, \xi_{j-1}$ which precede it.

For a proof, see Problem 3.

Theorem VI. Any non-empty subset of a linearly independent set of vectors is linearly independent.

Theorem VII. Any finite set S of vectors, not all the zero vector, contains a linearly independent subset U which spans the same vector space as S.

For a proof, see Problem 4.

Example 6: In the set $S = \{\xi_1, \xi_2, \xi_3\}$ of Example 5(*b*), ξ_1 and ξ_2 are linearly independent while $\xi_3 = 2\xi_2 - 3\xi_1$. Thus, $T_1 = \{\xi_1, \xi_2\}$ is a maximum linearly independent subset of S. But, since ξ_1 and ξ_3 are linearly independent (prove this) while $\xi_2 = \frac{1}{2}(\xi_3 + 3\xi_1)$, $T_2 = \{\xi_1, \xi_3\}$ is also a maximum linearly independent subset of S. Similarly, $T_3 = \{\xi_2, \xi_3\}$ is another. By Theorem VII each of the subsets T_1, T_2, T_3 spans the same space as does S.

The problem of determining whether a given set of vectors is linearly dependent or linearly independent (and, if linearly dependent, of selecting a maximum subset of linearly independent vectors) involves at most the study of certain systems of linear equations. While such investigations are not difficult, they can be extremely tedious. We shall postpone most of these problems until Chapter 14 where a neater procedure will be devised.

See Problem 5.

BASES OF A VECTOR SPACE

A set $S = \{\xi_1, \xi_2, \ldots, \xi_n\}$ of vectors of a vector space V over $\mathcal{F}$ is called a *basis* of V provided:

(i) S is a linearly independent set,

(ii) the vectors of S span V.

Define the *unit vectors* of $V_n(\mathcal{F})$ as follows:

$$\begin{aligned}
\epsilon_1 &= (u, 0, 0, 0, \ldots, 0, 0) \\
\epsilon_2 &= (0, u, 0, 0, \ldots, 0, 0) \\
\epsilon_3 &= (0, 0, u, 0, \ldots, 0, 0) \\
&\cdots\cdots\cdots\cdots\cdots\cdots \\
\epsilon_n &= (0, 0, 0, 0, \ldots, 0, u)
\end{aligned}$$

and consider the linear combination

$$\xi = a_1\epsilon_1 + a_2\epsilon_2 + \cdots + a_n\epsilon_n = (a_1, a_2, \ldots, a_n), \qquad a_i \in \mathcal{F} \tag{1}$$

If $\xi = \zeta$, then $a_1 = a_2 = \cdots = a_n = z$; hence, $E = \{\epsilon_1, \epsilon_2, \ldots, \epsilon_n\}$ is a linearly independent set. Also, if ξ is an arbitrary vector of $V_n(\mathcal{F})$, then (1) exhibits it as a linear combination of the unit vectors. Thus the set E spans $V_n(\mathcal{F})$ and is a basis.

Example 7: One basis of $V_4(R)$ is the unit basis

$$E = \{(1,0,0,0), (0,1,0,0), (0,0,1,0), (0,0,0,1)\}$$

Another basis is

$$F = \{(1,1,1,0), (0,1,1,1), (1,0,1,1), (1,1,0,1)\}$$

To prove this, consider the linear combination

$$\begin{aligned}
\xi &= a_1(1,1,1,0) + a_2(0,1,1,1) + a_3(1,0,1,1) + a_4(1,1,0,1) \\
&= (a_1 + a_3 + a_4,\ a_1 + a_2 + a_4,\ a_1 + a_2 + a_3,\ a_2 + a_3 + a_4); \qquad a_i \in R
\end{aligned}$$

If ξ is an arbitrary vector $(p, q, r, s) \in V_4(R)$, we find

$$\begin{aligned}
a_1 &= (p + q + r - 2s)/3 & a_3 &= (p + r + s - 2q)/3 \\
a_2 &= (q + r + s - 2p)/3 & a_4 &= (p + q + s - 2r)/3
\end{aligned}$$

Then F is a linearly independent set (prove this) and spans $V_4(R)$.

In Problem 6, we prove

Theorem VIII. If $S = \{\xi_1, \xi_2, \ldots, \xi_m\}$ is a basis of the vector space V over $\mathcal{F}$ and $T = \{\eta_1, \eta_2, \ldots, \eta_n\}$ is any linearly independent set of vectors of V, then $n \leq m$.

As consequences, we have

Theorem IX. If $S = \{\xi_1, \xi_2, \ldots, \xi_m\}$ is a basis of V over $\mathcal{F}$, then any $m + 1$ vectors of V necessarily form a linearly dependent set.

and

Theorem X. Every basis of a vector space V over $\mathcal{F}$ has the same number of elements.

The number defined in Theorem X is called the *dimension* of V. It is evident that dimension, as defined here, implies *finite* dimension. Not every vector space has finite dimension, as shown in

Example 8: (a) From Example 7, it follows that $V_4(R)$ has dimension 4.

(b) Consider $V = \{a_0 + a_1x + a_2x^2 + a_3x^3 + a_4x^4 : a_i \in R\}$
Clearly, $B = \{1, x, x^2, x^3, x^4\}$ is a basis and V has dimension 5.

(c) The vector space V of all polynomials in x over R has no finite basis and, hence, is without dimension. For, assume B, consisting of p linearly independent polynomials of V with degrees $\leq q$, to be a basis. Since no polynomial of V of degree $> q$ can be generated by B, it is not a basis.

See Problem 7.

SUBSPACES OF A VECTOR SPACE

Let V, of dimension n, be a vector space over $\mathcal{F}$ and U, of dimension $m < n$ having $B = \{\xi_1, \xi_2, \ldots, \xi_m\}$ as a basis, be a subspace of V. By Theorem VIII, only m of the unit vectors of V can be written as linear combinations of the elements of B; hence there exist vectors of V which are not in U. Let η_1 be such a vector and consider

$$k_1\xi_1 + k_2\xi_2 + \cdots + k_m\xi_m + k\eta_1 = \zeta, \qquad k_i, k \in \mathcal{F} \tag{2}$$

Now $k = z$, since otherwise $k^{-1} \in \mathcal{F}$,

$$\eta_1 = k^{-1}(-k_1\xi_1 - k_2\xi_2 - \cdots - k_m\xi_m)$$

and $\eta_1 \in U$, contrary to the definition of η_1. With $k = z$, (2) requires each $k_i = z$ since B is a basis, and we have proved

Theorem XI. If $B = \{\xi_1, \xi_2, \ldots, \xi_m\}$ is a basis of $U \subset V$ and if $\eta_1 \in V$ but $\eta_1 \notin U$, then $B \cup \{\eta_1\}$ is a linearly independent set.

When, in Theorem XI, $m + 1 = n$, the dimension of V, $B_1 = B \cup \{\eta_1\}$ is a basis of V; when $m + 1 < n$, B_1 is a basis of some subspace U_1 of V. In the latter case there exists a vector η_2 in V but not in U_1 such that the space U_2, having $B \cup \{\eta_1, \eta_2\}$ as basis, is either V or is properly contained in V, ... Thus we eventually obtain

Theorem XII. If $B = \{\xi_1, \xi_2, \ldots, \xi_m\}$ is a basis of $U \subset V$ having dimension n, there exist vectors $\eta_1, \eta_2, \ldots, \eta_{n-m}$ in V such that $B \cup \{\eta_1, \eta_2, \ldots, \eta_{n-m}\}$ is a basis of V.

See Problem 8.

Let U and W be subspaces of V. We define

$$U \cap W = \{\xi : \xi \in U, \xi \in W\}$$

$$U + W = \{\xi + \eta : \xi \in U, \eta \in W\}$$

and leave for the reader to prove that each is a subspace of V.

Example 9: Consider $V = V_4(R)$ over R with unit vectors $\epsilon_1, \epsilon_2, \epsilon_3, \epsilon_4$ as in Example 7. Let

$$U = \{a_1\epsilon_1 + a_2\epsilon_2 + a_3\epsilon_3 : a_i \in R\}$$

and

$$W = \{b_1\epsilon_2 + b_2\epsilon_3 + b_3\epsilon_4 : b_i \in R\}$$

be subspaces of dimensions 3 of V. Clearly,

$$U \cap W = \{c_1\epsilon_2 + c_2\epsilon_3 : c_i \in R\}, \quad \text{of dimension 2}$$

and

$$U + W = \{a_1\epsilon_1 + a_2\epsilon_2 + a_3\epsilon_3 + b_1\epsilon_2 + b_2\epsilon_3 + b_3\epsilon_4 : a_i, b_i \in R\}$$
$$= \{d_1\epsilon_1 + d_2\epsilon_2 + d_3\epsilon_3 + d_4\epsilon_4 : d_i \in R\} = V$$

Example 9 illustrates

Theorem XIII. If U and W, of dimension $r \leq n$ and $s \leq n$ respectively, are subspaces of a vector space V of dimension n and if $U \cap W$ and $U + W$ are of dimensions p and t respectively, then $t = r + s - p$.

For a proof, see Problem 9.

VECTOR SPACES OVER R

In this section, we shall restrict our attention to vector spaces $V = V_n(R)$ over R. This is done for two reasons: (1) our study will have applications in geometry and (2) of all possible fields, R will present a minimum in difficulties.

Consider in $V = V_2(R)$ the vectors $\xi = (a_1, a_2)$ and $\eta = (b_1, b_2)$ of Fig. 13-2. The length $|\xi|$ of ξ is given by $|\xi| = \sqrt{a_1^2 + a_2^2}$ and the length of η is given by $|\eta| = \sqrt{b_1^2 + b_2^2}$. By the law of cosines we have

$$|\xi - \eta|^2 \;=\; |\xi|^2 + |\eta|^2 - 2\,|\xi|\cdot|\eta| \cos\theta$$

so that

$$\cos\theta \;=\; \frac{(a_1^2 + a_2^2) + (b_1^2 + b_2^2) - [(a_1 - b_1)^2 + (a_2 - b_2)^2]}{2\,|\xi|\cdot|\eta|}$$

$$=\; \frac{a_1b_1 + a_2b_2}{|\xi|\cdot|\eta|}$$

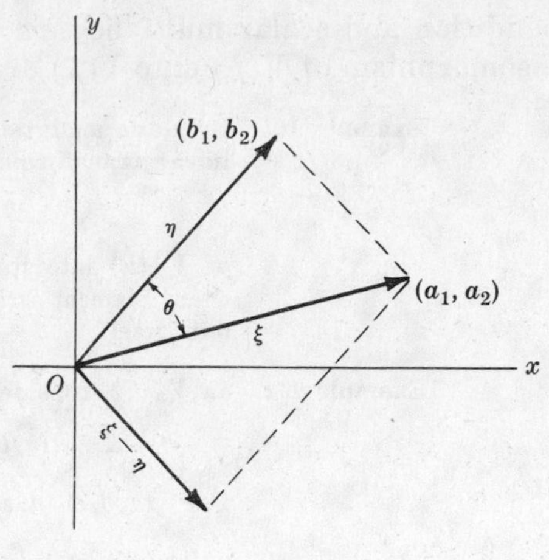

The expression for $\cos\theta$ suggests the definition of the *scalar product* (also called *dot product* and *inner product*) of ξ and η by

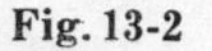
Fig. 13-2

$$\xi\cdot\eta \;=\; a_1b_1 + a_2b_2$$

Then $|\xi| = \sqrt{\xi\cdot\xi}$, $\cos\theta = \dfrac{\xi\cdot\eta}{|\xi|\cdot|\eta|}$, and the vectors ξ and η are *orthogonal* (i.e., mutually perpendicular, so that $\cos\theta = 0$) if and only if $\xi\cdot\eta = 0$.

In the vector space $V = V_n(R)$ we define for all $\xi = (a_1, a_2, \ldots, a_n)$ and $\eta = (b_1, b_2, \ldots, b_n)$,

$$\xi\cdot\eta \;=\; \sum a_ib_i \;=\; a_1b_1 + a_2b_2 + \cdots + a_nb_n$$

$$|\xi| \;=\; \sqrt{\xi\cdot\xi} \;=\; \sqrt{a_1^2 + a_2^2 + \cdots + a_n^2}$$

There follow

(1) $|s\xi| = |s|\cdot|\xi|$ for all $\xi \in V$ and $s \in R$

(2) $|\xi| \geqq 0$, the equality holding only when $\xi = 0$

(3) $|\xi\cdot\eta| \leqq |\xi|\cdot|\eta|$ (Schwarz inequality)

(4) $|\xi + \eta| \leqq |\xi| + |\eta|$ (Triangle inequality)

(5) ξ and η are orthogonal if and only if $\xi\cdot\eta = 0$.

For a proof of (3), see Problem 10.
See also Problems 11-13.

Suppose in $V_n(R)$ the vector η is orthogonal to each of $\xi_1, \xi_2, \ldots, \xi_m$. Then, since $\eta\cdot\xi_1 = \eta\cdot\xi_2 = \cdots = \eta\cdot\xi_m = 0$, we have $\eta\cdot(c_1\xi_1 + c_2\xi_2 + \cdots + c_m\xi_m) = 0$ for any $c_i \in R$ and have proved

Theorem XIV. If, in $V_n(R)$, a vector η is orthogonal to each vector of the set $\{\xi_1, \xi_2, \ldots, \xi_m\}$, then η is orthogonal to every vector of the space spanned by this set.

See Problem 14.

LINEAR TRANSFORMATIONS

A linear transformation of a vector space $V(\mathcal{F})$ into a vector space $W(\mathcal{F})$ over the same field $\mathcal{F}$ is a mapping T of $V(\mathcal{F})$ into $W(\mathcal{F})$ for which

(i) $(\xi_i + \xi_j)T \;=\; \xi_iT + \xi_jT$ for all $\xi_i, \xi_j \in V(\mathcal{F})$

(ii) $(s\xi_i)T \;=\; s(\xi_iT)$ for all $\xi_i \in V(\mathcal{F})$ and $s \in \mathcal{F}$

We shall restrict our attention here to the case $W(\mathcal{F}) = V(\mathcal{F})$, i.e. to the case when T is a mapping of $V(\mathcal{F})$ into itself. Since the mapping preserves the operations of vector addition and scalar multiplication, a linear transformation of $V(\mathcal{F})$ into itself is either an isomorphism of $V(\mathcal{F})$ onto $V(\mathcal{F})$ or a homomorphism of $V(\mathcal{F})$ into $V(\mathcal{F})$.

Example 10: In plane analytic geometry the familiar rotation of axes through an angle α is a linear transformation

$$T: \quad (x, y) \rightarrow (x \cos \alpha - y \sin \alpha,\ x \sin \alpha + y \cos \alpha)$$

of $V_2(R)$ into itself. Since distinct elements of $V_2(R)$ have distinct images and every element is an image (prove this), T is an example of an isomorphism of $V(R)$ onto itself.

Example 11: In $V_3(Q)$, consider the mapping

$$T: \quad (a, b, c) \rightarrow (a + b + 5c,\ a + 2c;\ 2b + 6c), \qquad (a, b, c) \in V_3(Q)$$

For $(a, b, c), (d, e, f) \in V_3(Q)$ and $s \in Q$, we have readily

(i) $(a, b, c) + (d, e, f) = (a + d,\ b + e,\ c + f) \rightarrow$

$$(a + d + b + e + 5c + 5f,\ a + d + 2c + 2f,\ 2b + 2e + 6c + 6f)$$
$$= (a + b + 5c,\ a + 2c,\ 2b + 6c) + (d + e + 5f,\ d + 2f,\ 2e + 6f)$$

i.e., $$[(a, b, c) + (d, e, f)]T = (a, b, c)T + (d, e, f)T$$

and

(ii) $s(a, b, c) = (sa, sb, sc) \rightarrow (sa + sb + 5sc,\ sa + 2sc,\ 2sb + 6sc)$
$$= s(a + b + 5c,\ a + 2c,\ 2b + 6c)$$

i.e., $$[s(a, b, c)]T = s[(a, b, c)T]$$

Thus T is a linear transformation on $V_3(Q)$.

Since $(0, 0, 1)$ and $(2, 3, 0)$ have the same image $(5, 2, 6)$, this linear transformation is an example of a homomorphism of $V_3(Q)$ into itself.

The linear transformation T of Example 10 may be written as

$$x(1, 0) + y(0, 1) \rightarrow x(\cos \alpha, \sin \alpha) + y(-\sin \alpha, \cos \alpha)$$

suggesting that T may be given as

$$T: \quad (1, 0) \rightarrow (\cos \alpha, \sin \alpha),\ (0, 1) \rightarrow (-\sin \alpha, \cos \alpha)$$

Likewise, T of Example 11 may be written as

$$a(1, 0, 0) + b(0, 1, 0) + c(0, 0, 1) \rightarrow a(1, 1, 0) + b(1, 0, 2) + c(5, 2, 6)$$

suggesting that T may be given as

$$T: \quad (1, 0, 0) \rightarrow (1, 1, 0),\ (0, 1, 0) \rightarrow (1, 0, 2),\ (0, 0, 1) \rightarrow (5, 2, 6)$$

Thus we infer:

Any linear transformation of a vector space into itself can be described completely by exhibiting its effect on the unit basis vectors of the space.

See Problem 15.

In Problem 16, we prove the more general

Theorem XV. If $\{\xi_1, \xi_2, \ldots, \xi_n\}$ is *any* basis of $V = V(\mathcal{F})$ and if $\{\eta_1, \eta_2, \ldots, \eta_n\}$ is any set of n elements of V, the mapping

$$T: \quad \xi_i \rightarrow \eta_i, \qquad (i = 1, 2, \ldots, n)$$

defines a linear transformation of V into itself.

In Problem 17, we prove

Theorem XVI. If T is a linear transformation of $V(\mathcal{F})$ into itself and if W is a subspace of $V(\mathcal{F})$, then $W_T = \{\xi T : \xi \in W\}$, the image of W under T, is also a subspace of $V(\mathcal{F})$.

Returning now to Example 11, we note that the images of the unit basis vectors of $V_3(Q)$ are linearly dependent, i.e., $2\epsilon_1 T + 3\epsilon_2 T - \epsilon_3 T = (0,0,0)$. Thus, $V_T \subset V$; in fact, since $(1,1,0)$ and $(1,0,2)$ are linearly independent, V_T has dimension 2. Defining the *rank* of a linear transformation T of a vector space V to be the dimension of the image space V_T of V under T, we have by Theorem XVI,

$$r_T = \text{rank of } T \leqq \text{dimension of } V$$

When the equality holds, we shall call the linear transformation T *non-singular*; otherwise we shall call T *singular*. Thus T of Example 11 is singular of rank 2.

Consider next the linear transformation T of Theorem XV and suppose T is singular. Since the image vectors η_i are then linearly dependent, there exist elements $s_i \in \mathcal{F}$, not all z, such that $\sum s_i\eta_i = \zeta$. Then, for $\xi = \sum s_i\xi_i$, we have $\xi T = \zeta$. Conversely, suppose $\eta = \sum t_i\xi_i \neq \zeta$ and

$$\eta T = \sum (t_i\xi_i)T = t_1(\xi_1 T) + t_2(\xi_2 T) + \cdots + t_n(\xi_n T) = \zeta$$

Then the image vectors $\xi_i T$, $(i = 1, 2, \ldots, n)$ must be linearly dependent. We have proved

Theorem XVII. A linear transformation T of a vector space $V(\mathcal{F})$ is singular if and only if there exists a non-zero vector $\xi \in V(\mathcal{F})$ such that $\xi T = \zeta$.

Example 12: Determine whether each of the following linear transformations of $V_4(Q)$ into itself is singular or non-singular:

$$(a)\quad A: \begin{cases} \epsilon_1 \to (1,1,0,0) \\ \epsilon_2 \to (0,1,1,0) \\ \epsilon_3 \to (0,0,1,1) \\ \epsilon_4 \to (0,1,0,1) \end{cases}, \qquad (b)\quad B: \begin{cases} \epsilon_1 \to (1,1,0,0) \\ \epsilon_2 \to (0,1,1,0) \\ \epsilon_3 \to (0,0,1,1) \\ \epsilon_4 \to (1,1,1,1) \end{cases}$$

Let $\xi = (a,b,c,d)$ be an arbitrary vector of $V_4(Q)$.

(a) Set $\xi A = (a\epsilon_1 + b\epsilon_2 + c\epsilon_3 + d\epsilon_4)A = (a,\, a+b+d,\, b+c,\, c+d) = 0$. Since this requires $a = b = c = d = 0$, that is, $\xi = 0$, A is non-singular.

(b) Set $\xi B = (a+d,\, a+b+d,\, b+c+d,\, c+d) = 0$. Since this is satisfied when $a = c = 1$, $b = 0$, $d = -1$, we have $(1,0,1,-1)B = 0$ and B is singular. This is evident by inspection, i.e., $\epsilon_1 B + \epsilon_3 B = \epsilon_4 B$. Then, since $\epsilon_1 B, \epsilon_2 B, \epsilon_3 B$ are clearly linearly independent, B has rank 3 and is singular.

We shall again (see the paragraph following Example 6) postpone additional examples and problems until Chapter 14.

THE ALGEBRA OF LINEAR TRANSFORMATIONS

Denote by $\mathcal{A}$ the set of all linear transformations of a given vector space $V(\mathcal{F})$ over $\mathcal{F}$ into itself and by $\mathcal{M}$ the set of all non-singular linear transformations in $\mathcal{A}$. Let addition $(+)$ and multiplication $(\cdot)$ on $\mathcal{A}$ be defined by

$$A + B: \quad \xi(A+B) = \xi A + \xi B, \qquad \xi \in V(\mathcal{F})$$

and

$$A \cdot B: \quad \xi(A \cdot B) = (\xi A)B, \qquad \xi \in V(\mathcal{F})$$

for all $A, B \in \mathcal{A}$. Let scalar multiplication be defined on $\mathcal{A}$ by

$$kA: \quad \xi(kA) = (k\xi)A, \qquad \xi \in V(\mathcal{F})$$

for all $A \in \mathcal{A}$ and $k \in \mathcal{F}$.

Example 13: Let

$$A: \begin{cases} \epsilon_1 \to (a, b) \\ \epsilon_2 \to (c, d) \end{cases} \qquad \text{and} \qquad B: \begin{cases} \epsilon_1 \to (e, f) \\ \epsilon_2 \to (g, h) \end{cases}$$

be linear transformations of $V_2(R)$ into itself. For any vector $\xi = (s, t) \in V_2(R)$, we find

$$\xi A \;=\; (s, t)A \;=\; (s\epsilon_1 + t\epsilon_2)A \;=\; s(a, b) + t(c, d)$$
$$= \; (sa + tc, sb + td)$$

$$\xi B \;=\; (se + tg, sf + th)$$

and

$$\xi(A + B) \;=\; (s, t)A + (s, t)B \;=\; \big(s(a + e) + t(c + g), s(b + f) + t(d + h)\big)$$

Thus we have

$$A + B: \begin{cases} \epsilon_1 \to (a + e,\ b + f) \\ \epsilon_2 \to (c + g, d + h) \end{cases}$$

Also,

$$\xi(A \cdot B) \;=\; \big((s, t)A\big)B \;=\; (sa + tc, sb + td)B$$
$$= \; (sa + tc) \cdot (e, f) \;+\; (sb + td) \cdot (g, h)$$
$$= \; \big(s(ae + bg) + t(ce + dg),\ s(af + bh) + t(cf + dh)\big)$$

and

$$A \cdot B: \begin{cases} \epsilon_1 \to (ae + bg,\ af + bh) \\ \epsilon_2 \to (ce + dg,\ cf + dh) \end{cases}$$

Finally, for any $k \in R$, we find

$$(k\xi)A \;=\; (ks, kt)A \;=\; \big(k(sa + tc),\ k(sb + td)\big)$$

and

$$kA: \begin{cases} \epsilon_1 \to (ka, kb) \\ \epsilon_2 \to (kc, kd) \end{cases}$$

In Problem 18, we prove

Theorem XVIII. The set $\mathcal{A}$ of all linear transformations of a vector space into itself forms a ring with respect to addition and multiplication as defined above.

In Problem 19, we prove

Theorem XIX. The set $\mathcal{M}$ of all non-singular linear transformations of a vector space into itself forms a group under multiplication.

We leave for the reader to prove

Theorem XX. If $\mathcal{A}$ is the set of all linear transformations of a vector space $V(\mathcal{F})$ over $\mathcal{F}$ into itself, then $\mathcal{A}$ itself is also a vector space over $\mathcal{F}$.

Let $A, B \in \mathcal{A}$. Since, for every $\xi \in V$,

$$\xi(A + B) \;=\; \xi A \;+\; \xi B$$

it is evident that

$$V_{(A+B)} \;\subseteq\; V_A + V_B$$

Then

$$\text{dimension of } V_{(A+B)} \;\leq\; \text{dimension of } V_A \;+\; \text{dimension of } V_B$$

and

$$r_{(A+B)} \;\leq\; r_A + r_B$$

Since for any linear transformation $T \in \mathcal{A}$,

$$\text{dimension of } V_T \leq \text{dimension of } V$$

we have $$\text{dimension of } V_{(A\cdot B)} \leq \text{dimension of } V_A$$

Also, since $V_A \subseteq V$, $$\text{dimension of } V_{(A\cdot B)} \leq \text{dimension of } V_B$$

Thus $$r_{(A\cdot B)} \leq r_A, \quad r_{(A\cdot B)} \leq r_B$$

Solved Problems

1. Prove: A non-empty subset U of a vector space V over $\mathcal{F}$ is a subspace of V if and only if U is closed with respect to scalar multiplication and vector addition as defined on V.

Suppose U is a subspace of V; then U is closed with respect to scalar multiplication and vector addition. Conversely, suppose U is a non-empty subset of V which is closed with respect to scalar multiplication and vector addition. Let $\xi \in U$; then $(-u)\xi = -(u\xi) = -\xi \in U$ and $\xi + (-\xi) = \zeta \in U$. Thus U is an abelian additive group. Since the properties (i)-(iv), Page 144, hold in V, they also hold in U. Thus U is a vector space over $\mathcal{F}$ and, hence, is a subspace of V.

2. In the vector space $V_3(R)$ over R (Example 3, Page 145), let U be spanned by $\xi_1 = (1, 2, -1)$ and $\xi_2 = (2, -3, 2)$ and W be spanned by $\xi_3 = (4, 1, 3)$ and $\xi_4 = (-3, 1, 2)$. Are U and W identical subspaces of V?

First, consider the vector $\xi = \xi_3 - \xi_4 = (7, 0, 1) \in W$ and the vector

$$\eta = x\xi_1 + y\xi_2 = (x + 2y,\ 2x - 3y,\ -x + 2y) \in U$$

Now ξ and η are the same provided $x, y \in R$ exist such that

$$(x + 2y,\ 2x - 3y,\ -x + 2y) = (7, 0, 1)$$

We find $x = 3$, $y = 2$. To be sure, this does not prove U and W identical; for that we need to be able to produce $x, y \in R$ such that

$$x\xi_1 + y\xi_2 = a\xi_3 + b\xi_4$$

for arbitrary $a, b \in R$.

From $\begin{cases} x + 2y = 4a - 3b \\ -x + 2y = 3a + 2b \end{cases}$ we find $x = \frac{1}{2}(a - 5b)$, $y = \frac{1}{4}(7a - b)$. Now $2x - 3y \neq a + b$; hence U and W are not identical.

Geometrically, U and W are distinct planes through O, the origin of coordinates, in ordinary space. They have, of course, a line of vectors in common; one of these common vectors is $(7, 0, 1)$.

3. Prove: The set of non-zero vectors $S = \{\xi_1, \xi_2, \ldots, \xi_m\}$ of V over $\mathcal{F}$ is linearly dependent if and only if some one of them, say ξ_j, can be expressed as a linear combination of the vectors $\xi_1, \xi_2, \ldots, \xi_{j-1}$ which precede it.

Suppose the m vectors are linearly dependent so that there exist scalars $k_1, k_2, \ldots, k_m$, not all equal to z, such that $\sum k_i\xi_i = \zeta$. Suppose further that the coefficient k_j is not z while the coefficients $k_{j+1}, k_{j+2}, \ldots, k_m$ are z (but not excluding, of course, the extreme case $j = m$). Then in effect

$$k_1\xi_1 + k_2\xi_2 + \cdots + k_j\xi_j = \zeta \tag{1}$$

and, since $k_j\xi_j \neq \zeta$, we have

$$k_j\xi_j = -k_1\xi_1 - k_2\xi_2 - \cdots - k_{j-1}\xi_{j-1}$$

or

$$\xi_j = q_1\xi_1 + q_2\xi_2 + \cdots + q_{j-1}\xi_{j-1} \tag{2}$$

with some of the $q_i \neq z$. Thus, ξ_j is a linear combination of the vectors which precede it.

Conversely, suppose (2) holds. Then

$$k_1\xi_1 + k_2\xi_2 + \cdots + k_j\xi_j + z\xi_{j+1} + z\xi_{j+2} + \cdots + z\xi_m = \zeta$$

with $k_j \neq z$ and the vectors $\xi_1, \xi_2, \ldots, \xi_m$ are linearly dependent.

4. Prove: Any finite set S of vectors, not all the zero vector, contains a linearly independent subset U which spans the same space as S.

Let $S = \{\xi_1, \xi_2, \ldots, \xi_m\}$. The discussion thus far indicates that U exists while Theorem V suggests the following procedure for extracting it from S. Considering each vector in turn from left to right, let us agree to exclude the vector in question if (1) it is the zero vector or (2) it can be written as a linear combination of all the vectors which precede it. Suppose there are $n \leq m$ vectors remaining which, after relabeling, we denote by $U = \{\eta_1, \eta_2, \ldots, \eta_n\}$. By its construction, U is a linearly independent subset of S which spans the same space as S.

Example 6, Page 146, shows, as might have been anticipated, that there will usually be linearly independent subsets of S, other than U, which will span the same space as S. An advantage of the procedure used above lies in the fact that, once the elements of S have been set down in some order, only one linearly independent subset, namely U, can be found.

5. Find a linearly independent subset U of the set $S = \{\xi_1, \xi_2, \xi_3, \xi_4\}$, where

$$\xi_1 = (1,2,-1),\ \xi_2 = (-3,-6,3),\ \xi_3 = (2,1,3),\ \xi_4 = (8,7,7) \quad \in \quad R,$$

which spans the same space as S.

First, we note that $\xi_1 \neq \zeta$ and move to ξ_2. Since $\xi_2 = -3\xi_1$, (i.e. ξ_2 is a linear combination of ξ_1), we exclude ξ_2 and move to ξ_3. Now $\xi_3 \neq s\xi_1$, for any $s \in R$; we move to ξ_4. Since ξ_4 is neither a scalar multiple of ξ_1 nor of ξ_3 (and, hence, is not automatically excluded), we write

$$s\xi_1 + t\xi_3 = s(1,2,-1) + t(2,1,3) = (8,7,7) = \xi_4$$

and seek a solution, if any, in R of the system

$$s + 2t = 8, \quad 2s + t = 7, \quad -s + 3t = 7$$

The reader will verify that $\xi_4 = 2\xi_1 + 3\xi_3$; hence, $U = \{\xi_1, \xi_3\}$ is the required subset.

6. Prove: If $S = \{\xi_1, \xi_2, \ldots, \xi_m\}$ is a basis of a vector space V over $\mathcal{F}$ and if $T = \{\eta_1, \eta_2, \ldots, \eta_n\}$ is a linearly independent set of vectors of V, then $n \leq m$.

Since each element of T can be written as a linear combination of the basis elements, the set

$$S' = \{\eta_1, \xi_1, \xi_2, \ldots, \xi_m\}$$

spans V and is linearly dependent. Now $\eta_1 \neq \zeta$; hence some one of the ξ's must be a linear combination of the elements which precede it in S'. Examining the ξ's in turn, suppose we find ξ_i satisfies this condition. Excluding ξ_i from S', we have the set

$$S_1 = \{\eta_1, \xi_1, \xi_2, \ldots, \xi_{i-1}, \xi_{i+1}, \ldots, \xi_m\}$$

which we shall now prove to be a basis of V. It is clear that S_1 spans the same space as S', i.e. S_1 spans V. Thus we need only to show that S_1 is a linearly independent set. Write

$$\eta_1 = a_1\xi_1 + a_2\xi_2 + \cdots + a_i\xi_i, \qquad a_j \in \mathcal{F},\ a_i \neq z$$

If S_1 were a linearly dependent set there would be some ξ_j, $j > i$, which could be expressed as

$$\xi_j = b_1\eta_1 + b_2\xi_1 + b_3\xi_2 + \cdots + b_i\xi_{i-1} + b_{i+1}\xi_{i+1} + \cdots + b_{j-1}\xi_{j-1}, \quad b_1 \neq z$$

whence, upon substituting for η_1,

$$\xi_j = c_1\xi_1 + c_2\xi_2 + \cdots + c_{j-1}\xi_{j-1}$$

contrary to the assumption that S is a linearly independent set.

Similarly, $$S_1' = \{\eta_2, \eta_1, \xi_1, \xi_2, \ldots, \xi_{i-1}, \xi_{i+1}, \ldots, \xi_m\}$$

is a linearly dependent set which spans V. Since η_1 and η_2 are linearly independent, some one of the ξ's in S_1', say ξ_j, is a linear combination of all the vectors which precede it. Repeating the argument above, we obtain (assuming $j > i$) the set

$$S_2 = \{\eta_2, \eta_1, \xi_1, \xi_2, \ldots, \xi_{i-1}, \xi_{i+1}, \ldots, \xi_{j-1}, \xi_{j+1}, \ldots, \xi_m\}$$

as a basis of V.

Now this procedure may be repeated until T is exhausted provided $n \leqq m$. Suppose $n > m$ and we have obtained

$$S_m = \{\eta_m, \eta_{m-1}, \ldots, \eta_2, \eta_1\}$$

as a basis of V. Consider $S_m' = S \cup \{\eta_{m+1}\}$. Since S_m is a basis of V and $\eta_{m+1} \in V$, then η_{m+1} is a linear combination of the vectors of S_m. But this contradicts the assumption on T. Hence $n \leqq m$, as required.

7. (*a*) Select a basis of $V_3(R)$ from the set

$$S = \{\xi_1, \xi_2, \xi_3, \xi_4\} = \{(1, -3, 2), (2, 4, 1), (3, 1, 3), (1, 1, 1)\}$$

(*b*) Express each of the unit vectors of $V_3(R)$ as linear combinations of the basis vectors found in (*a*).

(*a*) If the problem has a solution, some one of the ξ's must be a linear combination of those which precede it. By inspection, we find $\xi_3 = \xi_1 + \xi_2$. To prove $\{\xi_1, \xi_2, \xi_4\}$ a linearly independent set and, hence, the required basis, we need to show that

$$a\xi_1 + b\xi_2 + c\xi_4 = (a + 2b + c, -3a + 4b + c, 2a + b + c) = (0, 0, 0)$$

requires $a = b = c = 0$. We leave this for the reader.

The same result is obtained by showing that

$$s\xi_1 + t\xi_2 = \xi_4, \qquad s, t \in R$$

i.e., $\begin{cases} s + 2t = 1 \\ -3s + 4t = 1 \\ 2s + t = 1 \end{cases}$ is impossible. Finally any reader acquainted with determinants will recall that these equations have a solution if and only if $\begin{vmatrix} 1 & 2 & 1 \\ -3 & 4 & 1 \\ 2 & 1 & 1 \end{vmatrix} = 0.$

(*b*) Set $a\xi_1 + b\xi_2 + c\xi_4$ equal to the unit vectors $\epsilon_1, \epsilon_2, \epsilon_3$ in turn and obtain

$$\begin{cases} a + 2b + c = 1 \\ -3a + 4b + c = 0 \\ 2a + b + c = 0 \end{cases} \qquad \begin{cases} a + 2b + c = 0 \\ -3a + 4b + c = 1 \\ 2a + b + c = 0 \end{cases} \qquad \begin{cases} a + 2b + c = 0 \\ -3a + 4b + c = 0 \\ 2a + b + c = 1 \end{cases}$$

having solutions:

$$a = 3/2,\ b = 5/2,\ c = -11/2 \qquad a = b = -1/2,\ c = 3/2 \qquad a = -1,\ b = -2,\ c = 5$$

Thus $\epsilon_1 = \frac{1}{2}(3\xi_1 + 5\xi_2 - 11\xi_4)$, $\epsilon_2 = \frac{1}{2}(-\xi_1 - \xi_2 + 3\xi_4)$, and $\epsilon_3 = -\xi_1 - 2\xi_2 + 5\xi_4$.

8. For the vector space $V_4(Q)$ determine, if possible, a basis which includes the vectors $\xi_1 = (3, -2, 0, 0)$ and $\xi_2 = (0, 1, 0, 1)$.

Since $\xi_1 \neq \zeta$, $\xi_2 \neq \zeta$, and $\xi_2 \neq s\xi_1$ for any $s \in Q$, we know that ξ_1 and ξ_2 can be elements of a basis of $V_4(Q)$. Since the unit vectors $\epsilon_1, \epsilon_2, \epsilon_3, \epsilon_4$ (see Example 7, Page 147) are a basis of $V_4(Q)$, the set $S = \{\xi_1, \xi_2, \epsilon_1, \epsilon_2, \epsilon_3, \epsilon_4\}$ spans $V_4(Q)$ and surely contains a basis of the type we seek. Now ϵ_1 will be an element of this basis if and only if

$$a\xi_1 + b\xi_2 + c\epsilon_1 \;=\; (3a+c,\ -2a+b,\ 0,\ b) \;=\; (0, 0, 0, 0)$$

requires $a = b = c = 0$. Clearly, ϵ_1 can serve as an element of the basis. Again, ϵ_2 will be an element if and only if

$$a\xi_1 + b\xi_2 + c\epsilon_1 + d\epsilon_2 \;=\; (3a+c,\ -2a+b+d,\ 0,\ b) \;=\; (0, 0, 0, 0) \qquad (1)$$

requires $a = b = c = d = 0$. We have $b = 0 = 3a + c = -2a + d$; then (1) is satisfied by $a = 1$, $b = 0$, $c = -3$, $d = 2$ and so $\{\xi_1, \xi_2, \epsilon_1, \epsilon_2\}$ is not a basis. We leave for the reader to verify that $\{\xi_1, \xi_2, \epsilon_1, \epsilon_3\}$ is a basis.

9. Prove: If U and W, of dimensions $r \leq n$ and $s \leq n$ respectively, are subspaces of a vector space V of dimension n and if $U \cap W$ and $U + W$ are of dimensions p and t respectively, then $t = r + s - p$.

Take $A = \{\xi_1, \xi_2, \ldots, \xi_p\}$ as a basis of $U \cap W$ and, in agreement with Theorem XII, Page 148, take $B = A \cup \{\lambda_1, \lambda_2, \ldots, \lambda_{r-p}\}$ as a basis of U and $C = A \cup \{\mu_1, \mu_2, \ldots, \mu_{s-p}\}$ as a basis of W. Then, by definition, any vector of $U + W$ can be expressed as a linear combination of the vectors of

$$D \;=\; \{\xi_1, \xi_2, \ldots, \xi_p, \lambda_1, \lambda_2, \ldots, \lambda_{r-p}, \mu_1, \mu_2, \ldots, \mu_{s-p}\}$$

To show that D is a linearly independent set and, hence, is a basis of $U + W$, consider

$$a_1\xi_1 + a_2\xi_2 + \cdots + a_p\xi_p + b_1\lambda_1 + b_2\lambda_2 + \cdots + b_{r-p}\lambda_{r-p} + c_1\mu_1 + c_2\mu_2 + \cdots + c_{s-p}\mu_{s-p} = \zeta \qquad (1)$$

where $a_i, b_j, c_k \in \mathcal{F}$.

Set $\pi = c_1\mu_1 + c_2\mu_2 + \cdots + c_{s-p}\mu_{s-p}$. Now $\pi \in W$ and by (1), $\pi \in U$; thus, $\pi \in U \cap W$ and is a linear combination of the vectors of A, say, $\pi = d_1\xi_1 + d_2\xi_2 + \cdots + d_p\xi_p$. Then

$$c_1\mu_1 + c_2\mu_2 + \cdots + c_{s-p}\mu_{s-p} - d_1\xi_1 - d_2\xi_2 - \cdots - d_p\xi_p = \zeta$$

and, since C is a basis of W, each $c_i = z$ and each $d_i = z$. With each $c_i = z$, (1) becomes

$$a_1\xi_1 + a_2\xi_2 + \cdots + a_p\xi_p + b_1\lambda_1 + b_2\lambda_2 + \cdots + b_{r-p}\lambda_{r-p} = \zeta \qquad (1')$$

Since B is a basis of U, each $a_i = z$ and each $b_i = z$ in (1′). Then D is a linearly independent set and, hence, is a basis of $U + W$ of dimension $t = r + s - p$.

10. Prove: $|\xi \cdot \eta| \leq |\xi| \cdot |\eta|$ for all $\xi, \eta \in V_n(R)$.

For $\xi = 0$ or $\eta = 0$, we have $0 \leq 0$. Suppose $\xi \neq 0$ and $\eta \neq 0$; then $|\eta| = k|\xi|$ for some $k \in R^+$, and we have

$$\eta \cdot \eta = |\eta|^2 = [k \cdot |\xi|]^2 = k \cdot |\xi| \cdot |\eta| = k^2 \cdot |\xi|^2 = k^2(\xi \cdot \xi)$$

and

$$0 \leq (k\xi \pm \eta) \cdot (k\xi \pm \eta) = k^2(\xi \cdot \xi) \pm 2k(\xi \cdot \eta) + \eta \cdot \eta$$
$$= 2k \cdot |\xi| \cdot |\eta| \pm 2k(\xi \cdot \eta)$$

Hence

$$\pm 2k(\xi \cdot \eta) \leq 2k\,|\xi| \cdot |\eta|$$
$$\pm(\xi \cdot \eta) \leq |\xi| \cdot |\eta|$$

and

$$|\xi \cdot \eta| \leq |\xi| \cdot |\eta|$$

11. Given $\xi = (1, 2, 3, 4)$ and $\eta = (2, 0, -3, 1)$, find
(a) $\xi \cdot \eta$, (b) $|\xi|$ and $|\eta|$, (c) $|5\xi|$ and $|-3\eta|$, (d) $|\xi + \eta|$

(a) $\xi \cdot \eta = 1 \cdot 2 + 2 \cdot 0 + 3(-3) + 4 \cdot 1 = -3$

(b) $|\xi| = \sqrt{1 \cdot 1 + 2 \cdot 2 + 3 \cdot 3 + 4 \cdot 4} = \sqrt{30}$, $\quad |\eta| = \sqrt{4 + 9 + 1} = \sqrt{14}$

(c) $|5\xi| = \sqrt{25 + 25 \cdot 4 + 25 \cdot 9 + 25 \cdot 16} = 5\sqrt{30}$, $\quad |-3\eta| = \sqrt{9 \cdot 4 + 9 \cdot 9 + 9 \cdot 1} = 3\sqrt{14}$

(d) $\xi + \eta = (3, 2, 0, 5)$ and $|\xi + \eta| = \sqrt{9 + 4 + 25} = \sqrt{38}$

12. Given $\xi = (1,1,1)$ and $\eta = (3,4,5)$ in $V_3(R)$, find the shortest vector of the form $\lambda = \xi + s\eta$.

Here, $\lambda = (1+3s,\ 1+4s,\ 1+5s)$

and $|\lambda|^2 = 3 + 24s + 50s^2$

Now $|\lambda|$ is minimum when $24 + 100s = 0$. Hence, $\lambda = (7/25, 1/25, -1/5)$. We find easily that λ and η are orthogonal. We have thus solved the problem: Find the shortest distance from the point $P(1,1,1)$ in ordinary space to the line joining the origin and $Q(3,4,5)$.

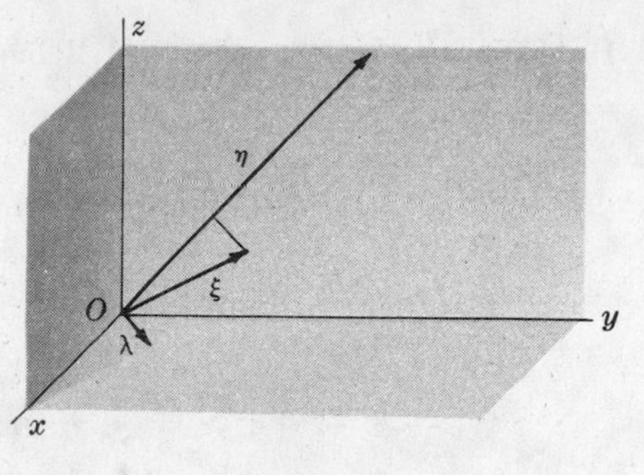

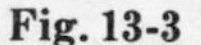
Fig. 13-3

13. For $\xi = (a_1, a_2, a_3)$, $\eta = (b_1, b_2, b_3) \in V_3(R)$, define

$$\xi \times \eta = (a_2b_3 - a_3b_2,\ a_3b_1 - a_1b_3,\ a_1b_2 - a_2b_1)$$

(a) Show that $\xi \times \eta$ is orthogonal to both ξ and η.

(b) Find a vector λ orthogonal to each of $\xi = (1,1,1)$ and $\eta = (1,2,-3)$.

(a) $\xi \cdot (\xi \times \eta) = a_1(a_2b_3 - a_3b_2) + a_2(a_3b_1 - a_1b_3) + a_3(a_1b_2 - a_2b_1) = 0$ and similarly for $\eta \cdot (\xi \times \eta)$.

(b) $\lambda = \xi \times \eta = \big(1(-3) - 2\cdot 1,\ 1\cdot 1 - 1(-3),\ 1\cdot 2 - 1\cdot 1\big) = (-5, 4, 1)$

14. Let $\xi = (1,1,1,1)$ and $\eta = (1,2,-3,0)$ be given vectors in $V_4(R)$.

(a) Show that they are orthogonal.

(b) Find two linearly independent vectors λ and μ which are orthogonal to both ξ and η.

(c) Find a non-zero vector ν orthogonal to each of ξ, η, λ and show that it is a linear combination of λ and μ.

(a) $\xi \cdot \eta = 1\cdot 1 + 1\cdot 2 + 1(-3) = 0$; thus ξ and η are orthogonal.

(b) Assume $(a,b,c,d) \in V_4(R)$ orthogonal to both ξ and η; then

$$\text{(i)} \qquad a+b+c+d = 0 \quad \text{and} \quad a+2b-3c = 0$$

First, take $c = 0$. Then $a + 2b = 0$ is satisfied by $a = 2$, $b = -1$; and $a+b+c+d = 0$ now gives $d = -1$. We have $\lambda = (2,-1,0,-1)$.

Next, take $b = 0$. Then $a - 3c = 0$ is satisfied by $a = 3$, $c = 1$; and $a+b+c+d = 0$ now gives $d = -4$. We have $\mu = (3,0,1,-4)$. Clearly, λ and μ are linearly independent.

Since an obvious solution of the equations (i) is $a = b = c = d = 0$, why not take $\lambda = (0,0,0,0)$?

(c) If $\nu = (a,b,c,d)$ is orthogonal to ξ, η and λ, then

$$a+b+c+d = 0, \quad a+2b-3c = 0 \quad \text{and} \quad 2a-b-d = 0$$

Adding the first and last equation, we have $3a + c = 0$ which is satisfied by $a = 1$, $c = -3$. Now $b = -5$, $d = 7$, and $\nu = (1,-5,-3,7)$. Finally, $\nu = 5\lambda - 3\mu$.

Note. It should be clear that the solutions obtained here are not unique. First of all, any non-zero scalar multiple of any or all of λ, μ, ν is also a solution. Moreover, in finding λ (also, μ) we arbitrarily chose $c = 0$ (also, $b = 0$). However, examine the solution in (c) and verify that ν is unique up to a scalar multiplier.

15. Find the image of $\xi = (1, 2, 3, 4)$ under the linear transformation

$$A: \begin{cases} \epsilon_1 \to (1, -2, 0, 4) \\ \epsilon_2 \to (2, 4, 1, -2) \\ \epsilon_3 \to (0, -1, 5, -1) \\ \epsilon_4 \to (1, 3, 2, 0) \end{cases}$$

of $V_4(Q)$ into itself.

We have

$$\xi = \epsilon_1 + 2\epsilon_2 + 3\epsilon_3 + 4\epsilon_4 \to$$
$$(1, -2, 0, 4) + 2(2, 4, 1, -2) + 3(0, -1, 5, -1) + 4(1, 3, 2, 0) = (9, 15, 25, -3)$$

16. Prove: If $\{\xi_1, \xi_2, \ldots, \xi_n\}$ is any basis of $V = V(\mathcal{F})$ and if $\{\eta_1, \eta_2, \ldots, \eta_n\}$ is any set of n elements of V, the mapping

$$T: \quad \xi_i \to \eta_i, \qquad (i = 1, 2, \ldots, n)$$

defines a linear transformation of V into itself.

Let $\xi = \sum s_i\xi_i$ and $\eta = \sum t_i\xi_i$ be any two vectors in V. Then

$$\xi + \eta = \sum(s_i + t_i)\xi_i \to \sum(s_i + t_i)\eta_i = \sum s_i\eta_i + \sum t_i\eta_i$$

so that (i) $(\xi + \eta)T = \xi T + \eta T$

Also, for any $s \in \mathcal{F}$ and any $\xi \in V$,

$$s\xi = s\sum s_i\xi_i \to s\sum s_i\eta_i$$

so that (ii) $(s\xi)T = s(\xi T)$

as required.

17. Prove: If T is a linear transformation of $V(\mathcal{F})$ into itself and if W is a subspace of $V(\mathcal{F})$, then $W_T = \{\xi T : \xi \in W\}$, the image of W under T, is also a subspace of $V(\mathcal{F})$.

For any $\xi T, \eta T \in W_T$, we have $\xi T + \eta T = (\xi + \eta)T$. Since $\xi, \eta \in W$ implies $\xi + \eta \in W$, then $(\xi + \eta)T \in W_T$. Thus W_T is closed under addition. Similarly, for any $\xi T \in W_T$, $s \in \mathcal{F}$, we have $s(\xi T) = (s\xi)T \in W_T$ since $\xi \in W$ implies $s\xi \in W$. Thus W_T is closed under scalar multiplication. This completes the proof.

18. Prove: The set $\mathcal{A}$ of all linear transformations of a vector space $V(\mathcal{F})$ into itself forms a ring with respect to addition and multiplication defined by

$$A + B: \quad \xi(A + B) = \xi A + \xi B, \qquad \xi \in V(\mathcal{F})$$
$$A \cdot B: \quad \xi(A \cdot B) = (\xi A)B, \qquad \xi \in V(\mathcal{F})$$

for all $A, B \in \mathcal{A}$.

Let $\xi, \eta \in V(\mathcal{F})$, $k \in \mathcal{F}$, and $A, B, C \in \mathcal{A}$. Then

$$(\xi + \eta)(A + B) = (\xi + \eta)A + (\xi + \eta)B = \xi A + \eta A + \xi B + \eta B$$
$$= \xi(A + B) + \eta(A + B)$$

and

$$(k\xi)(A + B) = (k\xi)A + (k\xi)B = k(\xi A) + k(\xi B)$$
$$= k(\xi A + \xi B) = k\xi(A + B)$$

Also,

$$(\xi + \eta)(A \cdot B) = [(\xi + \eta)A]B = (\xi A + \eta A)B$$
$$= (\xi A)B + (\eta A)B = \xi(A \cdot B) + \eta(A \cdot B)$$

Thus $A+B, A\cdot B \in \mathcal{A}$ and $\mathcal{A}$ is closed with respect to addition and multiplication.

Addition is both commutative and associative since

$$\xi(A+B) = \xi A + \xi B = \xi B + \xi A = \xi(B+A)$$

and

$$\xi[(A+B)+C] = \xi(A+B) + \xi C = \xi A + \xi B + \xi C$$
$$= \xi A + \xi(B+C) = \xi[A+(B+C)]$$

Let the mapping which carries each element of $V(\mathcal{F})$ into ζ be denoted by 0; i.e.,

$$0: \quad \xi 0 = \zeta, \quad \xi \in V(\mathcal{F})$$

Then $0 \in \mathcal{A}$ (show this), $\quad \xi(A+0) = \xi A + \xi 0 = \xi A + \zeta = \xi A$

and 0 is the additive identity element of $\mathcal{A}$.

For each $A \in \mathcal{A}$, let $-A$ be defined by

$$-A: \quad \xi(-A) = -(\xi A), \quad \xi \in V(\mathcal{F})$$

It follows readily that $-A \in \mathcal{A}$ and is the additive inverse of A since

$$\zeta = \zeta A = (\xi - \xi)A = \xi A + [-(\xi A)] = \xi[A+(-A)] = \xi \cdot 0$$

We have proved that $\mathcal{A}$ is an abelian additive group.

Multiplication is clearly associative but, in general, is not commutative (see Problem 55). To complete the proof that $\mathcal{A}$ is a ring, we prove one of the distributive laws

$$A \cdot (B+C) = A \cdot B + A \cdot C$$

and leave the other for the reader. We have

$$\xi[A \cdot (B+C)] = (\xi A)(B+C) = (\xi A)B + (\xi A)C$$
$$= \xi(A \cdot B) + \xi(A \cdot C) = \xi(A \cdot B + A \cdot C)$$

19. Prove: The set $\mathcal{M}$ of all non-singular linear transformations of a vector space $V(\mathcal{F})$ into itself forms a group under multiplication.

Let $A, B \in \mathcal{M}$. Since A and B are non-singular they map $V(\mathcal{F})$ onto $V(\mathcal{F})$, i.e., $V_A = V$ and $V_B = V$. Then $V_{(A\cdot B)} = (V_A)_B = V_B = V$ and $A \cdot B$ is non-singular. Thus $A \cdot B \in \mathcal{M}$ and $\mathcal{M}$ is closed under multiplication. The associative law holds in $\mathcal{M}$ since it holds in $\mathcal{A}$.

Let the mapping which carries each element of $V(\mathcal{F})$ into itself be denoted by I, i.e.,

$$I: \quad \xi I = \zeta, \quad \xi \in V(\mathcal{F})$$

Evidently, I is non-singular, belongs to $\mathcal{M}$, and since

$$\xi(I \cdot A) = (\xi I)A = \xi A = (\xi A)I = \xi(A \cdot I)$$

is the identity element in multiplication.

Now A is a one-to-one mapping of $V(\mathcal{F})$ onto itself; hence it has an inverse A^{-1} defined by

$$A^{-1}: \quad (\xi A)A^{-1} = \xi, \quad \xi \in V(\mathcal{F})$$

For any $\xi, \eta \in V(\mathcal{F})$, $A \in \mathcal{M}$, and $k \in \mathcal{F}$, we have $\xi A, \eta A \in V(\mathcal{F})$. Then, since

$$(\xi A + \eta A)A^{-1} = (\xi + \eta)A \cdot A^{-1} = \xi + \eta = (\xi A)A^{-1} + (\eta A)A^{-1}$$

and

$$[k(\xi A)]A^{-1} = [(k\xi)A]A^{-1} = k\xi = k[(\xi A)A^{-1}]$$

it follows that $A^{-1} \in \mathcal{A}$. But, by definition, A^{-1} is non-singular; hence $A^{-1} \in \mathcal{M}$. Thus each element of $\mathcal{M}$ has a multiplicative inverse and $\mathcal{M}$ is a multiplicative group.

Supplementary Problems

20. Using Fig. 13-1, Page 143:

(a) Identify the vectors $(1,0)$ and $(0,1)$; also $(a,0)$ and $(0,b)$.

(b) Verify $(a,b) = (a,0)+(0,b) = a(1,0)+b(0,1)$.

21. Using natural definitions for scalar multiplication and vector addition, show that each of the following are vector spaces over the indicated field:

(a) $V = R$; $\mathcal{F} = Q$ (b) $V = C$; $\mathcal{F} = R$

(c) $V = \{a + b\sqrt[3]{2} + c\sqrt{3} : a,b,c \in Q\}$; $\mathcal{F} = Q$

(d) V = all polynomials of degree ≤ 4 over R, including the zero polynomial; $\mathcal{F} = Q$

(e) $V = \{c_1e^x + c_2e^{3x} : c_1, c_2 \in R\}$; $\mathcal{F} = R$

(f) $V = \{(a_1, a_2, a_3) : a_i \in Q,\ a_1 + 2a_2 = 3a_3\}$; $\mathcal{F} = Q$

(g) $V = \{a + bx : a,b \in I/(3)\}$; $\mathcal{F} = I/(3)$

22. (a) Why is the set of all polynomials in x of degree > 4 over R not a vector space over R?

(b) Is the set of all polynomials $R[x]$ a vector space over Q? over C?

23. Let $\xi, \eta \in V$ over $\mathcal{F}$ and $s,t \in \mathcal{F}$. Prove:

(a) When $\xi \neq \zeta$, then $s\xi = t\xi$ implies $s = t$.

(b) When $s \neq z$, then $s\xi = s\eta$ implies $\xi = \eta$.

24. Let $\xi, \eta \neq \zeta \in V$ over $\mathcal{F}$. Show that ξ and η are linearly dependent if and only if $\xi = s\eta$ for some $s \in \mathcal{F}$.

25. (a) Let $\xi, \eta \in V(R)$. If ξ and η are linearly dependent over R, are they necessarily linearly dependent over Q? over C?

(b) Consider (a) with linearly dependent replaced by linearly independent.

26. Prove Theorem IV and Theorem VI, Page 146.

27. For the vector space $V = V_4(R) = \{(a,b,c,d) : a,b,c,d \in R\}$ over R, which of the following subsets are subspaces of V?

(a) $U = \{(a,a,a,a) : a \in R\}$

(b) $U = \{(a,b,a,b) : a,b \in I\}$

(c) $U = \{(a,2a,b,a+b) : a,b \in R\}$

(d) $U = \{(a_1,a_2,a_3,a_4) : a_i \in R,\ 2a_2 + 3a_3 = 0\}$

(e) $U = \{(a_1,a_2,a_3,a_4) : a_i \in R,\ 2a_2 + 3a_3 = 5\}$

28. Determine whether the following sets of vectors in $V_3(Q)$ are linearly dependent or independent over Q:

(a) $\{(0,0,0), (1,1,1)\}$ (d) $\{(0,1,-2), (1,-1,1), (1,2,1)\}$

(b) $\{(1,-2,3), (3,-6,9)\}$ (e) $\{(0,2,-4), (1,-2,-1), (1,-4,3)\}$

(c) $\{(1,-2,-3), (3,2,1)\}$ (f) $\{(1,-1,-1), (2,3,1), (-1,4,-2), (3,10,8)\}$

Ans. (c), (d) are linearly independent.

29. Determine whether the following sets of vectors in $V_3(I/(5))$ are linearly dependent or independent over $I/(5)$.

(a) $\{(1,2,4), (2,4,1)\}$ (c) $\{(0,1,1), (1,0,1), (3,3,2)\}$

(b) $\{(2,3,4), (3,2,1)\}$ (d) $\{(4,1,3), (2,3,1), (4,1,0)\}$

Ans. (a), (c) are linearly independent.

30. For the set $S = \{(1,2,1), (2,3,2), (3,2,3), (1,1,1)\}$ of vectors in $V(I/(5))$ find a maximum linearly independent subset T and express each of the remaining vectors as linear combinations of the elements of T.

31. Find the dimension of the subset of $V_3(Q)$ spanned by each set of vectors in Problem 28.
Ans. (*a*), (*b*) 1; (*c*), (*e*) 2; (*d*), (*f*) 3

32. For the vector space C over R, show
(*a*) $\{1, i\}$ is a basis.
(*b*) $\{a + bi, c + di\}$ is a basis if and only if $ad - bc \neq 0$.

33. In each of the following, find a basis of the vector space which includes the given set of vectors:
(*a*) $\{(1,1,0), (0,1,1)\}$ in $V_3(Q)$.
(*b*) $\{(2,1,-1,-2), (2,3,-2,1), (4,2,-1,3)\}$ in $V_4(Q)$.
(*c*) $\{(2,1,1,0), (1,2,0,1)\}$ in $V_4(I/(3))$.
(*d*) $\{(i,0,1,0), (0,i,1,0)\}$ in $V_4(C)$.

34. Show that $S = \{\xi_1, \xi_2, \xi_3\} = \{(i, 1+i, 2), (2+i, i, 1), (3, 3+2i, -1)\}$ is a basis of $V_3(C)$ and express each of the unit vectors of $V_3(C)$ as a linear combination of the elements of S.
Ans.
$$\epsilon_1 = [(-39-6i)\xi_1 + (80-i)\xi_2 + (2-13i)\xi_3]/173$$
$$\epsilon_2 = [(86-40i)\xi_1 + (-101+51i)\xi_2 + (71-29i)\xi_3]/346$$
$$\epsilon_3 = [(104+16i)\xi_1 + (75-55i)\xi_2 + (-63-23i)\xi_3]/346$$

35. Prove: If $k_1\xi_1 + k_2\xi_2 + k_3\xi_3 = \zeta$, where $k_1k_2 \neq z$, then $\{\xi_1, \xi_3\}$ and $\{\xi_2, \xi_3\}$ generate the same space.

36. Prove Theorem IX, Page 147.

37. Prove: If $\{\xi_1, \xi_2, \xi_3\}$ is a basis of $V_3(Q)$, so also is $\{\xi_1 + \xi_2, \xi_2 + \xi_3, \xi_3 + \xi_1\}$. Is this true in $V_3(I/(2))$? In $V_3(I/(3))$?

38. Prove Theorem X, Page 147. *Hint.* Assume A and B, containing respectively m and n elements, are bases of V. First, associate S with A and T with B, then S with B and T with A, and apply Theorem VIII, Page 147.

39. Prove: If V is a vector space of dimension $n \geqq 0$, any linearly independent set of n vectors of V is a basis.

40. Let $\xi_1, \xi_2, \ldots, \xi_m \in V$ and $S = \{\xi_1, \xi_2, \ldots, \xi_m\}$ span a subspace $U \subset V$. Show that the minimum number of vectors of V necessary to span U is the maximum number of linearly independent vectors in S.

41. Let $\{\xi_1, \xi_2, \ldots, \xi_n\}$ be a basis of V. Show that every vector $\xi \in V$ has a unique representation as a linear combination of the basis vectors.
Hint. Suppose $\xi = \sum c_i\xi_i = \sum d_i\xi_i$; then $\sum c_i\xi_i - \sum d_i\xi_i = \zeta$.

42. Prove: If U and W are subspaces of a vector space V, so also are $U \cap W$ and $U + W$.

43. Let the subspaces U and W of $V_4(Q)$ be spanned by
$$A = \{(2,-1,1,0), (1,0,2,1)\} \quad \text{and} \quad B = \{(0,0,1,0), (0,1,0,1), (4,-1,5,2)\}$$
respectively. Verify Theorem XIII, Page 148. Find a basis of $U + W$ which includes the vectors of A; also a basis which includes the vectors of B.

44. Show that $P = \{(a, b, -b, a) : a, b \in R\}$ with addition defined by
$$(a, b, -b, a) + (c, d, -d, c) = \big(a+c, b+d, -(b+d), a+c\big)$$
and scalar multiplication defined by $k(a, b, -b, a) = (ka, kb, -kb, ka)$, for all $(a, b, -b, a), (c, d, -d, c) \in P$ and $k \in R$, is a vector space of dimension two.

45. Let the prime p be a prime element of G of Problem 8, Chapter 10. Denote by $\mathcal{F}$ the field $G/(p)$ and by $\mathcal{F}'$ the prime field $I/(p)$ of $\mathcal{F}$. The field $\mathcal{F}$, considered as a vector space over $\mathcal{F}'$, has as basis $\{1, i\}$; hence, $\mathcal{F} = \{a_1 \cdot 1 + a_2 \cdot i : a_1, a_2 \in \mathcal{F}'\}$. (*a*) Show that $\mathcal{F}$ has at most p^2 elements. (*b*) Show that $\mathcal{F}$ has at least p^2 elements, (that is, $a_1 \cdot 1 + a_2 \cdot i = b_1 \cdot 1 + b_2 \cdot i$ if and only if $a_1 - b_1 = a_2 - b_2 = 0$) and hence exactly p^2 elements.

46. Generalize Problem 45 to a finite field $\mathcal{F}$ of characteristic p, a prime, over its prime field having n elements as basis.

47. For $\xi, \eta, \mu \in V_n(R)$ and $k \in R$, prove:
(a) $\xi \cdot \eta = \eta \cdot \xi$, (b) $(\xi + \eta) \cdot \mu = \xi \cdot \mu + \eta \cdot \mu$, (c) $(k\xi \cdot \eta) = k(\xi \cdot \eta)$.

48. Obtain from the Schwarz inequality $-1 \leq (\xi \cdot \eta)/(|\xi| \cdot |\eta|) \leq 1$ to show that $\cos\theta = \dfrac{\xi \cdot \eta}{|\xi| \cdot |\eta|}$ determines one and only one angle between 0° and 180°.

49. Let length be defined as in $V_n(R)$. Show that $(1,1) \in V_2(Q)$ is without length while $(1,i) \in V_2(C)$ is of length 0. Can you suggest a definition of $\xi \cdot \eta$ so that each non-zero vector of $V_2(C)$ will have length different from 0?

50. Let $\xi, \eta \in V_n(R)$ such that $|\xi| = |\eta|$. Show that $\xi - \eta$ and $\xi + \eta$ are orthogonal. What is the geometric interpretation?

51. For the vector space V of all polynomials in x over a field $\mathcal{F}$, verify that each of the following mappings of V into itself is a linear transformation of V.

(a) $\alpha(x) \to \alpha(x)$ (c) $\alpha(x) \to \alpha(-x)$

(b) $\alpha(x) \to -\alpha(x)$ (d) $\alpha(x) \to \alpha(0)$

52. Show that the mapping $T: (a, b) \to (a+1, b+1)$ of $V_2(R)$ into itself is not a linear transformation.
Hint. Compare $(\epsilon_1 + \epsilon_2)T$ and $(\epsilon_1 T + \epsilon_2 T)$.

53. For each of the linear transformations A, examine the image of an arbitrary vector ξ to determine the rank of A and, if A is singular, to determine a non-zero vector whose image is 0.

(a) $A: \quad \epsilon_1 \to (2,1), \quad \epsilon_2 \to (1,2)$

(b) $A: \quad \epsilon_1 \to (3,-4), \quad \epsilon_2 \to (-3,4)$

(c) $A: \quad \epsilon_1 \to (1,1,2), \quad \epsilon_2 \to (2,1,3), \quad \epsilon_3 \to (1,0,-2)$

(d) $A: \quad \epsilon_1 \to (1,-1,1), \quad \epsilon_2 \to (-3,3,-3), \quad \epsilon_3 \to (2,3,4)$

(e) $A: \quad \epsilon_1 \to (0,1,-1), \quad \epsilon_2 \to (-1,1,1), \quad \epsilon_3 \to (1,0,-2)$

(f) $A: \quad \epsilon_1 \to (1,0,3), \quad \epsilon_2 \to (0,1,1), \quad \epsilon_3 \to (2,2,8)$

Ans. (a) non-singular; (b) singular, $(1,1)$; (c) non-singular; (d) singular, $(3,1,0)$; (e) singular, $(-1,1,1)$; (f) singular, $(2,2,-1)$

54. For all $A, B \in \mathcal{A}$ and $k, l \in \mathcal{F}$, prove

(a) $kA \in \mathcal{A}$

(b) $k(A+B) = kA + kB$; $(k+l)A = kA + lA$

(c) $k(A \cdot B) = (kA)B = A(kB)$; $(k \cdot l)A = k(lA)$

(d) $0 \cdot A = k0 = 0$; $uA = A$

with 0 defined on Page 159. Together with Problem 18, this completes the proof of Theorem XX, Page 152.

55. Compute $B \cdot A$ for the linear transformations of Example 13, Page 152, to show that, in general, $A \cdot B \neq B \cdot A$.

56. For the linear transformations on $V_3(R)$

$$A: \begin{cases} \epsilon_1 \to (a, b, c) \\ \epsilon_2 \to (d, e, f) \\ \epsilon_3 \to (g, h, i) \end{cases} \quad \text{and} \quad B: \begin{cases} \epsilon_1 \to (j, k, l) \\ \epsilon_2 \to (m, n, p) \\ \epsilon_3 \to (q, r, s) \end{cases}$$

obtain

$$A + B: \begin{cases} \epsilon_1 \to (a+j,\ b+k,\ c+l) \\ \epsilon_2 \to (d+m,\ e+n,\ f+p) \\ \epsilon_3 \to (g+q,\ h+r,\ i+s) \end{cases}$$

$$A \cdot B : \begin{cases} \epsilon_1 \rightarrow (aj+bm+cq,\ ak+bn+cr,\ al+bp+cs) \\ \epsilon_2 \rightarrow (dj+em+fq,\ dk+en+fr,\ dl+ep+fs) \\ \epsilon_3 \rightarrow (gj+hm+iq,\ gk+hn+ir,\ gl+hp+is) \end{cases}$$

and, for any $k \in R$,

$$kA : \begin{cases} \epsilon_1 \rightarrow (ka, kb, kc) \\ \epsilon_2 \rightarrow (kd, ke, kf) \\ \epsilon_3 \rightarrow (kg, kh, ki) \end{cases}$$

57. Compute the inverse A^{-1} of $A : \begin{cases} \epsilon_1 \rightarrow (1,1) \\ \epsilon_2 \rightarrow (2,3) \end{cases}$ of $V_2(R)$.

Hint. Take $A^{-1} : \begin{cases} \epsilon_1 \rightarrow (p,q) \\ \epsilon_2 \rightarrow (r,s) \end{cases}$ and consider $(\xi A)A^{-1} = \xi$ where $\xi = (a, b)$.

Ans. $\begin{cases} \epsilon_1 \rightarrow (3,-1) \\ \epsilon_2 \rightarrow (-2,1) \end{cases}$

58. For the mapping

$$T_1 : \begin{cases} \epsilon_1 = (1,0) \rightarrow (1,1,1) \\ \epsilon_2 = (0,1) \rightarrow (0,1,2) \end{cases} \quad \text{of } V = V_2(R) \text{ into } W = V_3(R)$$

verify:

(*a*) T_1 is a linear transformation of V into W.

(*b*) The image of $\xi = (2,1) \in V$ is $(2,3,4) \in W$.

(*c*) The vector $(1,2,2) \in W$ is not an image.

(*d*) V_{T_1} has dimension 2.

59. For the mapping

$$T_2 : \begin{cases} \epsilon_1 = (1,0,0) \rightarrow (1,0,1,1) \\ \epsilon_2 = (0,1,0) \rightarrow (0,1,1,1) \\ \epsilon_3 = (0,0,1) \rightarrow (1,-1,0,0) \end{cases} \quad \text{of } V = V_3(R) \text{ into } W = V_4(R)$$

verify:

(*a*) T_2 is a linear transformation of V into W.

(*b*) The image of $\xi = (1,-1,-1) \in V$ is $(0,0,0,0) \in W$.

(*c*) V_{T_2} has dimension 2 and $r_{T_2} = 2$.

60. For T_1 of Problem 58 and T_2 of Problem 59, verify

$$T_1 \cdot T_2 : \begin{cases} \epsilon_1 = (1,0) \rightarrow (2,0,2,2) \\ \epsilon_2 = (0,1) \rightarrow (2,-1,1,1) \end{cases}$$

What is the rank of $T_1 \cdot T_2$?

Chapter 14

Matrices

INTRODUCTION

Consider again the linear transformations on $V_3(R)$

$$A: \begin{cases} \epsilon_1 \rightarrow (a, b, c) \\ \epsilon_2 \rightarrow (d, e, f) \\ \epsilon_3 \rightarrow (g, h, i) \end{cases} \quad \text{and} \quad B: \begin{cases} \epsilon_1 \rightarrow (j, k, l) \\ \epsilon_2 \rightarrow (m, n, p) \\ \epsilon_3 \rightarrow (q, r, s) \end{cases} \tag{1}$$

for which (see Problem 56, Chapter 13, Page 162)

$$A + B: \begin{cases} \epsilon_1 \rightarrow (a+j,\ b+k,\ c+l) \\ \epsilon_2 \rightarrow (d+m,\ e+n,\ f+p) \\ \epsilon_3 \rightarrow (g+q,\ h+r,\ i+s) \end{cases}$$

$$A \cdot B: \begin{cases} \epsilon_1 \rightarrow (aj+bm+cq,\ ak+bn+cr,\ al+bp+cs) \\ \epsilon_2 \rightarrow (dj+em+fq,\ dk+en+fr,\ dl+ep+fs) \\ \epsilon_3 \rightarrow (gj+hm+iq,\ gk+hn+ir,\ gl+hp+is) \end{cases}$$

$$kA: \begin{cases} \epsilon_1 \rightarrow (ka, kb, kc) \\ \epsilon_2 \rightarrow (kd, ke, kf) \\ \epsilon_3 \rightarrow (kg, kh, ki) \end{cases}, \qquad k \in R$$

As a step toward simplifying matters, let the present notation for linear transformations A and B be replaced by the arrays

$$A = \begin{bmatrix} a & b & c \\ d & e & f \\ g & h & i \end{bmatrix} \quad \text{and} \quad B = \begin{bmatrix} j & k & l \\ m & n & p \\ q & r & s \end{bmatrix} \tag{2}$$

effected by enclosing the image vectors in brackets and then removing the excess parentheses and commas. Our problem is to translate the operations on linear transformations into corresponding operations on their arrays. We have:

The *sum* $A + B$ of the arrays A and B is the array whose elements are the sums of the corresponding elements of A and B.

The *scalar product* kA of k, any scalar, and A is the array whose elements are k times the corresponding elements of A.

In forming the product $A \cdot B$ think of A as consisting of the vectors ρ_1, ρ_2, ρ_3 (the *row vectors* of A) whose components are the elements of the rows of A and think of B as consisting of the vectors $\gamma_1, \gamma_2, \gamma_3$ (the *column vectors* of B) whose components are the elements of the columns of B. Then

$$A \cdot B = \begin{bmatrix} \rho_1 \\ \rho_2 \\ \rho_3 \end{bmatrix} \cdot [\gamma_1 \ \gamma_2 \ \gamma_3] = \begin{bmatrix} \rho_1 \cdot \gamma_1 & \rho_1 \cdot \gamma_2 & \rho_1 \cdot \gamma_3 \\ \rho_2 \cdot \gamma_1 & \rho_2 \cdot \gamma_2 & \rho_2 \cdot \gamma_3 \\ \rho_3 \cdot \gamma_1 & \rho_3 \cdot \gamma_2 & \rho_3 \cdot \gamma_3 \end{bmatrix}$$

where $\rho_i \cdot \gamma_j$ is the inner product of ρ_i and γ_j. Note carefully that in $A \cdot B$ the inner product of every row vector of A with every column vector of B appears; also, the elements of any row of $A \cdot B$ are those inner products whose first factor is the corresponding row vector of A.

Example 1:

(*a*) When $A = \begin{bmatrix} 1 & 2 \\ 3 & 4 \end{bmatrix}$ and $B = \begin{bmatrix} 5 & 6 \\ 7 & 8 \end{bmatrix}$ over Q, we have

$$A + B = \begin{bmatrix} 1+5 & 2+6 \\ 3+7 & 4+8 \end{bmatrix} = \begin{bmatrix} 6 & 8 \\ 10 & 12 \end{bmatrix}, \qquad 10A = \begin{bmatrix} 10 \cdot 1 & 10 \cdot 2 \\ 10 \cdot 3 & 10 \cdot 4 \end{bmatrix} = \begin{bmatrix} 10 & 20 \\ 30 & 40 \end{bmatrix}$$

$$A \cdot B = \begin{bmatrix} 1 & 2 \\ 3 & 4 \end{bmatrix} \cdot \begin{bmatrix} 5 & 6 \\ 7 & 8 \end{bmatrix} = \begin{bmatrix} 1 \cdot 5 + 2 \cdot 7 & 1 \cdot 6 + 2 \cdot 8 \\ 3 \cdot 5 + 4 \cdot 7 & 3 \cdot 6 + 4 \cdot 8 \end{bmatrix} = \begin{bmatrix} 19 & 22 \\ 43 & 50 \end{bmatrix}$$

and
$$B \cdot A = \begin{bmatrix} 5 & 6 \\ 7 & 8 \end{bmatrix} \cdot \begin{bmatrix} 1 & 2 \\ 3 & 4 \end{bmatrix} = \begin{bmatrix} 5+18 & 10+24 \\ 7+24 & 14+32 \end{bmatrix} = \begin{bmatrix} 23 & 34 \\ 31 & 46 \end{bmatrix}$$

(*b*) When $A = \begin{bmatrix} 1 & 0 & -2 \\ 0 & -3 & 1 \\ 2 & 1 & 0 \end{bmatrix}$ and $B = \begin{bmatrix} 3 & -1 & 0 \\ -2 & 0 & 3 \\ 0 & 2 & 1 \end{bmatrix}$ over Q, we have

$$A \cdot B = \begin{bmatrix} 3 & -1-4 & 2 \\ 6 & 2 & -9-1 \\ 6-2 & -2 & 3 \end{bmatrix} = \begin{bmatrix} 3 & -5 & 2 \\ 6 & 2 & -10 \\ 4 & -2 & 3 \end{bmatrix} \quad \text{and} \quad B \cdot A = \begin{bmatrix} 3 & 3 & -7 \\ 4 & 3 & 4 \\ -2 & -7 & 2 \end{bmatrix}$$

SQUARE MATRICES

The arrays of the preceding section, on which addition, multiplication, and scalar multiplication have been defined, are called *square matrices*; more precisely, they are *square matrices of order* 3 since they have 3^2 elements. (In Example 1(*a*), the square matrices are of order 2).

In order to permit the writing in full of square matrices of higher orders, we now introduce a more uniform notation. Hereafter, the elements of an arbitrary square matrix will be denoted by a single letter with varying subscripts; for example,

$$A = \begin{bmatrix} a_{11} & a_{12} & a_{13} \\ a_{21} & a_{22} & a_{23} \\ a_{31} & a_{32} & a_{33} \end{bmatrix} \quad \text{and} \quad B = \begin{bmatrix} b_{11} & b_{12} & b_{13} & b_{14} \\ b_{21} & b_{22} & b_{23} & b_{24} \\ b_{31} & b_{32} & b_{33} & b_{34} \\ b_{41} & b_{42} & b_{43} & b_{44} \end{bmatrix}$$

Any element, say, b_{24} is to be thought of as $b_{2,4}$ although unless necessary (e.g., b_{121} which could be $b_{12,1}$ or $b_{1,21}$) we shall not print the comma. One advantage of this notation is that each element discloses its exact position in the matrix. For example, the element b_{24} stands in the second row and fourth column, the element b_{32} stands in the third row and second column, and so on. Another advantage is that we may indicate the matrices A and B above by writing

$$A = [a_{ij}], \qquad (i = 1,2,3;\ j = 1,2,3)$$

and
$$B = [b_{ij}], \qquad (i = 1,2,3,4;\ j = 1,2,3,4)$$

Then with A defined above and $C = [c_{ij}]$, $(i, j = 1, 2, 3)$, the product

$$A \cdot C = \begin{bmatrix} \sum a_{1j}c_{j1} & \sum a_{1j}c_{j2} & \sum a_{1j}c_{j3} \\ \sum a_{2j}c_{j1} & \sum a_{2j}c_{j2} & \sum a_{2j}c_{j3} \\ \sum a_{3j}c_{j1} & \sum a_{3j}c_{j2} & \sum a_{3j}c_{j3} \end{bmatrix}$$

where in each case the summation extends over all values of j; for example,

$$\sum a_{2j}c_{j3} = a_{21}c_{13} + a_{22}c_{23} + a_{23}c_{33}, \text{ etc.}$$

Two square matrices L and M will be called equal, $L = M$, if and only if one is the duplicate of the other; i.e., if and only if L and M are the same linear transformation. Thus two equal matrices necessarily have the same order.

In any square matrix the diagonal drawn from the upper left corner to the lower right corner is called the *principal diagonal* of the matrix. The elements standing in a principal diagonal are those and only those (a_{11}, a_{22}, a_{33} of A, for example) whose row and column indices are equal.

By definition, there is a one-to-one correspondence between the set of all linear transformations of a vector space over $\mathcal{F}$ of dimension n into itself and the set of all square matrices over $\mathcal{F}$ of order n (set of all n-square matrices over $\mathcal{F}$). Moreover, we have defined addition and multiplication on these matrices so that this correspondence is an isomorphism. Hence, by Theorems XVIII and XIX of Chapter 13, Page 152, we have

Theorem I. With respect to addition and multiplication, the set of all n-square matrices over $\mathcal{F}$ is a ring $\mathcal{R}$ with unity.

As a consequence:

Addition is both associative and commutative on $\mathcal{R}$.

Multiplication is associative but, in general, not commutative on $\mathcal{R}$.

Multiplication is both left and right distributive with respect to addition.

There exists a matrix 0_n or 0, the zero element of $\mathcal{R}$, each of whose elements is the zero element of $\mathcal{F}$. For example, $0_2 = \begin{bmatrix} 0 & 0 \\ 0 & 0 \end{bmatrix}$ and $0_3 = \begin{bmatrix} 0 & 0 & 0 \\ 0 & 0 & 0 \\ 0 & 0 & 0 \end{bmatrix}$ are *zero matrices* over R of orders 2 and 3 respectively.

There exists a matrix I_n or I, the unity of $\mathcal{R}$, having the unity of $\mathcal{F}$ as all elements along the principal diagonal and the zero element of $\mathcal{F}$ elsewhere. For example, $I_2 = \begin{bmatrix} 1 & 0 \\ 0 & 1 \end{bmatrix}$ and $I_3 = \begin{bmatrix} 1 & 0 & 0 \\ 0 & 1 & 0 \\ 0 & 0 & 1 \end{bmatrix}$ are *identity matrices* over R of orders 2 and 3 respectively.

For each $A = [a_{ij}] \in \mathcal{R}$, there exists an additive inverse $-A = (-1)[a_{ij}] = [-a_{ij}]$ such that $A + (-A) = 0$.

Throughout the remainder of this book we shall use 0 and 1 respectively to denote the zero element and unity of any field (see Page 118). Whenever z and u, originally reserved to denote these elements, appear they will have quite different connotations. Also, 0 and I as defined above over R will be used as the zero and identity matrices over any field $\mathcal{F}$.

By Theorem XX, Chapter 13, Page 152, we have

Theorem II. The set of all n-square matrices over $\mathcal{F}$ is itself a vector space.

A set of basis elements for this vector space consists of the n^2 matrices

$$E_{ij},\ (i,j = 1,2,3,\ldots,n)$$

of order n having 1 as element in the (i,j) position and 0's elsewhere. For example,

$$\{E_{11}, E_{12}, E_{21}, E_{22}\} = \left\{\begin{bmatrix}1 & 0\\0 & 0\end{bmatrix}, \begin{bmatrix}0 & 1\\0 & 0\end{bmatrix}, \begin{bmatrix}0 & 0\\1 & 0\end{bmatrix}, \begin{bmatrix}0 & 0\\0 & 1\end{bmatrix}\right\}$$

is a basis of the vector space of all 2-square matrices over $\mathcal{F}$; and for any $A = \begin{bmatrix}a & b\\c & d\end{bmatrix}$, we have $A = aE_{11} + bE_{12} + cE_{21} + dE_{22}$.

TOTAL MATRIX ALGEBRA

The set of all n-square matrices over $\mathcal{F}$ with the operations of addition, multiplication and scalar multiplication by elements of $\mathcal{F}$ is called the *total matrix algebra* $\mathcal{M}_n(\mathcal{F})$. Now just as there are subgroups of groups, subrings of rings, ..., so there are subalgebras of $\mathcal{M}_n(\mathcal{F})$. For example, the set of all matrices $\mathcal{M}$ of the form $\begin{bmatrix}a & b & c\\2c & a & b\\2b & 2c & a\end{bmatrix}$, where $a,b,c \in Q$, is a subalgebra of $\mathcal{M}_3(Q)$. All that is needed to prove this is to show that addition, multiplication and scalar multiplication by elements of Q on elements of $\mathcal{M}$ invariably yield elements of $\mathcal{M}$. Addition and scalar multiplication give no trouble, and so $\mathcal{M}$ is a subspace of the vector space $\mathcal{M}_3(Q)$. For multiplication, note that a basis of $\mathcal{M}$ is the set $\left\{I,\ X = \begin{bmatrix}0 & 1 & 0\\0 & 0 & 1\\2 & 0 & 0\end{bmatrix},\ Y = \begin{bmatrix}0 & 0 & 1\\2 & 0 & 0\\0 & 2 & 0\end{bmatrix}\right\}$. We leave for the reader to show that for $A, B \in \mathcal{M}$,

$$\begin{aligned} A\cdot B &= (aI + bX + cY)(xI + yX + zY) \\ &= (ax + 2bz + 2cy)I + (ay + bx + 2cz)X + (az + by + cx)Y \ \in\ \mathcal{M}; \end{aligned}$$

also, that multiplication is commutative on $\mathcal{M}$.

A MATRIX OF ORDER $m \times n$

By a matrix over $\mathcal{F}$ we shall mean any rectangular array of elements of $\mathcal{F}$; for example,

$$A = \begin{bmatrix}a_{11} & a_{12} & a_{13}\\a_{21} & a_{22} & a_{23}\\a_{31} & a_{32} & a_{33}\end{bmatrix}, \quad B = \begin{bmatrix}b_{11} & b_{12} & b_{13} & b_{14}\\b_{21} & b_{22} & b_{23} & b_{24}\\b_{31} & b_{32} & b_{33} & b_{34}\end{bmatrix}, \quad C = \begin{bmatrix}c_{11} & c_{12}\\c_{21} & c_{22}\\c_{31} & c_{32}\\c_{41} & c_{42}\end{bmatrix}$$

or $\quad A = [a_{ij}],\ (i,j = 1,2,3) \qquad B = [b_{ij}],\ (i = 1,2,3;\ j = 1,2,3,4)$

$$C = [c_{ij}],\ (i = 1,2,3,4;\ j = 1,2)$$

Any such matrix of m rows and n columns will be called a *matrix of order* $m \times n$.

For fixed m and n, consider the set of all matrices over $\mathcal{F}$ of order $m \times n$. With addition and scalar multiplication defined exactly as for square matrices, we have

Theorem II′. The set of all matrices over $\mathcal{F}$ of order $m \times n$ is a vector space over $\mathcal{F}$.

Multiplication cannot be defined on this set unless $m = n$. However, we may, as Problem 60, Chapter 13, Page 163, suggests, define the product of *certain* rectangular arrays. For example, using the matrices A, B, C above, we can form $A\cdot B$ but not $B\cdot A$; $B\cdot C$ but not $C\cdot B$; and neither $A\cdot C$ nor $C\cdot A$. The reason is clear; in order to form $L\cdot M$ the number of columns of L must equal the number of rows of M. For B and C above, we have

$$B \cdot C = \begin{bmatrix} \rho_1 \\ \rho_2 \\ \rho_3 \end{bmatrix} \cdot [\gamma_1 \ \gamma_2] = \begin{bmatrix} \rho_1 \cdot \gamma_1 & \rho_1 \cdot \gamma_2 \\ \rho_2 \cdot \gamma_1 & \rho_2 \cdot \gamma_2 \\ \rho_3 \cdot \gamma_1 & \rho_3 \cdot \gamma_2 \end{bmatrix} = \begin{bmatrix} \sum b_{1j}c_{j1} & \sum b_{1j}c_{j2} \\ \sum b_{2j}c_{j1} & \sum b_{2j}c_{j2} \\ \sum b_{3j}c_{j1} & \sum b_{3j}c_{j2} \end{bmatrix}$$

Thus the product of a matrix of order $m \times n$ and a matrix of order $n \times p$, both over the same field $\mathcal{F}$, is a matrix of order $m \times p$.

See Problems 2-3.

SOLUTIONS OF A SYSTEM OF LINEAR EQUATIONS

The study of matrices, thus far, has been dominated by our previous study of linear transformations of vector spaces. We might, however, have begun our study of matrices by noting the one-to-one correspondence between all systems of homogeneous linear equations over R and the set of arrays of their coefficients; for example

$$\begin{array}{ll} \text{(i)} & 2x + 3y + z = 0 \\ \text{(ii)} & x - y + 4z = 0 \\ \text{(iii)} & 4x + 11y - 5z = 0 \end{array} \quad \text{and} \quad \begin{bmatrix} 2 & 3 & 1 \\ 1 & -1 & 4 \\ 4 & 11 & -5 \end{bmatrix} \tag{3}$$

What we plan to do here is to show that a matrix, considered as the coefficient matrix of a system of homogeneous equations, can be used (in place of the actual equations) to obtain solutions of the system. In each of the steps below, we state our "moves", the result of these "moves" in terms of the equations, and finally in terms of the matrix.

The given system (*3*) has the trivial solution $x = y = z = 0$; it will have non-trivial solutions if and only if one of the equations is a linear combination of the other two, i.e. if and only if the row vectors of the coefficient matrix are linearly dependent. The procedure for finding the non-trivial solutions, if any, is well known to the reader. The set of "moves" is never unique; we shall proceed as follows: multiply the second equation by 2 and subtract from the first equation, then multiply the second equation by 4 and subtract from the third equation to obtain

$$\begin{array}{ll} \text{(i)} - 2\text{(ii)} & 0x + 5y - 7z = 0 \\ \text{(ii)} & x - y + 4z = 0 \\ \text{(iii)} - 4\text{(ii)} & 0x + 15y - 21z = 0 \end{array} \qquad \begin{bmatrix} 0 & 5 & -7 \\ 1 & -1 & 4 \\ 0 & 15 & -21 \end{bmatrix} \tag{4}$$

In (*4*), multiply the first equation by 3 and subtract from the third equation to obtain

$$\begin{array}{ll} \text{(i)} - 2\text{(ii)} & 0x + 5y - 7z = 0 \\ \text{(ii)} & x - y + 4z = 0 \\ \text{(iii)} + 2\text{(ii)} - 3\text{(i)} & 0x + 0y + 0z = 0 \end{array} \qquad \begin{bmatrix} 0 & 5 & -7 \\ 1 & -1 & 4 \\ 0 & 0 & 0 \end{bmatrix} \tag{5}$$

Finally, in (*5*), multiply the first equation by 1/5 and, after entering it as the first equation in (*6*), add it to the second equation. We have

$$\begin{array}{ll} \frac{1}{5}[\text{(i)} - 2\text{(ii)}] & 0x + y - 7z/5 = 0 \\ \frac{1}{5}[3\text{(ii)} + \text{(i)}] & x + 0y + 13z/5 = 0 \\ \text{(iii)} + 2\text{(ii)} - 3\text{(i)} & 0x + 0y + 0z = 0 \end{array} \qquad \begin{bmatrix} 0 & 1 & -7/5 \\ 1 & 0 & 13/5 \\ 0 & 0 & 0 \end{bmatrix} \tag{6}$$

Now if we take for z any arbitrary $r \in R$, we have as solutions of the system: $x = -13r/5$, $y = 7r/5$, $z = r$.

We summarize: from the given system of equations (3), we extracted the matrix $A = \begin{bmatrix} 2 & 3 & 1 \\ 1 & -1 & 4 \\ 4 & 11 & -5 \end{bmatrix}$; by operating on the rows of A we obtained the matrix $B = \begin{bmatrix} 0 & 1 & -7/5 \\ 1 & 0 & 13/5 \\ 0 & 0 & 0 \end{bmatrix}$; considering B as a coefficient matrix in the same unknowns, we read out the solutions of the original system. We give now three problems from vector spaces whose solutions follow easily.

Example 2: Is $\xi_1 = (2,1,4)$, $\xi_2 = (3,-1,11)$, $\xi_3 = (1,4,-5)$ a basis of $V_3(R)$?

We set $x\xi_1 + y\xi_2 + z\xi_3 = (2x+3y+z,\ x-y+4z,\ 4x+11y-5z) = 0 = (0,0,0)$ and obtain the system of equations (3). Using the solution $x = -13/5$, $y = 7/5$, $z = 1$, we find $\xi_3 = (13/5)\xi_1 - (7/5)\xi_2$. Thus the given set is not a basis. This, of course, is implied by the matrix (5) having a row of zeros.

Example 3: Is the set $\rho_1 = (2,3,1)$, $\rho_2 = (1,-1,4)$, $\rho_3 = (4,11,-5)$ a basis of $V_3(R)$?

The given vectors are the row vectors of (3). From the record of moves in (5), we extract (iii) + 2(ii) − 3(i) = 0 or $\rho_3 + 2\rho_2 - 3\rho_1 = 0$. Thus the set is not a basis.

Note. The problems solved in Examples 2 and 3 are of the same type and the computations are identical; the initial procedures, however, are quite different. In Example 2, the given vectors constitute the columns of the matrix and the operations on the matrix involve linear combinations of the corresponding components of these vectors. In Example 3, the given vectors constitute the rows of the matrix and the operations on this matrix involve linear combinations of the vectors themselves. We shall continue to use the notation of Chapter 13 in which a vector of $V_n(\mathcal{F})$ is written as a row of elements and thus hereafter use the procedure of Example 3.

Example 4: Show that the linear transformation

$$T: \begin{cases} \epsilon_1 \rightarrow (2,3,1) = \rho_1 \\ \epsilon_2 \rightarrow (1,-1,4) = \rho_2 \\ \epsilon_3 \rightarrow (4,11,-5) = \rho_3 \end{cases}$$

of $V = V_3(R)$ is singular and find a vector of V whose image is 0.

We write $T = \begin{bmatrix} 2 & 3 & 1 \\ 1 & -1 & 4 \\ 4 & 11 & -5 \end{bmatrix} = \begin{bmatrix} \rho_1 \\ \rho_2 \\ \rho_3 \end{bmatrix}$. By Example 3, $\rho_3 + 2\rho_2 - 3\rho_1 = 0$; thus the image of any vector of V is a linear combination of the vectors ρ_1 and ρ_2. Hence V_T has dimension 2 and T is singular. This, of course, is implied by the matrix (5) having a single row of zeros.

Since $3\rho_1 - 2\rho_2 - \rho_3 = 0$, the image of $\eta = (3,-2,-1)$ is 0, i.e.

$$(3,-2,-1) \cdot \begin{bmatrix} 2 & 3 & 1 \\ 1 & -1 & 4 \\ 4 & 11 & -5 \end{bmatrix} = 0$$

Note. The vector $(3,-2,-1)$ may be considered as a 1×3-matrix; hence the indicated product above is a valid one.

See Problem 4.

ELEMENTARY TRANSFORMATIONS ON A MATRIX

In solving a system of homogeneous linear equations with coefficients in $\mathcal{F}$ certain operations may be performed on the elements (equations) of the system without changing its solution or solutions:

Any two equations may be interchanged.

Any equation may be multiplied by any scalar $k \neq 0$ of $\mathcal{F}$.

Any equation may be multiplied by any scalar and added to any other equation.

The operations, called *elementary row transformations,* thereby induced on the coefficient matrix of the system are:

The interchange of the ith and jth rows, denoted by H_{ij}.

The multiplication of every element of the ith row by a non-zero scalar k, denoted by $H_i(k)$.

The addition to the elements of the ith row of k (a scalar) times the corresponding elements of the jth row, denoted by $H_{ij}(k)$.

Later we shall find useful the *elementary column transformations* on a matrix which we now list:

The interchange of the ith and jth columns, denoted by K_{ij}.

The multiplication of every element of the ith column by a non-zero scalar k, denoted by $K_i(k)$.

The addition to the elements of the ith column of k (a scalar) times the corresponding elements of the jth column, denoted by $K_{ij}(k)$.

Two matrices A and B will be called *row (column) equivalent* if B can be obtained from A by a sequence of elementary row (column) transformations. Similarly, two matrices A and B will be called *equivalent* if B can be obtained from A by a sequence of row *and* column transformations. When B is row equivalent, column equivalent or equivalent to A, we shall write $B \sim A$. We leave for the reader to show that $\sim$ is an equivalence relation.

Example 5: (*a*) Show that the set $\{(1,2,1,2), (2,4,3,4), (1,3,2,3), (0,3,1,3)\}$ is not a basis of $V_4(Q)$. (*b*) If T is the linear transformation having the vectors of (*a*) in order as images of $\epsilon_1, \epsilon_2, \epsilon_3, \epsilon_4$, what is the rank of T? (*c*) Find a basis of $V_4(Q)$ containing a maximum linearly independent subset of the vectors of (*a*).

(*a*) Using in turn $H_{21}(-2), H_{31}(-1);\ H_{13}(-2), H_{43}(-3);\ H_{42}(2),$ we have

$$A = \begin{bmatrix} 1 & 2 & 1 & 2 \\ 2 & 4 & 3 & 4 \\ 1 & 3 & 2 & 3 \\ 0 & 3 & 1 & 3 \end{bmatrix} \sim \begin{bmatrix} 1 & 2 & 1 & 2 \\ 0 & 0 & 1 & 0 \\ 0 & 1 & 1 & 1 \\ 0 & 3 & 1 & 3 \end{bmatrix} \sim \begin{bmatrix} 1 & 0 & -1 & 0 \\ 0 & 0 & 1 & 0 \\ 0 & 1 & 1 & 1 \\ 0 & 0 & -2 & 0 \end{bmatrix} \sim \begin{bmatrix} 1 & 0 & -1 & 0 \\ 0 & 0 & 1 & 0 \\ 0 & 1 & 1 & 1 \\ 0 & 0 & 0 & 0 \end{bmatrix}$$

The set is not a basis.

(*b*) Using $H_{12}(1), H_{32}(-1)$ on the final matrix obtained in (*a*), we have

$$A \sim \begin{bmatrix} 1 & 0 & -1 & 0 \\ 0 & 0 & 1 & 0 \\ 0 & 1 & 1 & 1 \\ 0 & 0 & 0 & 0 \end{bmatrix} \sim \begin{bmatrix} 1 & 0 & 0 & 0 \\ 0 & 0 & 1 & 0 \\ 0 & 1 & 0 & 1 \\ 0 & 0 & 0 & 0 \end{bmatrix} = B$$

Now B has the maximum number of zeros possible in any matrix row equivalent to A (verify this). Since B has 3 non-zero row vectors, V_T is of dimension 3 and $r_T = 3$.

(*c*) A check of the moves will show that a multiple of the fourth row was never added to any of the other three. Thus the first three vectors of the given set are linear combinations of the non-zero row vectors of B. The first three vectors of the given set together with any vector not a linear combination of the non-zero row vectors of B, for example ϵ_2 or ϵ_4, is a basis of $V_4(Q)$.

Consider the rows of a given matrix A as a set S of row vectors of $V_n(\mathcal{F})$ and interpret the elementary row transformations on A in terms of the vectors of S as:

The interchange of any two vectors in S.

The replacement of any vector $\xi \in S$ by a non-zero scalar multiple $a\xi$.

The replacement of any vector $\xi \in S$ by a linear combination $\xi + b\eta$ of ξ and any other vector $\eta \in S$.

The foregoing examples illustrate

Theorem III. The above operations on a set S of vectors of $V_n(\mathcal{F})$ neither increase nor decrease the number of linearly independent vectors in S.

See Problems 5-7.

UPPER TRIANGULAR, LOWER TRIANGULAR AND DIAGONAL MATRICES

A square matrix $A = [a_{ij}]$ is called *upper triangular* if $a_{ij} = 0$ whenever $i > j$, and is called *lower triangular* if $a_{ij} = 0$ whenever $i < j$. A square matrix which is both upper and lower triangular is called a *diagonal* matrix. For example, $\begin{bmatrix} 1 & 2 & 3 \\ 0 & 0 & 4 \\ 0 & 0 & 2 \end{bmatrix}$ is upper triangular, $\begin{bmatrix} 1 & 0 & 0 \\ 2 & 3 & 0 \\ 3 & 4 & 5 \end{bmatrix}$ is lower triangular, while $\begin{bmatrix} 1 & 0 & 0 \\ 0 & 0 & 0 \\ 0 & 0 & 2 \end{bmatrix}$ and $\begin{bmatrix} -2 & 0 & 0 \\ 0 & 3 & 0 \\ 0 & 0 & 1 \end{bmatrix}$ are diagonal.

By means of elementary transformations, any square matrix can be reduced to upper triangular, lower triangular, and diagonal form.

Example 6:

Reduce $A = \begin{bmatrix} 1 & 2 & 3 \\ 4 & 5 & 6 \\ 5 & 7 & 8 \end{bmatrix}$ over Q to upper triangular, lower triangular, and diagonal form.

(*a*) Using $H_{21}(-4)$, $H_{31}(-5)$; $H_{32}(-1)$, we obtain

$$A = \begin{bmatrix} 1 & 2 & 3 \\ 4 & 5 & 6 \\ 5 & 7 & 8 \end{bmatrix} \sim \begin{bmatrix} 1 & 2 & 3 \\ 0 & -3 & -6 \\ 0 & -3 & -7 \end{bmatrix} \sim \begin{bmatrix} 1 & 2 & 3 \\ 0 & -3 & -6 \\ 0 & 0 & -1 \end{bmatrix} \quad \text{which is upper triangular.}$$

(*b*) Using $H_{12}(-2/5)$, $H_{23}(-5/7)$; $H_{12}(-21/10)$, $H_{23}(-1/28)$

$$A = \begin{bmatrix} 1 & 2 & 3 \\ 4 & 5 & 6 \\ 5 & 7 & 8 \end{bmatrix} \sim \begin{bmatrix} -3/5 & 0 & 3/5 \\ 3/7 & 0 & 2/7 \\ 5 & 7 & 8 \end{bmatrix} \sim \begin{bmatrix} -3/2 & 0 & 0 \\ 1/4 & -1/4 & 0 \\ 5 & 7 & 8 \end{bmatrix} \quad \text{which is lower triangular.}$$

(*c*) Using $H_{21}(-4)$, $H_{31}(-5)$, $H_{32}(-1)$; $H_{12}(2/3)$; $H_{13}(-1)$, $H_{23}(-6)$

$$A \sim \begin{bmatrix} 1 & 2 & 3 \\ 0 & -3 & -6 \\ 0 & 0 & -1 \end{bmatrix} \sim \begin{bmatrix} 1 & 0 & -1 \\ 0 & -3 & -6 \\ 0 & 0 & -1 \end{bmatrix} \sim \begin{bmatrix} 1 & 0 & 0 \\ 0 & -3 & 0 \\ 0 & 0 & -1 \end{bmatrix} \quad \text{which is diagonal.}$$

See also Problem 8.

A CANONICAL FORM

In Problem 9, we prove

Theorem IV. Any non-zero matrix A over $\mathcal{F}$ can be reduced by a sequence of elementary row transformations to a *row canonical matrix* (*echelon matrix*) C having the properties:

(i) Each of the first r rows of C has at least one non-zero element; the remaining rows, if any, consist entirely of zero elements.

(ii) In the ith row $(i = 1, 2, \ldots, r)$ of C, its first non-zero element is 1. Let the column in which this element stands be numbered j_i.

(iii) The only non-zero element in the column numbered j_i, $(i = 1, 2, \ldots, r)$ is the element 1 of the ith row.

(iv) $j_1 < j_2 < \cdots < j_r$.

Example 7: (a) The matrix B of Problem 6, Page 187, is a row canonical matrix. The first non-zero element of the first row is 1 and stands in the first column, the first non-zero element of the second row is 1 and stands in the second column, the first non-zero element of the third row is 1 and stands in the fifth column. Thus, $j_1 = 1$, $j_2 = 2$, $j_3 = 5$ and $j_1 < j_2 < j_3$ is satisfied.

(b) The matrix B of Problem 7, Page 187, fails to meet condition (iv) and is not a row canonical matrix. It may, however, be reduced to $\begin{bmatrix} 1 & 0 & 0 & 1 \\ 0 & 1 & 0 & -1 \\ 0 & 0 & 1 & 1 \\ 0 & 0 & 0 & 0 \\ 0 & 0 & 0 & 0 \end{bmatrix} = C$, a row canonical matrix, by the elementary row transformations H_{12}, H_{13}.

In Problem 5, Page 186, the matrix B is a row canonical matrix; it is also the identity matrix of order 3. The linear transformation A is non-singular; we shall also call the matrix A non-singular. Thus,

> An n-square matrix is *non-singular* if and only if it is row equivalent to the identity matrix I_n.

Any n-square matrix which is not non-singular is called *singular*. The terms singular and non-singular are never used when the matrix is of order $m \times n$ with $m \neq n$.

The rank of a linear transformation A is the number of linearly independent vectors in the set of image vectors. We shall call the rank of the linear transformation A the row rank of the matrix A. Thus,

> The *row rank* of an $m \times n$ matrix is the number of non-zero rows in its row equivalent canonical matrix.

It is not necessary, of course, to reduce a matrix to row canonical form to determine its rank. For example, the rank of the matrix A in Problem 7 can be obtained as readily from B as from the row canonical matrix C of Example 7(b).

ELEMENTARY COLUMN TRANSFORMATIONS

Beginning with a matrix A and using only elementary column transformations, we may obtain matrices called column equivalent to A. Among these is a *column canonical matrix* D whose properties are precisely those obtained by interchanging "row" and "column" in the list of properties of the row canonical matrix C. We define the *column rank* of A to be the number of columns of D having at least one non-zero element. Our only interest in all of this is

Theorem V. The row rank and the column rank of any matrix A are equal.

For a proof, see Problem 10.

As a consequence, we define

> The *rank* of a matrix is its row (column) rank.

Let a matrix A over $\mathcal{F}$ of order $m \times n$ and rank r be reduced to its row canonical form C. Then using the element 1 which appears in each of the first r rows of C and appropriate transformations of the type $K_{ij}(k)$, C may be reduced to a matrix whose only non-zero elements are these 1's. Finally, using transformations of the type K_{ij}, these 1's can be made to occupy the diagonal positions in the first r rows and first r columns. The resulting matrix, denoted by N, is called the *normal form* of A.

Example 8:

(*a*) In Problem 4 we have

$$A = \begin{bmatrix} 1 & 2 & 2 & 0 \\ 2 & 5 & 3 & 1 \\ 3 & 8 & 4 & 2 \\ 2 & 7 & 1 & 3 \end{bmatrix} \sim \begin{bmatrix} 1 & 0 & 4 & -2 \\ 0 & 1 & -1 & 1 \\ 0 & 0 & 0 & 0 \\ 0 & 0 & 0 & 0 \end{bmatrix} = C$$

Using $K_{31}(-4)$, $K_{41}(2)$; $K_{32}(1)$, $K_{42}(-1)$ on C, we obtain

$$A \sim \begin{bmatrix} 1 & 0 & 4 & -2 \\ 0 & 1 & -1 & 1 \\ 0 & 0 & 0 & 0 \\ 0 & 0 & 0 & 0 \end{bmatrix} \sim \begin{bmatrix} 1 & 0 & 0 & 0 \\ 0 & 1 & -1 & 1 \\ 0 & 0 & 0 & 0 \\ 0 & 0 & 0 & 0 \end{bmatrix} \sim \begin{bmatrix} 1 & 0 & 0 & 0 \\ 0 & 1 & 0 & 0 \\ 0 & 0 & 0 & 0 \\ 0 & 0 & 0 & 0 \end{bmatrix} = \begin{bmatrix} I_2 & 0 \\ 0 & 0 \end{bmatrix}, \quad \text{the normal form.}$$

(*b*) The matrix B is the normal form of A in Problem 5.

(*c*) For the matrix of Problem 6, we obtain using on B the elementary column transformations $K_{31}(-4)$, $K_{32}(1)$, $K_{42}(-2)$; K_{35}

$$A \sim \begin{bmatrix} 1 & 0 & 4 & 0 & 0 \\ 0 & 1 & -1 & 2 & 0 \\ 0 & 0 & 0 & 0 & 1 \end{bmatrix} \sim \begin{bmatrix} 1 & 0 & 0 & 0 & 0 \\ 0 & 1 & 0 & 0 & 0 \\ 0 & 0 & 1 & 0 & 0 \end{bmatrix} = [I_3 \;\; 0]$$

Note. From these examples it might be thought that, in reducing A to its normal form, one first works with row transformations and then exclusively with column transformations. This order is not necessary. See Problem 11.

ELEMENTARY MATRICES

The matrix which results when an elementary row (column) transformation is applied to the identity matrix I_n is called an *elementary row (column) matrix*. Any elementary row (column) matrix will be denoted by the same symbol used to denote the elementary transformation which produces the matrix.

Example 9:

When $I = \begin{bmatrix} 1 & 0 & 0 \\ 0 & 1 & 0 \\ 0 & 0 & 1 \end{bmatrix}$, we have

$$H_{13} = \begin{bmatrix} 0 & 0 & 1 \\ 0 & 1 & 0 \\ 1 & 0 & 0 \end{bmatrix} = K_{13}, \quad H_2(k) = \begin{bmatrix} 1 & 0 & 0 \\ 0 & k & 0 \\ 0 & 0 & 1 \end{bmatrix} = K_2(k), \quad H_{23}(k) = \begin{bmatrix} 1 & 0 & 0 \\ 0 & 1 & k \\ 0 & 0 & 1 \end{bmatrix} = K_{32}(k)$$

By Theorem III, we have

Theorem VI. Every elementary matrix is non-singular.

and

Theorem VII. The product of two or more elementary matrices is non-singular.

There follows easily

Theorem VIII. To perform an elementary row (column) transformation H (K) on a matrix A of order $m \times n$, form the product $H \cdot A$ ($A \cdot K$) where H (K) is the matrix obtained by performing the transformation H (K) on I.

The matrices H and K of Theorem VIII will carry no indicator of their orders. If A is of order $m \times n$, then a product such as $H_{13} \cdot A \cdot K_{23}(k)$ must imply that H_{13} is of order m while $K_{23}(k)$ is of order n since otherwise the indicated product is meaningless.

Example 10: Given $A = \begin{bmatrix} 1 & 2 & 3 & 4 \\ 5 & 6 & 7 & 8 \\ 2 & 4 & 6 & 8 \end{bmatrix}$ over Q, calculate

$$(a)\quad H_{13} \cdot A = \begin{bmatrix} 0 & 0 & 1 \\ 0 & 1 & 0 \\ 1 & 0 & 0 \end{bmatrix} \cdot \begin{bmatrix} 1 & 2 & 3 & 4 \\ 5 & 6 & 7 & 8 \\ 2 & 4 & 6 & 8 \end{bmatrix} = \begin{bmatrix} 2 & 4 & 6 & 8 \\ 5 & 6 & 7 & 8 \\ 1 & 2 & 3 & 4 \end{bmatrix}$$

$$(b)\quad H_1(-3) \cdot A = \begin{bmatrix} -3 & 0 & 0 \\ 0 & 1 & 0 \\ 0 & 0 & 1 \end{bmatrix} \cdot \begin{bmatrix} 1 & 2 & 3 & 4 \\ 5 & 6 & 7 & 8 \\ 2 & 4 & 6 & 8 \end{bmatrix} = \begin{bmatrix} -3 & -6 & -9 & -12 \\ 5 & 6 & 7 & 8 \\ 2 & 4 & 6 & 8 \end{bmatrix}$$

$$(c)\quad A \cdot K_{41}(-4) = \begin{bmatrix} 1 & 2 & 3 & 4 \\ 5 & 6 & 7 & 8 \\ 2 & 4 & 6 & 8 \end{bmatrix} \cdot \begin{bmatrix} 1 & 0 & 0 & -4 \\ 0 & 1 & 0 & 0 \\ 0 & 0 & 1 & 0 \\ 0 & 0 & 0 & 1 \end{bmatrix} = \begin{bmatrix} 1 & 2 & 3 & 0 \\ 5 & 6 & 7 & -12 \\ 2 & 4 & 6 & 0 \end{bmatrix}$$

Suppose now that $H_1, H_2, \ldots, H_s$ and $K_1, K_2, \ldots, K_t$ are sequences of elementary transformations which when performed, in the order of their subscripts, on a matrix A reduce it to B, i.e.,

$$H_s \cdot \ldots \cdot H_2 \cdot H_1 \cdot A \cdot K_1 \cdot K_2 \cdot \ldots \cdot K_t = B$$

Then, defining $S = H_s \cdot \ldots \cdot H_2 \cdot H_1$ and $T = K_1 \cdot K_2 \cdot \ldots \cdot K_t$, we have

$$S \cdot A \cdot T = B$$

Now A and B are equivalent matrices. The proof of the converse

Theorem IX. If A and B are equivalent matrices there exist non-singular matrices S and T such that $S \cdot A \cdot T = B$.

will be given in the next section.

As a consequence of Theorem IX, we have

Theorem IX′. For any matrix A there exist non-singular matrices S and T such that $S \cdot A \cdot T = N$, the normal form of A.

Example 11: Find non-singular matrices S and T over Q such that

$$S \cdot A \cdot T = S \cdot \begin{bmatrix} 1 & 2 & -1 \\ 3 & 8 & 2 \\ 4 & 9 & -1 \end{bmatrix} \cdot T = N, \text{ the normal form of } A.$$

Using $H_{21}(-3), H_{31}(-4), K_{21}(-2), K_{31}(1)$, we find $A = \begin{bmatrix} 1 & 2 & -1 \\ 3 & 8 & 2 \\ 4 & 9 & -1 \end{bmatrix} \sim \begin{bmatrix} 1 & 0 & 0 \\ 0 & 2 & 5 \\ 0 & 1 & 3 \end{bmatrix}$. Then, using $H_{23}(-1)$, we obtain $A \sim \begin{bmatrix} 1 & 0 & 0 \\ 0 & 1 & 2 \\ 0 & 1 & 3 \end{bmatrix}$ and finally $H_{32}(-1), K_{32}(-2)$ yield the normal form $\begin{bmatrix} 1 & 0 & 0 \\ 0 & 1 & 0 \\ 0 & 0 & 1 \end{bmatrix}$. Thus,

$$H_{32}(-1) \cdot H_{23}(-1) \cdot H_{31}(-4) \cdot H_{21}(-3) \cdot A \cdot K_{21}(-2) \cdot K_{31}(1) \cdot K_{32}(-2)$$

$$= \begin{bmatrix} 1 & 0 & 0 \\ 0 & 1 & 0 \\ 0 & -1 & 1 \end{bmatrix} \cdot \begin{bmatrix} 1 & 0 & 0 \\ 0 & 1 & -1 \\ 0 & 0 & 1 \end{bmatrix} \cdot \begin{bmatrix} 1 & 0 & 0 \\ 0 & 1 & 0 \\ -4 & 0 & 1 \end{bmatrix} \cdot \begin{bmatrix} 1 & 0 & 0 \\ -3 & 1 & 0 \\ 0 & 0 & 1 \end{bmatrix} \cdot A \cdot \begin{bmatrix} 1 & -2 & 0 \\ 0 & 1 & 0 \\ 0 & 0 & 1 \end{bmatrix} \cdot \begin{bmatrix} 1 & 0 & 1 \\ 0 & 1 & 0 \\ 0 & 0 & 1 \end{bmatrix} \cdot \begin{bmatrix} 1 & 0 & 0 \\ 0 & 1 & -2 \\ 0 & 0 & 1 \end{bmatrix}$$

$$= \begin{bmatrix} 1 & 0 & 0 \\ 1 & 1 & -1 \\ -5 & -1 & 2 \end{bmatrix} \cdot \begin{bmatrix} 1 & 2 & -1 \\ 3 & 8 & 2 \\ 4 & 9 & -1 \end{bmatrix} \cdot \begin{bmatrix} 1 & -2 & 5 \\ 0 & 1 & -2 \\ 0 & 0 & 1 \end{bmatrix} = \begin{bmatrix} 1 & 0 & 0 \\ 0 & 1 & 0 \\ 0 & 0 & 1 \end{bmatrix}$$

An alternate procedure is as follows: We begin with the array

$$\begin{array}{ccc} & & \\ & & \\ & & \\ & & \\ & I_3 & \\ A & I_3 & = \end{array} \begin{array}{cccccc} 1 & 0 & 0 & & & \\ 0 & 1 & 0 & & & \\ 0 & 0 & 1 & & & \\ 1 & 2 & -1 & 1 & 0 & 0 \\ 3 & 8 & 2 & 0 & 1 & 0 \\ 4 & 9 & -1 & 0 & 0 & 1 \end{array}$$

and proceed to reduce A to I. In doing so, each row transformation is performed on the rows of six elements and each column transformation is performed on the columns of six elements. Using $H_{21}(-3)$, $H_{31}(-4)$; $K_{21}(-2)$, $K_{31}(1)$; $H_{23}(-1)$; $H_{32}(-1)$; $K_{32}(-2)$ we obtain

$$\begin{array}{cccccc} 1 & 0 & 0 & & & \\ 0 & 1 & 0 & & & \\ 0 & 0 & 1 & & & \\ 1 & 2 & -1 & 1 & 0 & 0 \\ 3 & 8 & 2 & 0 & 1 & 0 \\ 4 & 9 & -1 & 0 & 0 & 1 \end{array} \rightarrow \begin{array}{cccccc} 1 & 0 & 0 & & & \\ 0 & 1 & 0 & & & \\ 0 & 0 & 1 & & & \\ 1 & 2 & -1 & 1 & 0 & 0 \\ 0 & 2 & 5 & -3 & 1 & 0 \\ 0 & 1 & 3 & -4 & 0 & 1 \end{array} \rightarrow \begin{array}{cccccc} 1 & -2 & 1 & & & \\ 0 & 1 & 0 & & & \\ 0 & 0 & 1 & & & \\ 1 & 0 & 0 & 1 & 0 & 0 \\ 0 & 2 & 5 & -3 & 1 & 0 \\ 0 & 1 & 3 & -4 & 0 & 1 \end{array}$$

$$\rightarrow \begin{array}{cccccc} 1 & -2 & 1 & & & \\ 0 & 1 & 0 & & & \\ 0 & 0 & 1 & & & \\ 1 & 0 & 0 & 1 & 0 & 0 \\ 0 & 1 & 2 & 1 & 1 & -1 \\ 0 & 1 & 3 & -4 & 0 & 1 \end{array} \rightarrow \begin{array}{cccccc} 1 & -2 & 1 & & & \\ 0 & 1 & 0 & & & \\ 0 & 0 & 1 & & & \\ 1 & 0 & 0 & 1 & 0 & 0 \\ 0 & 1 & 2 & 1 & 1 & -1 \\ 0 & 0 & 1 & -5 & -1 & 2 \end{array} \rightarrow \begin{array}{cccccc} 1 & -2 & 5 & & & \\ 0 & 1 & -2 & & & \\ 0 & 0 & 1 & & & \\ 1 & 0 & 0 & 1 & 0 & 0 \\ 0 & 1 & 0 & 1 & 1 & -1 \\ 0 & 0 & 1 & -5 & -1 & 2 \end{array} = \begin{array}{cc} & \\ & \\ & \\ & \\ T & \\ I & S \end{array}$$

and $S \cdot A \cdot T = I$, as before.

See also Problem 12.

INVERSES OF ELEMENTARY MATRICES

For each of the elementary transformations there is an inverse transformation, that is, a transformation which undoes whatever the elementary transformation does. In terms of elementary transformations or of elementary matrices, we find readily

$$H_{ij}^{-1} = H_{ij} \qquad K_{ij}^{-1} = K_{ij}$$

$$H_i^{-1}(k) = H_i(1/k) \qquad K_i^{-1}(k) = K_i(1/k)$$

$$H_{ij}^{-1}(k) = H_{ij}(-k) \qquad K_{ij}^{-1}(k) = K_{ij}(-k)$$

Thus,

Theorem X. The inverse of an elementary row (column) transformation is an elementary row (column) transformation of the same order.

and

Theorem XI. The inverse of an elementary row (column) matrix is non-singular.

In Problem 13, we prove

Theorem XII. The inverse of the product of two matrices A and B, each of which has an inverse, is the product of the inverses in reverse order, that is,

$$(A \cdot B)^{-1} = B^{-1} \cdot A^{-1}$$

Theorem XII may be extended readily to the inverse of the product of any number of matrices. In particular, we have:

$$\text{if } S = H_s \cdot \ldots \cdot H_2 \cdot H_1 \text{ then } S^{-1} = H_1^{-1} \cdot H_2^{-1} \cdot \ldots \cdot H_s^{-1},$$
$$\text{if } T = K_1 \cdot K_2 \cdot \ldots \cdot K_t \text{ then } T^{-1} = K_t^{-1} \cdot \ldots \cdot K_2^{-1} \cdot K_1^{-1}.$$

Suppose A of order $m \times n$. By Theorem IX′, there exist non-singular matrices S of order m and T of order n such that $S \cdot A \cdot T = N$, the normal form of A. Then

$$A = S^{-1}(S \cdot A \cdot T)T^{-1} = S^{-1} \cdot N \cdot T^{-1}$$

In particular, we have

Theorem XIII. If A is non-singular and if $S \cdot A \cdot T = I$, then

$$A = S^{-1} \cdot T^{-1}$$

that is, every non-singular matrix of order n can be expressed as a product of elementary matrices of the same order.

Example 12: In Example 11, we have

$$S = H_{32}(-1) \cdot H_{23}(-1) \cdot H_{31}(-4) \cdot H_{21}(-3) \quad \text{and} \quad T = K_{21}(-2) \cdot K_{31}(1) \cdot K_{32}(-2)$$

Then

$$S^{-1} = H_{21}^{-1}(-3) \cdot H_{31}^{-1}(-4) \cdot H_{23}^{-1}(-1) \cdot H_{32}^{-1}(-1) = H_{21}(3) \cdot H_{31}(4) \cdot H_{23}(1) \cdot H_{32}(1)$$

$$= \begin{bmatrix} 1 & 0 & 0 \\ 3 & 1 & 0 \\ 0 & 0 & 1 \end{bmatrix} \cdot \begin{bmatrix} 1 & 0 & 0 \\ 0 & 1 & 0 \\ 4 & 0 & 1 \end{bmatrix} \cdot \begin{bmatrix} 1 & 0 & 0 \\ 0 & 1 & 1 \\ 0 & 0 & 1 \end{bmatrix} \cdot \begin{bmatrix} 1 & 0 & 0 \\ 0 & 1 & 0 \\ 0 & 1 & 1 \end{bmatrix} = \begin{bmatrix} 1 & 0 & 0 \\ 3 & 2 & 1 \\ 4 & 1 & 1 \end{bmatrix},$$

$$T^{-1} = K_{32}(2) \cdot K_{31}(-1) \cdot K_{21}(2) = \begin{bmatrix} 1 & 0 & 0 \\ 0 & 1 & 2 \\ 0 & 0 & 1 \end{bmatrix} \cdot \begin{bmatrix} 1 & 0 & -1 \\ 0 & 1 & 0 \\ 0 & 0 & 1 \end{bmatrix} \cdot \begin{bmatrix} 1 & 2 & 0 \\ 0 & 1 & 0 \\ 0 & 0 & 1 \end{bmatrix} = \begin{bmatrix} 1 & 2 & -1 \\ 0 & 1 & 2 \\ 0 & 0 & 1 \end{bmatrix}$$

$$\text{and} \quad A = S^{-1} \cdot T^{-1} = \begin{bmatrix} 1 & 0 & 0 \\ 3 & 2 & 1 \\ 4 & 1 & 1 \end{bmatrix} \cdot \begin{bmatrix} 1 & 2 & -1 \\ 0 & 1 & 2 \\ 0 & 0 & 1 \end{bmatrix} = \begin{bmatrix} 1 & 2 & -1 \\ 3 & 8 & 2 \\ 4 & 9 & -1 \end{bmatrix}.$$

Suppose A and B over $\mathcal{F}$ of order $m \times n$ have the same rank. Then they have the same normal form N and there exist non-singular matrices $S_1, T_1; S_2, T_2$ such that $S_1AT_1 = N = S_2BT_2$. Using the inverse S_1^{-1} and T_1^{-1} of S_1 and T_1, we obtain

$$A = S_1^{-1} \cdot S_1AT_1 \cdot T_1^{-1} = S_1^{-1} \cdot S_2BT_2 \cdot T_1^{-1} = (S_1^{-1} \cdot S_2)B(T_2 \cdot T_1^{-1}) = S \cdot B \cdot T$$

Thus A and B are equivalent. We leave the converse for the reader and state

Theorem XIV. Two $m \times n$ matrices A and B over $\mathcal{F}$ are equivalent if and only if they have the same rank.

THE INVERSE OF A NON-SINGULAR MATRIX

The inverse A^{-1}, if it exists, of a square matrix A has the property

$$A \cdot A^{-1} = A^{-1} \cdot A = I$$

Since the rank of a product of two matrices cannot exceed the rank of either factor (see Chapter 13), we have

Theorem XV. The inverse of a matrix A exists if and only if A is non-singular.

Let A be non-singular. By Theorem IX′ there exist non-singular matrices S and T such that $S \cdot A \cdot T = I$. Then $A = S^{-1} \cdot T^{-1}$ and, by Theorem XII,

$$A^{-1} = (S^{-1} \cdot T^{-1})^{-1} = T \cdot S$$

Example 13:

Using the results of Example 11, we find

$$A^{-1} = T \cdot S = \begin{bmatrix} 1 & -2 & 5 \\ 0 & 1 & -2 \\ 0 & 0 & 1 \end{bmatrix} \cdot \begin{bmatrix} 1 & 0 & 0 \\ 1 & 1 & -1 \\ -5 & -1 & 2 \end{bmatrix} = \begin{bmatrix} -26 & -7 & 12 \\ 11 & 3 & -5 \\ -5 & -1 & 2 \end{bmatrix}$$

In computing the inverse of a non-singular matrix it will be found simpler to use elementary row transformations alone.

Example 14:

Find the inverse of $A = \begin{bmatrix} 1 & 2 & -1 \\ 3 & 8 & 2 \\ 4 & 9 & -1 \end{bmatrix}$ of Example 11 using only elementary row transformations.

We have

$$[A\ I] = \begin{bmatrix} 1 & 2 & -1 & 1 & 0 & 0 \\ 3 & 8 & 2 & 0 & 1 & 0 \\ 4 & 9 & -1 & 0 & 0 & 1 \end{bmatrix} \sim \begin{bmatrix} 1 & 2 & -1 & 1 & 0 & 0 \\ 0 & 2 & 5 & -3 & 1 & 0 \\ 0 & 1 & 3 & -4 & 0 & 1 \end{bmatrix} \sim \begin{bmatrix} 1 & 2 & -1 & 1 & 0 & 0 \\ 0 & 1 & 3 & -4 & 0 & 1 \\ 0 & 2 & 5 & -3 & 1 & 0 \end{bmatrix}$$

$$\sim \begin{bmatrix} 1 & 0 & -7 & 9 & 0 & -2 \\ 0 & 1 & 3 & -4 & 0 & 1 \\ 0 & 0 & -1 & 5 & 1 & -2 \end{bmatrix} \sim \begin{bmatrix} 1 & 0 & 0 & -26 & -7 & 12 \\ 0 & 1 & 0 & 11 & 3 & -5 \\ 0 & 0 & 1 & -5 & -1 & 2 \end{bmatrix} = [I\ A^{-1}].$$

See also Problem 14.

MINIMUM POLYNOMIAL OF A SQUARE MATRIX

Let $A \neq 0$ be an n-square matrix over $\mathcal{F}$. Since $A \in \mathcal{M}_n(\mathcal{F})$, the set $\{I, A, A^2, \ldots, A^{n^2}\}$ is linearly dependent and there exist scalars $a_0, a_1, a_2, \ldots, a_{n^2}$ not all 0 such that

$$\phi(A) = a_0 I + a_1 A + a_2 A^2 + \cdots + a_{n^2} A^{n^2} = 0$$

In this section we shall be concerned with that monic polynomial $m(\lambda) \in \mathcal{F}[\lambda]$ of minimum degree such that $m(A) = 0$. Clearly, either $m(\lambda) = \phi(\lambda)$ or $m(\lambda)$ is a proper divisor of $\phi(\lambda)$. In either case, $m(\lambda)$ will be called the *minimum polynomial* of A.

The most elementary procedure for obtaining the minimum polynomial of $A \neq 0$ involves the following routine:

(1) If $A = a_0 I$, $a_0 \in \mathcal{F}$, then $m(\lambda) = \lambda - a_0$.

(2) If $A \neq aI$ for all $a \in \mathcal{F}$ but $A^2 = a_1 A + a_0 I$ with $a_0, a_1 \in \mathcal{F}$, then $m(\lambda) = \lambda^2 - a_1\lambda - a_0$.

(3) If $A^2 \neq aA + bI$ for all $a, b \in \mathcal{F}$ but $A^3 = a_2 A^2 + a_1 A + a_0 I$ with $a_0, a_1, a_2 \in \mathcal{F}$, then $m(\lambda) = \lambda^3 - a_2\lambda^2 - a_1\lambda - a_0$

and so on.

Example 15:

Find the minimum polynomial of $A = \begin{bmatrix} 1 & 2 & 2 \\ 2 & 1 & 2 \\ 2 & 2 & 1 \end{bmatrix}$ over Q.

Since $A \neq a_0 I$ for all $a_0 \in Q$, set

$$A^2 = \begin{bmatrix} 9 & 8 & 8 \\ 8 & 9 & 8 \\ 8 & 8 & 9 \end{bmatrix} = a_1 \begin{bmatrix} 1 & 2 & 2 \\ 2 & 1 & 2 \\ 2 & 2 & 1 \end{bmatrix} + a_0 \begin{bmatrix} 1 & 0 & 0 \\ 0 & 1 & 0 \\ 0 & 0 & 1 \end{bmatrix} = \begin{bmatrix} a_1 + a_0 & 2a_1 & 2a_1 \\ 2a_1 & a_1 + a_0 & 2a_1 \\ 2a_1 & 2a_1 & a_1 + a_0 \end{bmatrix}$$

After checking every entry, we conclude that $A^2 = 4A + 5I$ and the minimum polynomial is $\lambda^2 - 4\lambda - 5$.

See also Problem 15.

Example 15 and Problem 15 suggest that the constant term of the minimum polynomial of $A \neq 0$ is different from zero if and only if A is non-singular. A second procedure for computing the inverse of a non-singular matrix follows.

Example 16: Find the inverse A^{-1}, given that the minimum polynomial of $A = \begin{bmatrix} 1 & 2 & 2 \\ 2 & 1 & 2 \\ 2 & 2 & 1 \end{bmatrix}$ (see Example 15) is $\lambda^2 - 4\lambda - 5$.

Since $A^2 - 4A - 5I = 0$ we have, after multiplying by A^{-1}, $A - 4I - 5A^{-1} = 0$;

$$\text{hence,} \quad A^{-1} = \tfrac{1}{5}(A - 4I) = \begin{bmatrix} -3/5 & 2/5 & 2/5 \\ 2/5 & -3/5 & 2/5 \\ 2/5 & 2/5 & -3/5 \end{bmatrix}.$$

SYSTEMS OF LINEAR EQUATIONS

Let $\mathcal{F}$ be a given field and $x_1, x_2, \ldots, x_n$ be indeterminates. By a *linear form* over $\mathcal{F}$ in the n indeterminates we shall mean a polynomial of the type

$$ax_1 + bx_2 + \cdots + px_n$$

in which $a, b, \ldots, p \in \mathcal{F}$. Consider now a system of m linear equations

$$\left\{\begin{array}{l} a_{11}x_1 + a_{12}x_2 + \cdots + a_{1n}x_n = h_1 \\ a_{21}x_1 + a_{22}x_2 + \cdots + a_{2n}x_n = h_2 \\ \cdots\cdots\cdots\cdots\cdots\cdots\cdots\cdots\cdots \\ a_{m1}x_1 + a_{m2}x_2 + \cdots + a_{mn}x_n = h_m \end{array}\right. \tag{7}$$

in which both the coefficients a_{ij} and the constant terms h_i are elements of $\mathcal{F}$. It is to be noted that the equality sign in (7) cannot be interpreted as in previous chapters since in each equation the right member is in $\mathcal{F}$ but the left member is not. Following common practice, we write (7) to indicate that elements $r_1, r_2, \ldots, r_n \in \mathcal{F}$ are sought such that when x_i is replaced by r_i, $(i = 1, 2, \ldots, n)$, the system will consist of true equalities of elements of $\mathcal{F}$. Any such set of elements r_i is called a *solution* of (7).

Denote by $$A = [a_{ij}], \quad (i = 1, 2, \ldots, m;\ j = 1, 2, \ldots, n)$$
the coefficient matrix of (7) and by $S = \{\xi_1, \xi_2, \ldots, \xi_m\}$ the set of row vectors of A. Since the ξ_i are vectors of $V_n(\mathcal{F})$, the number of linearly independent vectors in S is $r \leq n$. Without loss of generality, we can (and will) assume that these r linearly independent vectors constitute the first r rows of A since this at most requires the writing of the equations of (7) in some different order.

Suppose now that we have found a vector $\rho = (r_1, r_2, \ldots, r_n) \in V_n(\mathcal{F})$ such that

$$\xi_1 \cdot \rho = h_1, \quad \xi_2 \cdot \rho = h_2, \quad \ldots, \quad \xi_r \cdot \rho = h_r$$

Since each ξ_i, $(i = r+1, r+2, \ldots, m)$ is a linear combination with coefficients in $\mathcal{F}$ of the r linearly independent vectors of A, it follows that

$$\xi_{r+1} \cdot \rho = h_{r+1}, \quad \xi_{r+2} \cdot \rho = h_{r+2}, \quad \ldots, \quad \xi_m \cdot \rho = h_m \tag{8}$$

and $x_1 = r_1$, $x_2 = r_2$, $\ldots$, $x_n = r_n$ is a solution of (7) if and only if in (8) each h_i is the same linear combination of $h_1, h_2, \ldots, h_r$ as ξ_i is of the set $\xi_1, \xi_2, \ldots, \xi_r$, that is, if and only if

the row rank of the *augmented matrix* $$[A\ H] = \begin{bmatrix} a_{11} & a_{12} & \ldots & a_{1n} & h_1 \\ a_{21} & a_{22} & \ldots & a_{2n} & h_2 \\ \cdots & \cdots & \cdots & \cdots & \cdots \\ a_{m1} & a_{m2} & \ldots & a_{mn} & h_m \end{bmatrix}$$ is also r.

We have proved

Theorem XVI. A system (7) of m linear equations in n unknowns will have a solution if and only if the row rank of the coefficient matrix A and of the augmented matrix $[A\ H]$ of the system are equal.

Suppose A and $[A\ H]$ have common row rank $r<n$ and that $[A\ H]$ has been reduced to its row canonical form

$$\begin{bmatrix} 1 & 0 & 0 & \ldots & 0 & c_{1,r+1} & c_{1,r+2} & \ldots & c_{1n} & k_1 \\ 0 & 1 & 0 & \ldots & 0 & c_{2,r+1} & c_{2,r+2} & \ldots & c_{2n} & k_2 \\ \cdots & & & & & & & & & \\ 0 & 0 & 0 & \ldots & 1 & c_{r,r+1} & c_{r,r+2} & \ldots & c_{rn} & k_r \\ 0 & 0 & 0 & \ldots & 0 & 0 & 0 & \ldots & 0 & 0 \\ \cdots & & & & & & & & & \\ 0 & 0 & 0 & \ldots & 0 & 0 & 0 & \ldots & 0 & 0 \end{bmatrix}$$

Let arbitrary values $s_{r+1}, s_{r+2}, \ldots, s_n \in \mathcal{F}$ be assigned to $x_{r+1}, x_{r+2}, \ldots, x_n$; then

$$\begin{aligned} x_1 &= k_1 - c_{1,r+1} \cdot s_{r+1} - c_{1,r+2} \cdot s_{r+2} - \cdots - c_{1n} \cdot s_n \\ x_2 &= k_2 - c_{2,r+1} \cdot s_{r+1} - c_{2,r+2} \cdot s_{r+2} - \cdots - c_{2n} \cdot s_n \\ &\cdots \\ x_r &= k_r - c_{r,r+1} \cdot s_{r+1} - c_{r,r+2} \cdot s_{r+2} - \cdots - c_{rn} \cdot s_n \end{aligned}$$

are uniquely determined. We have

Theorem XVI′. In a system (7) in which the common row rank of A and $[A\ H]$ is $r<n$, certain $n-r$ of the unknowns may be assigned arbitrary values in $\mathcal{F}$ and then the remaining r unknowns are uniquely determined in terms of these.

Systems of Non-homogeneous Linear Equations

We call (7) a system of non-homogeneous linear equations over $\mathcal{F}$ provided not every $h_i = 0$. To discover whether or not such a system has a solution as well as to find the solution (solutions), if any, we proceed to reduce the augmented matrix $[A\ H]$ of the system to its row canonical form. The various possibilities are illustrated in the examples below.

Example 17:

Consider the system $\begin{cases} x_1 + 2x_2 - 3x_3 + x_4 = 1 \\ 2x_1 - x_2 + 2x_3 - x_4 = 1 \\ 4x_1 + 3x_2 - 4x_3 + x_4 = 2 \end{cases}$ over Q.

We have

$$[A\ H] = \begin{bmatrix} 1 & 2 & -3 & 1 & 1 \\ 2 & -1 & 2 & -1 & 1 \\ 4 & 3 & -4 & 1 & 2 \end{bmatrix} \sim \begin{bmatrix} 1 & 2 & -3 & 1 & 1 \\ 0 & -5 & 8 & -3 & -1 \\ 0 & -5 & 8 & -3 & -2 \end{bmatrix} \sim \begin{bmatrix} 1 & 2 & -3 & 1 & 1 \\ 0 & -5 & 8 & -3 & -1 \\ 0 & 0 & 0 & 0 & -1 \end{bmatrix}$$

Although this is not the row canonical form, we see readily that

$$r_A = 2 < 3 = r_{[A\ H]}$$

and the system is *incompatible*, i.e., has no solution.

Example 18:

Consider the system $\begin{cases} x_1 + 2x_2 - x_3 = -1 \\ 3x_1 + 8x_2 + 2x_3 = 28 \\ 4x_1 + 9x_2 - x_3 = 14 \end{cases}$ over Q.

We have

$$[A\ H] = \begin{bmatrix} 1 & 2 & -1 & -1 \\ 3 & 8 & 2 & 28 \\ 4 & 9 & -1 & 14 \end{bmatrix} \sim \begin{bmatrix} 1 & 2 & -1 & -1 \\ 0 & 2 & 5 & 31 \\ 0 & 1 & 3 & 18 \end{bmatrix} \sim \begin{bmatrix} 1 & 2 & -1 & -1 \\ 0 & 1 & 3 & 18 \\ 0 & 2 & 5 & 31 \end{bmatrix}$$

$$\sim \begin{bmatrix} 1 & 0 & -7 & -37 \\ 0 & 1 & 3 & 18 \\ 0 & 0 & -1 & -5 \end{bmatrix} \sim \begin{bmatrix} 1 & 0 & 0 & -2 \\ 0 & 1 & 0 & 3 \\ 0 & 0 & 1 & 5 \end{bmatrix}$$

Here, $r_A = r_{[A\ H]} = 3 =$ the number of unknowns. There is one and only one solution: $x_1 = -2$, $x_2 = 3$, $x_3 = 5$.

Example 19:

Consider the system $\begin{cases} x_1 + x_2 + x_3 + x_4 + x_5 = 3 \\ 2x_1 + 3x_2 + 3x_3 + x_4 - x_5 = 0 \\ -x_1 + 2x_2 - 5x_3 + 2x_4 - x_5 = 1 \\ 3x_1 - x_2 + 2x_3 - 3x_4 - 2x_5 = -1 \end{cases}$ over Q.

We have

$$[A\ H] = \begin{bmatrix} 1 & 1 & 1 & 1 & 1 & 3 \\ 2 & 3 & 3 & 1 & -1 & 0 \\ -1 & 2 & -5 & 2 & -1 & 1 \\ 3 & -1 & 2 & -3 & -2 & -1 \end{bmatrix} \sim \begin{bmatrix} 1 & 1 & 1 & 1 & 1 & 3 \\ 0 & 1 & 1 & -1 & -3 & -6 \\ 0 & 3 & -4 & 3 & 0 & 4 \\ 0 & -4 & -1 & -6 & -5 & -10 \end{bmatrix}$$

$$\sim \begin{bmatrix} 1 & 0 & 0 & 2 & 4 & 9 \\ 0 & 1 & 1 & -1 & -3 & -6 \\ 0 & 0 & -7 & 6 & 9 & 22 \\ 0 & 0 & 3 & -10 & -17 & -34 \end{bmatrix} \sim \begin{bmatrix} 1 & 0 & 0 & 2 & 4 & 9 \\ 0 & 1 & 1 & -1 & -3 & -6 \\ 0 & 0 & -1 & -14 & -25 & -46 \\ 0 & 0 & 3 & -10 & -17 & -34 \end{bmatrix}$$

$$\sim \begin{bmatrix} 1 & 0 & 0 & 2 & 4 & 9 \\ 0 & 1 & 1 & -1 & -3 & -6 \\ 0 & 0 & 1 & 14 & 25 & 46 \\ 0 & 0 & 3 & -10 & -17 & -34 \end{bmatrix} \sim \begin{bmatrix} 1 & 0 & 0 & 2 & 4 & 9 \\ 0 & 1 & 0 & -15 & -28 & -52 \\ 0 & 0 & 1 & 14 & 25 & 46 \\ 0 & 0 & 0 & -52 & -92 & -172 \end{bmatrix}$$

$$\sim \begin{bmatrix} 1 & 0 & 0 & 2 & 4 & 9 \\ 0 & 1 & 0 & -15 & -28 & -52 \\ 0 & 0 & 1 & 14 & 25 & 46 \\ 0 & 0 & 0 & 1 & 23/13 & 43/13 \end{bmatrix} \sim \begin{bmatrix} 1 & 0 & 0 & 0 & 6/13 & 31/13 \\ 0 & 1 & 0 & 0 & -19/13 & -31/13 \\ 0 & 0 & 1 & 0 & 3/13 & -4/13 \\ 0 & 0 & 0 & 1 & 23/13 & 43/13 \end{bmatrix}$$

Here both A and $[A\ H]$ are of rank 4; the system is *compatible*, i.e., has one or more solutions. Unlike the system of Example 18, the rank is less than the number of unknowns. Now the given system is equivalent to $\begin{cases} x_1 + \frac{6}{13}x_5 = 31/13 \\ x_2 - \frac{19}{13}x_5 = -31/13 \\ x_3 + \frac{3}{13}x_5 = -4/13 \\ x_4 + \frac{23}{13}x_5 = 43/13 \end{cases}$ and it is clear that if we assign to x_5 any value $r \in Q$ then $x_1 = (31 - 6r)/13$, $x_2 = (-31 + 19r)/13$, $x_3 = (-4 - 3r)/13$, $x_4 = (43 - 23r)/13$, $x_5 = r$ is a solution. For instance, $x_1 = 1$, $x_2 = 2$, $x_3 = -1$, $x_4 = -2$, $x_5 = 3$ and $x_1 = 31/13$, $x_2 = -31/13$, $x_3 = -4/13$, $x_4 = 43/13$, $x_5 = 0$ are particular solutions of the system.

See also Problems 16-18.

These examples and problems illustrate

Theorem XVII. A system of non-homogeneous linear equations over $\mathcal{F}$ in n unknowns has a solution in $\mathcal{F}$ if and only if the rank of its coefficient matrix is equal to the rank of its augmented matrix. When the common rank is n, the system has a unique solution. When the common rank is $r < n$, certain $n-r$ of the unknowns may be assigned arbitrary values in $\mathcal{F}$ and then the remaining r unknowns are uniquely determined in terms of these.

When $m=n$ in system (7), we may proceed as follows:

(i) Write the system in matrix form $\begin{bmatrix} a_{11} & a_{12} & \dots & a_{1n} \\ a_{21} & a_{22} & \dots & a_{2n} \\ \dots & \dots & \dots & \dots \\ a_{n1} & a_{n2} & \dots & a_{nn} \end{bmatrix} \cdot \begin{bmatrix} x_1 \\ x_2 \\ \cdot \\ x_n \end{bmatrix} = \begin{bmatrix} h_1 \\ h_2 \\ \cdot \\ h_n \end{bmatrix}$ or, more compactly, as $A \cdot X = H$ where X is the $n\times 1$ matrix of unknowns and H is the $n\times 1$ matrix of constant terms.

(ii) Proceed with the matrix A as in computing A^{-1}. If, along the way, a row or column of zero elements is obtained, A is singular and we must begin anew with the matrix $[A\ H]$ as in the first procedure. However, if A is non-singular with inverse A^{-1}, then $A^{-1}(A\cdot X) = A^{-1}\cdot H$ and $X = A^{-1}\cdot H$.

Example 20:

For the system of Example 18, we have from Example 14, $A^{-1} = \begin{bmatrix} -26 & -7 & 12 \\ 11 & 3 & -5 \\ -5 & -1 & 2 \end{bmatrix}$; then

$$X = \begin{bmatrix} x_1 \\ x_2 \\ x_3 \end{bmatrix} = A^{-1}\cdot H = \begin{bmatrix} -26 & -7 & 12 \\ 11 & 3 & -5 \\ -5 & -1 & 2 \end{bmatrix} \cdot \begin{bmatrix} -1 \\ 28 \\ 14 \end{bmatrix} = \begin{bmatrix} -2 \\ 3 \\ 5 \end{bmatrix}$$ and we obtain the unique solution as before.

Systems of Homogeneous Linear Equations

We call (7) a system of *homogeneous linear equations* provided each $h_i = 0$. Since then the rank of the coefficient matrix and the augmented matrix are the same, the system always has one or more solutions. If the rank is n, then the *trivial solution* $x_1 = x_2 = \cdots = x_n = 0$ is the unique solution; if the rank is $r<n$, Theorem XVI′ assures the existence of non-trivial solutions. We have

Theorem XVIII. A system of homogeneous linear equations over $\mathcal{F}$ in n unknowns always has the trivial solution $x_1 = x_2 = \cdots = x_n = 0$. If the rank of the coefficient matrix is n, the trivial solution is the only solution; if the rank is $r<n$, certain $n-r$ of the unknowns may be assigned arbitrary values in $\mathcal{F}$ and then the remaining r unknowns are uniquely determined in terms of these.

Example 21: Solve the system $\begin{cases} x_1 + 2x_2 - x_3 = 0 \\ 3x_1 + 8x_2 + 2x_3 = 0 \\ 4x_1 + 9x_2 - x_3 = 0 \end{cases}$ over Q.

By Example 18, $A \sim I_3$. Thus, $x_1 = x_2 = x_3 = 0$ is the only solution.

Example 22: Solve the system $\begin{cases} x_1 + x_2 + x_3 + x_4 = 0 \\ 2x_1 + 3x_2 + 2x_3 + x_4 = 0 \\ 3x_1 + 4x_2 + 3x_3 + 2x_4 = 0 \end{cases}$ over Q.

We have

$$A = \begin{bmatrix} 1 & 1 & 1 & 1 \\ 2 & 3 & 2 & 1 \\ 3 & 4 & 3 & 2 \end{bmatrix} \sim \begin{bmatrix} 1 & 1 & 1 & 1 \\ 0 & 1 & 0 & -1 \\ 0 & 1 & 0 & -1 \end{bmatrix} \sim \begin{bmatrix} 1 & 0 & 1 & 2 \\ 0 & 1 & 0 & -1 \\ 0 & 0 & 0 & 0 \end{bmatrix}$$ of rank 2

Setting $x_3 = s$, $x_4 = t$ where $s,t \in Q$, we obtain the required solutions as: $x_1 = -s-2t$, $x_2 = t$, $x_3 = s$, $x_4 = t$.

See also Problem 19.

DETERMINANT OF A SQUARE MATRIX

To each square matrix A over $\mathcal{F}$ there may be associated a unique element $a \in \mathcal{F}$. This element a, called the *determinant* of A is denoted either by $\det A$ or $|A|$. Consider the n-square matrix

$$A = \begin{bmatrix} a_{11} & a_{12} & a_{13} & \dots & a_{1n} \\ a_{21} & a_{22} & a_{23} & \dots & a_{2n} \\ a_{31} & a_{32} & a_{33} & \dots & a_{3n} \\ \dots & \dots & \dots & \dots & \dots \\ a_{n1} & a_{n2} & a_{n3} & \dots & a_{nn} \end{bmatrix}$$

and a product

$$a_{1j_1}\, a_{2j_2}\, a_{3j_3} \dots a_{nj_n}$$

of n of its elements selected so that one and only one element comes from any row and one and only one element comes from any column. Note that the factors in this product have been arranged so that the row indices (first subscripts) are in natural order $1, 2, 3, \dots, n$. The sequence of column indices (second subscripts) is some permutation

$$\rho = (j_1, j_2, j_3, \dots, j_n)$$

of the digits $1, 2, 3, \dots, n$. For this permutation, define $\epsilon_\rho = +1$ or -1 according as ρ is even or odd and form the signed product

$$(a) \qquad \epsilon_\rho\, a_{1j_1}\, a_{2j_2}\, a_{3j_3} \dots a_{nj_n}$$

The set S_n of all permutations of n symbols contains $n!$ elements; hence, $n!$ distinct signed products of the type (a) can be formed. The determinant of A is defined to be the sum of these $n!$ signed products (called terms of $|A|$), i.e.,

$$(b) \qquad |A| = \sum_{S_n} \epsilon_\rho\, a_{1j_1}\, a_{2j_2}\, a_{3j_3} \dots a_{nj_n}$$

Example 23:

(i) $$\begin{vmatrix} a_{11} & a_{12} \\ a_{21} & a_{22} \end{vmatrix} = \epsilon_{12}a_{11}a_{22} + \epsilon_{21}a_{12}a_{21} = a_{11}a_{22} - a_{12}a_{21}$$

Thus, the determinant of a matrix of order 2 is the product of the diagonal elements of the matrix minus the product of the off-diagonal elements.

(ii) $$\begin{aligned} \begin{vmatrix} a_{11} & a_{12} & a_{13} \\ a_{21} & a_{22} & a_{23} \\ a_{31} & a_{32} & a_{33} \end{vmatrix} &= \epsilon_{123}a_{11}a_{22}a_{33} + \epsilon_{132}a_{11}a_{23}a_{32} + \epsilon_{213}a_{12}a_{21}a_{33} \\ &\qquad + \epsilon_{231}a_{12}a_{23}a_{31} + \epsilon_{312}a_{13}a_{21}a_{32} + \epsilon_{321}a_{13}a_{22}a_{31} \\ &= a_{11}a_{22}a_{33} - a_{11}a_{23}a_{32} - a_{12}a_{21}a_{33} + a_{12}a_{23}a_{31} + a_{13}a_{21}a_{32} - a_{13}a_{22}a_{31} \\ &= a_{11}(a_{22}a_{33} - a_{23}a_{32}) - a_{12}(a_{21}a_{33} - a_{23}a_{31}) + a_{13}(a_{21}a_{32} - a_{22}a_{31}) \\ &= a_{11}\begin{vmatrix} a_{22} & a_{23} \\ a_{32} & a_{33} \end{vmatrix} - a_{12}\begin{vmatrix} a_{21} & a_{23} \\ a_{31} & a_{33} \end{vmatrix} + a_{13}\begin{vmatrix} a_{21} & a_{22} \\ a_{31} & a_{32} \end{vmatrix} \\ &= (-1)^{1+1}a_{11}\begin{vmatrix} a_{22} & a_{23} \\ a_{32} & a_{33} \end{vmatrix} + (-1)^{1+2}a_{12}\begin{vmatrix} a_{21} & a_{23} \\ a_{31} & a_{33} \end{vmatrix} \\ &\qquad + (-1)^{1+3}a_{13}\begin{vmatrix} a_{21} & a_{22} \\ a_{31} & a_{32} \end{vmatrix} \end{aligned}$$

called the expansion of the determinant along its first row. It will be left for the reader to work out an expansion along each row and along each column.

PROPERTIES OF DETERMINANTS

Throughout this section, A is the n-square matrix whose determinant $|A|$ is given by (b) of the preceding section.

From (b) there follow easily

Theorem XIX. If every element of a row (column) of a square matrix A is zero, then $|A| = 0$.

Theorem XX. If A is upper (lower) triangular or is diagonal, then $|A| = a_{11}a_{22}a_{33}\cdots a_{nn}$, the product of the diagonal elements.

Theorem XXI. If B is obtained from A by multiplying its ith row (ith column) by a non-zero scalar k, then $|B| = k|A|$.

Let us now take a closer look at (a). Since ρ is a mapping of $S = \{1, 2, 3, \ldots, n\}$ into itself, it may be given (see Chapter 1) as

$$\rho: \quad 1\rho = j_1,\ 2\rho = j_2,\ 3\rho = j_3,\ \ldots,\ n\rho = j_n$$

With this notation, (a) takes the form

$$(a') \qquad \epsilon_\rho\, a_{1,1\rho}\, a_{2,2\rho}\, a_{3,3\rho} \ldots a_{n,n\rho}$$

and (b) takes the form

$$(b') \qquad |A| = \sum_{S_n} \epsilon_\rho\, a_{1,1\rho}\, a_{2,2\rho}\, a_{3,3\rho} \ldots a_{n,n\rho}$$

Since S_n is a group, it contains the inverse

$$\rho^{-1}: \quad j_1\rho^{-1} = 1,\ j_2\rho^{-1} = 2,\ j_3\rho^{-1} = 3,\ \ldots,\ j_n\rho^{-1} = n$$

of ρ. Moreover, ρ and ρ^{-1} are either both odd or both even. Thus (a) may be written as

$$\epsilon_{\rho^{-1}}\, a_{j_1\rho^{-1},j_1}\, a_{j_2\rho^{-1},j_2}\, a_{j_3\rho^{-1},j_3} \cdots a_{j_n\rho^{-1},j_n}$$

and, after reordering the factors so that the column indices are in natural order, as

$$(a'') \qquad \epsilon_{\rho^{-1}}\, a_{1\rho^{-1},1}\, a_{2\rho^{-1},2}\, a_{3\rho^{-1},3} \ldots a_{n\rho^{-1},n}$$

and (b) as

$$(b'') \qquad |A| = \sum_{S_n} \epsilon_{\rho^{-1}}\, a_{1\rho^{-1},1}\, a_{2\rho^{-1},2}\, a_{3\rho^{-1},3} \ldots a_{n\rho^{-1},n}$$

For each square matrix $A = [a_{ij}]$ define *transpose* A, denoted by A^T, to be the matrix obtained by interchanging the rows and columns of A. For example,

$$\text{when} \quad A = \begin{bmatrix} a_{11} & a_{12} & a_{13} \\ a_{21} & a_{22} & a_{23} \\ a_{31} & a_{32} & a_{33} \end{bmatrix} \quad \text{then} \quad A^T = \begin{bmatrix} a_{11} & a_{21} & a_{31} \\ a_{12} & a_{22} & a_{32} \\ a_{13} & a_{23} & a_{33} \end{bmatrix}$$

Let us also write $A^T = [a_{ij}^T]$, where $a_{ij}^T = a_{ji}$ for all i and j. Then the term of $|A^T|$

$$\begin{aligned} \epsilon_\rho\, a_{1,j_1}^T\, a_{2,j_2}^T\, a_{3,j_3}^T \ldots a_{n,j_n}^T &= \epsilon_\rho\, a_{j_1,1}\, a_{j_2,2}\, a_{j_3,3} \ldots a_{j_n,n} \\ &= \epsilon_\rho\, a_{1\rho,1}\, a_{2\rho,2}\, a_{3\rho,3} \ldots a_{n\rho,n} \end{aligned}$$

is by (b'') a term of $|A|$. Since this is true for every $\rho \in S_n$, we have proved

Theorem XXII. If A^T is the transpose of the square matrix A, then $|A^T| = |A|$.

Now let A be any square matrix and B be the matrix obtained by multiplying the ith row of A by a non-zero scalar k. In terms of elementary matrices $B = H_i(k) \cdot A$; and, by Theorem XXI,

$$|B| = |H_i(k) \cdot A| = k|A|$$

But $|H_i(k)| = k$; hence, $|H_i(k)\cdot A| = |H_i(k)|\cdot|A|$. By an independent proof or by Theorem XXII, we have also

$$|A\cdot K_i(k)| \;=\; |A|\cdot|K_i(k)|$$

Next, denote by B the matrix obtained from A by interchanging its ith and jth columns and denote by τ the corresponding transposition (i, j). The effect of τ on (a') is to produce

$$(a''') \qquad \epsilon_{\rho\tau}\, a_{1,1\rho\tau}\, a_{2,2\rho\tau}\, a_{3,3\rho\tau} \ldots a_{n,n\rho\tau}$$

hence,

$$|B| \;=\; \sum_{S_n} \epsilon_{\rho\tau}\, a_{1,1\rho\tau}\, a_{2,2\rho\tau}\, a_{3,3\rho\tau} \ldots a_{n,n\rho\tau}$$

Now $\sigma = \rho\tau \in S_n$ is even when ρ is odd and is odd when ρ is even; hence, $\epsilon_\sigma = -\epsilon_\rho$. Moreover, with τ fixed, let ρ range over S_n; then σ ranges over S_n and so

$$|B| \;=\; \sum_{S_n} \epsilon_\sigma\, a_{1,1\sigma}\, a_{2,2\sigma}\, a_{3,3\sigma} \ldots a_{n,n\sigma} \;=\; -|A|$$

We have proved

Theorem XXIII. If B is obtained from A by interchanging any two of its rows (columns), then $|B| = -|A|$.

Since in Theorem XXIII $B = A\cdot K_{ij}$ and $|K_{ij}| = -1$, we have $|A\cdot K_{ij}| = |A|\cdot|K_{ij}|$ and, by symmetry, $|H_{ij}\cdot A| = |H_{ij}|\cdot|A|$.

There follows readily, excluding all fields of characteristic two,

Theorem XXIV. If two rows (columns) of A are identical, then $|A| = 0$.

Finally, let B be obtained from A by adding to its ith row the product of k (a scalar) and its jth row. Assuming $j < i$,

$$\begin{aligned}
|B| &= \sum_{S_n} \epsilon_\rho\, a_{1,1\rho} \ldots a_{j,j\rho} \ldots a_{i-1,(i-1)\rho}\,(a_{i,i\rho} + k a_{j,j\rho})\, a_{i+1,(i+1)\rho} \ldots a_{n,n\rho} \\
&= \sum_{S_n} \epsilon_\rho\, a_{1,1\rho}\, a_{2,2\rho}\, a_{3,3\rho} \ldots a_{n,n\rho} \\
&\qquad + \sum_{S_n} \epsilon_\rho\, a_{1,1\rho} \ldots a_{j,j\rho} \ldots a_{i-1,(i-1)\rho}\,(k a_{j,j\rho})\, a_{i+1,(i+1)\rho} \ldots a_{n,n\rho} \\
&= |A| + 0 \;=\; |A| \qquad \text{(using } (b') \text{ and Theorems XXI and XXIV)}
\end{aligned}$$

We have proved (the case $j > i$ being left for the reader)

Theorem XXV. If B is obtained from A by adding to its ith row the product of k (a scalar) and its jth row, then $|B| = |A|$. The theorem also holds when row is replaced by column throughout.

Since, in Theorem XXV, $B = H_{ij}(k)\cdot A$ and $|H_{ij}(k)| = |I| = 1$, we have

$$|H_{ij}(k)\cdot A| = |H_{ij}(k)|\cdot|A| \qquad \text{and} \qquad |A\cdot K_{ij}(k)| = |A|\cdot|K_{ij}(k)|$$

We have now also proved

Theorem XXVI. If A is an n-square matrix and $H(K)$ is any n-square elementary row (column) matrix, then

$$|H\cdot A| = |H|\cdot|A| \qquad \text{and} \qquad |A\cdot K| = |A|\cdot|K|$$

By Theorem IX′, any square matrix A can be expressed as

$$(c) \qquad A \;=\; H_1^{-1}\cdot H_2^{-1} \ldots H_s^{-1}\cdot N\cdot K_t^{-1} \ldots K_2^{-1}\cdot K_1^{-1}$$

Then, by repeated applications of Theorem XXVI, we obtain

$$\begin{aligned}|A| &= |H_1^{-1}| \cdot |H_2^{-1} \ldots H_s^{-1} \cdot N \cdot K_t^{-1} \ldots K_2^{-1} \cdot K_1^{-1}| \\ &= |H_1^{-1}| \cdot |H_2^{-1}| \cdot |H_3^{-1} \ldots H_s^{-1} \cdot N \cdot K_t^{-1} \ldots K_2^{-1} \cdot K_1^{-1}| \\ &= \ldots\ldots\ldots\ldots\ldots\ldots\ldots\ldots\ldots\ldots \\ &= |H_1^{-1}| \cdot |H_2^{-1}| \ldots |H_s^{-1}| \cdot |N| \cdot |K_t^{-1}| \ldots |K_2^{-1}| \cdot |K_1^{-1}|\end{aligned}$$

If A is non-singular, then $N = I$ and $|N| = 1$; if A is singular, then one or more of the diagonal elements of N is 0 and $|N| = 0$. Thus,

Theorem XXVII. A square matrix A is non-singular if and only if $|A| \neq 0$.

and

Theorem XXVIII. If A and B are n-square matrices, then $|A \cdot B| = |A| \cdot |B|$.

EVALUATION OF DETERMINANTS

Using the result of Example 23 (ii), we have

$$\begin{aligned}\begin{vmatrix} 1 & 2 & 3 \\ 4 & 5 & 6 \\ 5 & 7 & 8 \end{vmatrix} &= (1)\begin{vmatrix} 5 & 6 \\ 7 & 8 \end{vmatrix} - (2)\begin{vmatrix} 4 & 6 \\ 5 & 8 \end{vmatrix} + (3)\begin{vmatrix} 4 & 5 \\ 5 & 7 \end{vmatrix} \\ &= (40-42) - 2(32-30) + 3(28-25) \\ &= -2 - 4 + 9 = 3\end{aligned}$$

The most practical procedure for evaluating $|A|$ of order $n \geqq 3$, consists in reducing A to triangular form using elementary transformations of the types $H_{ij}(k)$ and $K_{ij}(k)$ exclusively (they do not disturb the value of $|A|$) and then applying Theorem XX. If other elementary transformations are used, careful records must be kept since the effect of H_{ij} or K_{ij} is to change the sign of $|A|$ while that of $H_i(k)$ or $K_j(k)$ is to multiply $|A|$ by k.

Example 24: An examination of the triangular forms obtained in Example 6 and Problem 8 shows that while the diagonal elements are not unique, the product of the diagonal elements is. From Example 6(a), Page 171, we have

$$|A| = \begin{vmatrix} 1 & 2 & 3 \\ 4 & 5 & 6 \\ 5 & 7 & 8 \end{vmatrix} = \begin{vmatrix} 1 & 2 & 3 \\ 0 & -3 & -6 \\ 0 & -3 & -7 \end{vmatrix} = \begin{vmatrix} 1 & 2 & 3 \\ 0 & -3 & -6 \\ 0 & 0 & -1 \end{vmatrix} = (1)(-3)(-1) = 3$$

See also Problem 20.

Solved Problems

1. Find the image of $\xi = (1,2,3,4)$ under the linear transformation $A = \begin{bmatrix} 1 & -2 & 0 & 4 \\ 2 & 4 & 1 & -2 \\ 0 & -1 & 5 & -1 \\ 1 & 3 & 2 & 0 \end{bmatrix}$ of $V_4(Q)$ into itself.

$$\begin{aligned}\xi A &= \xi[\gamma_1\ \gamma_2\ \gamma_3\ \gamma_4] = (\xi\cdot\gamma_1, \xi\cdot\gamma_2, \xi\cdot\gamma_3, \xi\cdot\gamma_4) \\ &= (9, 15, 25, -3)\end{aligned}$$

(See Problem 15, Chapter 13, Page 158.)

2. Compute $A \cdot B$ and $B \cdot A$, given $A = \begin{bmatrix} 1 \\ 2 \\ 3 \end{bmatrix}$ and $B = [4\ 5\ 6]$.

$$A \cdot B = \begin{bmatrix} 1 \\ 2 \\ 3 \end{bmatrix} \cdot [4\ 5\ 6] = \begin{bmatrix} 1\cdot 4 & 1\cdot 5 & 1\cdot 6 \\ 2\cdot 4 & 2\cdot 5 & 2\cdot 6 \\ 3\cdot 4 & 3\cdot 5 & 3\cdot 6 \end{bmatrix} = \begin{bmatrix} 4 & 5 & 6 \\ 8 & 10 & 12 \\ 12 & 15 & 18 \end{bmatrix}$$

and
$$B \cdot A = [4\ 5\ 6] \cdot \begin{bmatrix} 1 \\ 2 \\ 3 \end{bmatrix} = [4\cdot 1 + 5\cdot 2 + 6\cdot 3] = [32]$$

3. When $A = \begin{bmatrix} 1 & 2 & -2 \\ 3 & 0 & 1 \end{bmatrix}$ and $B = \begin{bmatrix} 2 & 1 & 0 & -1 \\ 1 & 3 & -2 & 0 \\ 0 & 1 & -1 & -1 \end{bmatrix}$, find $A \cdot B$.

$$A \cdot B = \begin{bmatrix} 2+2 & 1+6-2 & -4+2 & -1+2 \\ 6 & 3+1 & -1 & -3-1 \end{bmatrix} = \begin{bmatrix} 4 & 5 & -2 & 1 \\ 6 & 4 & -1 & -4 \end{bmatrix}$$

4. Show that the linear transformation $\begin{bmatrix} 1 & 2 & 2 & 0 \\ 2 & 5 & 3 & 1 \\ 3 & 8 & 4 & 2 \\ 2 & 7 & 1 & 3 \end{bmatrix}$ in $V_4(R)$ is singular and find a vector whose image is 0.

Using in turn $H_{21}(-2), H_{31}(-3), H_{41}(-2);\ H_{12}(-2), H_{32}(-2), H_{42}(-3),$ we have

$$\begin{bmatrix} 1 & 2 & 2 & 0 \\ 2 & 5 & 3 & 1 \\ 3 & 8 & 4 & 2 \\ 2 & 7 & 1 & 3 \end{bmatrix} \sim \begin{bmatrix} 1 & 2 & 2 & 0 \\ 0 & 1 & -1 & 1 \\ 0 & 2 & -2 & 2 \\ 0 & 3 & -3 & 3 \end{bmatrix} \sim \begin{bmatrix} 1 & 0 & 4 & -2 \\ 0 & 1 & -1 & 1 \\ 0 & 0 & 0 & 0 \\ 0 & 0 & 0 & 0 \end{bmatrix}$$

The transformation is singular, of rank 2.

Designate the equivalent matrices as A, B, C respectively and denote by $\rho_1, \rho_2, \rho_3, \rho_4$ the row vectors of A, by $\rho_1', \rho_2', \rho_3', \rho_4'$ the row vectors of B, and by $\rho_1'', \rho_2'', \rho_3'', \rho_4''$ the row vectors of C. Using the moves in order, we have

$$\rho_2' = \rho_2 - 2\rho_1, \quad \rho_3' = \rho_3 - 3\rho_1, \quad \rho_4' = \rho_4 - 2\rho_1$$

$$\rho_1'' = \rho_1 - 2\rho_2', \quad \rho_3'' = \rho_3' - 2\rho_2', \quad \rho_4'' = \rho_4' - 3\rho_2'$$

Now $\quad \rho_3'' = \rho_3' - 2\rho_2' = (\rho_3 - 3\rho_1) - 2(\rho_2 - 2\rho_1) = \rho_3 - 2\rho_2 + \rho_1 = 0$

while $\quad \rho_4'' = \rho_4' - 3\rho_2' = (\rho_4 - 2\rho_1) - 3(\rho_2 - 2\rho_1) = \rho_4 - 3\rho_2 + 4\rho_1 = 0$

Thus, the image of $\xi = (1, -2, 1, 0)$ is 0; also, the image of $\eta = (4, -3, 0, 1)$ is 0. Show that the vectors whose images are 0 fill a subspace of dimension 2 in $V_4(R)$.

5. Show that the linear transformation $A = \begin{bmatrix} 1 & 2 & 3 \\ 2 & 1 & 3 \\ 3 & 2 & 1 \end{bmatrix}$ is non-singular.

We find

$$A = \begin{bmatrix} 1 & 2 & 3 \\ 2 & 1 & 3 \\ 3 & 2 & 1 \end{bmatrix} \sim \begin{bmatrix} 1 & 2 & 3 \\ 0 & -3 & -3 \\ 0 & -4 & -8 \end{bmatrix} \sim \begin{bmatrix} 1 & 2 & 3 \\ 0 & 1 & 1 \\ 0 & -4 & -8 \end{bmatrix} \sim \begin{bmatrix} 1 & 0 & 1 \\ 0 & 1 & 1 \\ 0 & 0 & -4 \end{bmatrix} \sim \begin{bmatrix} 1 & 0 & 1 \\ 0 & 1 & 1 \\ 0 & 0 & 1 \end{bmatrix} \sim \begin{bmatrix} 1 & 0 & 0 \\ 0 & 1 & 0 \\ 0 & 0 & 1 \end{bmatrix} = B$$

The row vectors of A are linearly independent; the linear transformation A is non-singular.

6. Find the rank of the linear transformation $A = \begin{bmatrix} 1 & 2 & 2 & 4 & 2 \\ 2 & 5 & 3 & 10 & 7 \\ 3 & 5 & 7 & 10 & 4 \end{bmatrix}$ of $V_3(R)$ into $V_5(R)$.

We find

$$\begin{bmatrix} 1 & 2 & 2 & 4 & 2 \\ 2 & 5 & 3 & 10 & 7 \\ 3 & 5 & 7 & 10 & 4 \end{bmatrix} \sim \begin{bmatrix} 1 & 2 & 2 & 4 & 2 \\ 0 & 1 & -1 & 2 & 3 \\ 0 & -1 & 1 & -2 & -2 \end{bmatrix} \sim \begin{bmatrix} 1 & 0 & 4 & 0 & -4 \\ 0 & 1 & -1 & 2 & 3 \\ 0 & 0 & 0 & 0 & 1 \end{bmatrix} \sim \begin{bmatrix} 1 & 0 & 4 & 0 & 0 \\ 0 & 1 & -1 & 2 & 0 \\ 0 & 0 & 0 & 0 & 1 \end{bmatrix} = B$$

The image vectors are linearly independent; $r_A = 3$.

7. From the set $\{(2,5,0,-3), (3,2,1,2), (1,2,1,0), (5,6,3,2), (1,-2,-1,2)\}$ of vectors in $V_4(R)$, select a maximum linearly independent subset.

The given set is linearly dependent (why?). We find

$$A = \begin{bmatrix} 2 & 5 & 0 & -3 \\ 3 & 2 & 1 & 2 \\ 1 & 2 & 1 & 0 \\ 5 & 6 & 3 & 2 \\ 1 & -2 & -1 & 2 \end{bmatrix} \sim \begin{bmatrix} 0 & 1 & -2 & -3 \\ 0 & -4 & -2 & 2 \\ 1 & 2 & 1 & 0 \\ 0 & -4 & -2 & 2 \\ 0 & -4 & -2 & 2 \end{bmatrix} \sim \begin{bmatrix} 0 & 1 & -2 & -3 \\ 0 & 0 & -10 & -10 \\ 1 & 0 & 5 & 6 \\ 0 & 0 & -10 & -10 \\ 0 & 0 & -10 & -10 \end{bmatrix}$$

$$\sim \begin{bmatrix} 0 & 1 & -2 & -3 \\ 0 & 0 & 1 & 1 \\ 1 & 0 & 5 & 6 \\ 0 & 0 & -10 & -10 \\ 0 & 0 & -10 & -10 \end{bmatrix} \sim \begin{bmatrix} 0 & 1 & 0 & -1 \\ 0 & 0 & 1 & 1 \\ 1 & 0 & 0 & 1 \\ 0 & 0 & 0 & 0 \\ 0 & 0 & 0 & 0 \end{bmatrix} = B$$

From an examination of the moves, it is clear that the first three vectors of A are linear combinations of the three linearly independent vectors of B (check this). Thus, $\{(2,5,0,-3), (3,2,1,2), (1,2,1,0)\}$ is a maximum linearly independent subset of A. Can you conclude that any three vectors of A are necessarily linearly independent? Check your answer by considering the subset $(1,2,1,0)$, $(5,6,3,2)$, $(1,-2,-1,2)$.

8. By means of elementary column transformations, reduce $A = \begin{bmatrix} 1 & 2 & 3 \\ 4 & 5 & 6 \\ 5 & 7 & 8 \end{bmatrix}$ to upper triangular, lower triangular, and diagonal form.

Using $K_{13}(-2/3)$, $K_{23}(-5/6)$; $K_{12}(1)$, $K_{20}(-1/24)$, we obtain

$$A = \begin{bmatrix} 1 & 2 & 3 \\ 4 & 5 & 6 \\ 5 & 7 & 8 \end{bmatrix} \sim \begin{bmatrix} -1 & -1/2 & 3 \\ 0 & 0 & 6 \\ -1/3 & 1/3 & 8 \end{bmatrix} \sim \begin{bmatrix} -3/2 & -5/8 & 3 \\ 0 & -1/4 & 6 \\ 0 & 0 & 8 \end{bmatrix}$$ which is upper triangular.

Using $K_{21}(-2)$, $K_{31}(-3)$; $K_{32}(-2)$ we obtain

$$A = \begin{bmatrix} 1 & 2 & 3 \\ 4 & 5 & 6 \\ 5 & 7 & 8 \end{bmatrix} \sim \begin{bmatrix} 1 & 0 & 0 \\ 4 & -3 & -6 \\ 5 & -3 & -7 \end{bmatrix} \sim \begin{bmatrix} 1 & 0 & 0 \\ 4 & -3 & 0 \\ 5 & -3 & -1 \end{bmatrix}$$ which is lower triangular.

Using $K_{21}(-2)$, $K_{31}(-3)$, $K_{32}(-2)$; $K_{13}(5)$, $K_{23}(-3)$; $K_{12}(4/3)$ we obtain

$$A \sim \begin{bmatrix} 1 & 0 & 0 \\ 4 & -3 & 0 \\ 5 & -3 & -1 \end{bmatrix} \sim \begin{bmatrix} 1 & 0 & 0 \\ 4 & -3 & 0 \\ 0 & 0 & -1 \end{bmatrix} \sim \begin{bmatrix} 1 & 0 & 0 \\ 0 & -3 & 0 \\ 0 & 0 & -1 \end{bmatrix}$$ which is diagonal.

9. Prove: Any non-zero matrix A over $\mathcal{F}$ can be reduced by a sequence of elementary row transformations to a row canonical matrix (echelon matrix) C having the properties:

(i) Each of the first r rows of C has at least one non-zero element; the remaining rows, if any, consist entirely of zero elements.

(ii) In the ith row $(i = 1, 2, \ldots, r)$ of C, its first non-zero element is 1, the unity of $\mathcal{F}$. Let the column in which this element stands be numbered j_i.

(iii) The only non-zero element in the column numbered j_i $(i = 1, 2, \ldots, r)$ is the element 1 in the ith row.

(iv) $j_1 < j_2 < \cdots < j_r$.

Consider the first non-zero column, numbered j_1, of A;

(a) If $a_{1j_1} \neq 0$, use $H_1(a_{1j_1}^{-1})$ to reduce it to 1, if necessary.

(b) If $a_{1j_1} = 0$ but $a_{pj_1} \neq 0$, use H_{1p} and proceed as in (a).

(c) Use transformations of the type $H_{i1}(k)$ to obtain zeros elsewhere in the j_1 column when necessary.

If non-zero elements of the resulting matrix B occur only in the first row, then $B = C$; otherwise, there is a non-zero element elsewhere in the column numbered $j_2 > j_1$. If $b_{2j_2} \neq 0$, use $H_2(b_{2j_2}^{-1})$ as in (a) and proceed as in (c); if $b_{2j_2} = 0$ but $b_{qj_2} \neq 0$, use H_{2q} and proceed as in (a) and (c).

If non-zero elements of the resulting matrix occur only in the first two rows, we have reached C; otherwise, there is a column numbered $j_3 > j_2$ having non-zero elements elsewhere in the column. If ... and so on; ultimately, we must reach C.

10. Prove: The row rank and column rank of any matrix A over $\mathcal{F}$ are equal.

Consider any $m \times n$ matrix and suppose it has row rank r and column rank s. Now a maximum linearly independent subset of the column vectors of this matrix consists of s vectors. By interchanging columns, if necessary, let it be arranged that the first s columns are linearly independent. We leave for the reader to show that such interchanges of columns will neither increase nor decrease the row rank of the given matrix. Without loss in generality, we may suppose in

$$A = \begin{bmatrix} a_{11} & a_{12} & \cdots & a_{1s} & a_{1,s+1} & \cdots & a_{1n} \\ a_{21} & a_{22} & \cdots & a_{2s} & a_{2,s+1} & \cdots & a_{2n} \\ \cdots & \cdots & \cdots & \cdots & \cdots & \cdots & \cdots \\ a_{s1} & a_{s2} & \cdots & a_{ss} & a_{s,s+1} & \cdots & a_{sn} \\ \cdots & \cdots & \cdots & \cdots & \cdots & \cdots & \cdots \\ a_{m1} & a_{m2} & \cdots & a_{ms} & a_{m,s+1} & \cdots & a_{m,n} \end{bmatrix}$$

the first s column vectors $\gamma_1, \gamma_2, \ldots, \gamma_s$ are linearly independent while each of the remaining $n - s$ column vectors is some linear combination of these, say,

$$\gamma_{s+t} = c_{t1}\gamma_1 + c_{t2}\gamma_2 + \cdots + c_{ts}\gamma_s, \quad (t = 1, 2, \ldots, n-s)$$

with $c_{ij} \in \mathcal{F}$. Define the following vectors:

$$\rho_1 = (a_{11}, a_{12}, \ldots, a_{1s}), \quad \rho_2 = (a_{21}, a_{22}, \ldots, a_{2s}), \quad \ldots, \quad \rho_m = (a_{m1}, a_{m2}, \ldots, a_{ms})$$

and $$\sigma_1 = (a_{11}, a_{21}, \ldots, a_{s+1,1}), \quad \sigma_2 = (a_{12}, a_{22}, \ldots, a_{s+1,2}), \quad \ldots, \quad \sigma_n = (a_{1n}, a_{2n}, \ldots, a_{s+1,n})$$

Since the ρ's lie in a space $V_s(\mathcal{F})$, any $s+1$ of them forms a linearly dependent set. Thus there exist scalars $b_1, b_2, \ldots, b_{s+1}$ in $\mathcal{F}$ not all 0 such that

$$\begin{aligned} b_1\rho_1 + b_2\rho_2 + \cdots + b_{s+1}\rho_{s+1} &= (b_1a_{11} + b_2a_{21} + \cdots + b_{s+1}a_{s+1,1}, \ b_1a_{12} + b_2a_{22} + \cdots \\ &\qquad + b_{s+1}a_{s+1,2}, \ \ldots, \ b_1a_{1s} + b_2a_{2s} + \cdots + b_{s+1}a_{s+1,s}) \\ &= (\xi \cdot \sigma_1, \ \xi \cdot \sigma_2, \ \ldots, \ \xi \cdot \sigma_s) = \zeta \end{aligned}$$

where $\zeta = (0, 0, \ldots, 0) = 0$ is the zero vector of $V_s(\mathcal{F})$ and $\xi = (b_1, b_2, \ldots, b_{s+1})$. Then

$$\xi \cdot \sigma_1 = \xi \cdot \sigma_2 = \cdots = \xi \cdot \sigma_s = 0$$

Consider any one of the remaining σ's, say,

$$\begin{aligned} \sigma_{s+k} &= (a_{1,s+k}, a_{2,s+k}, \ldots, a_{s+1,s+k}) \\ &= (c_{k1}a_{11} + c_{k2}a_{12} + \cdots + c_{ks}a_{1s}, c_{k1}a_{21} + c_{k2}a_{22} + \cdots + c_{ks}a_{2s}, \ldots, \\ &\qquad c_{k1}a_{s+1,1} + c_{k2}a_{s+1,2} + \cdots + c_{ks}a_{s+1,s}) \end{aligned}$$

Then
$$\xi \cdot \sigma_{s+k} = c_{k1}(\xi \cdot \sigma_1) + c_{k2}(\xi \cdot \sigma_2) + \cdots + c_{ks}(\xi \cdot \sigma_s) = 0$$

Thus any set of $s+1$ rows of A is linearly dependent; hence $s \leq r$, that is,

the column rank of a matrix cannot exceed its row rank.

To complete the proof, we must show that $r \leq s$. This may be done in either of two ways:

(i) Repeat the above argument beginning with A, having its first r rows linearly independent, and concluding that its first $r+1$ columns are linearly dependent.

(ii) Consider the transpose of A

$$A^T = \begin{bmatrix} a_{11} & a_{21} & \cdots & a_{m1} \\ a_{12} & a_{22} & \cdots & a_{m2} \\ \cdots & \cdots & \cdots & \cdots \\ a_{1n} & a_{2n} & \cdots & a_{mn} \end{bmatrix}$$

whose rows are the corresponding columns of A. Now the row rank of A^T is s, the column rank of A, and the column rank of A^T is r, the row rank of A. By the argument above the column rank of A^T cannot exceed its row rank; i.e. $r \leq s$.

In either case we have $r = s$, as was to be proved.

11. Reduce $A = \begin{bmatrix} 3 & 2 & 3 & 4 & 5 \\ 2 & -1 & 4 & 5 & 1 \\ 4 & 5 & 1 & 2 & -3 \end{bmatrix}$ over R to normal form.

First we use $H_{12}(-1)$ to obtain the element 1 in the first row and first column; thus

$$A = \begin{bmatrix} 3 & 2 & 3 & 4 & 5 \\ 2 & -1 & 4 & 5 & 1 \\ 4 & 5 & 1 & 2 & -3 \end{bmatrix} \sim \begin{bmatrix} 1 & 3 & -1 & -1 & 4 \\ 2 & -1 & 4 & 5 & 1 \\ 4 & 5 & 1 & 2 & -3 \end{bmatrix}$$

Using $H_{21}(-2)$, $H_{31}(-4)$, $K_{21}(-3)$, $K_{31}(1)$, $K_{41}(1)$, $K_{51}(-4)$, we have

$$A \sim \begin{bmatrix} 1 & 0 & 0 & 0 & 0 \\ 0 & -7 & 6 & 7 & -7 \\ 0 & -7 & 5 & 6 & -19 \end{bmatrix}$$

Then using $H_{32}(-1)$, $K_2(-1/7)$, $K_{32}(-6)$, $K_{42}(-7)$, $K_{52}(7)$, we have

$$A \sim \begin{bmatrix} 1 & 0 & 0 & 0 & 0 \\ 0 & 1 & 0 & 0 & 0 \\ 0 & 0 & -1 & -1 & -12 \end{bmatrix}$$

and finally, using $H_3(-1)$, $K_{43}(-1)$, $K_{53}(-12)$,

$$A \sim \begin{bmatrix} 1 & 0 & 0 & 0 & 0 \\ 0 & 1 & 0 & 0 & 0 \\ 0 & 0 & 1 & 0 & 0 \end{bmatrix}$$

12. Reduce $A = \begin{bmatrix} 1 & 1 & 1 & 2 \\ 2 & 1 & -3 & -6 \\ 3 & 3 & 1 & 2 \end{bmatrix}$ over R to normal form N and find matrices S and T such that $S \cdot A \cdot T = N$.

We find

$$\begin{array}{c} I_4 \\ A \quad I_3 \end{array} = \begin{array}{ccccccc} 1 & 0 & 0 & 0 & & & \\ 0 & 1 & 0 & 0 & & & \\ 0 & 0 & 1 & 0 & & & \\ 0 & 0 & 0 & 1 & & & \\ 1 & 1 & 1 & 2 & 1 & 0 & 0 \\ 2 & 1 & -3 & -6 & 0 & 1 & 0 \\ 3 & 3 & 1 & 2 & 0 & 0 & 1 \end{array} \rightarrow \begin{array}{ccccccc} 1 & 0 & 0 & 0 & & & \\ 0 & 1 & 0 & 0 & & & \\ 0 & 0 & 1 & 0 & & & \\ 0 & 0 & 0 & 1 & & & \\ 1 & 1 & 1 & 2 & 1 & 0 & 0 \\ 0 & -1 & -5 & -10 & -2 & 1 & 0 \\ 0 & 0 & -2 & -4 & -3 & 0 & 1 \end{array}$$

$$\rightarrow \begin{array}{ccccccc} 1 & -1 & -1 & -2 & & & \\ 0 & 1 & 0 & 0 & & & \\ 0 & 0 & 1 & 0 & & & \\ 0 & 0 & 0 & 1 & & & \\ 1 & 0 & 0 & 0 & 1 & 0 & 0 \\ 0 & -1 & -5 & -10 & -2 & 1 & 0 \\ 0 & 0 & -2 & -4 & -3 & 0 & 1 \end{array} \rightarrow \begin{array}{ccccccc} 1 & -1 & -1 & -2 & & & \\ 0 & 1 & 0 & 0 & & & \\ 0 & 0 & 1 & 0 & & & \\ 0 & 0 & 0 & 1 & & & \\ 1 & 0 & 0 & 0 & 1 & 0 & 0 \\ 0 & 1 & 5 & 10 & 2 & -1 & 0 \\ 0 & 0 & 1 & 2 & 3/2 & 0 & -1/2 \end{array}$$

$$\rightarrow \begin{array}{ccccccc} 1 & -1 & -1 & -2 & & & \\ 0 & 1 & 0 & 0 & & & \\ 0 & 0 & 1 & 0 & & & \\ 0 & 0 & 0 & 1 & & & \\ 1 & 0 & 0 & 0 & 1 & 0 & 0 \\ 0 & 1 & 0 & 0 & -11/2 & -1 & 5/2 \\ 0 & 0 & 1 & 2 & 3/2 & 0 & -1/2 \end{array} \rightarrow \begin{array}{ccccccc} 1 & -1 & -1 & 0 & & & \\ 0 & 1 & 0 & 0 & & & \\ 0 & 0 & 1 & -2 & & & \\ 0 & 0 & 0 & 1 & & & \\ 1 & 0 & 0 & 0 & 1 & 0 & 0 \\ 0 & 1 & 0 & 0 & -11/2 & -1 & 5/2 \\ 0 & 0 & 1 & 0 & 3/2 & 0 & -1/2 \end{array} = \begin{array}{cc} T & \\ N & S \end{array}$$

Hence $S = \begin{bmatrix} 1 & 0 & 0 \\ -11/2 & -1 & 5/2 \\ 3/2 & 0 & -1/2 \end{bmatrix}$ and $T = \begin{bmatrix} 1 & -1 & -1 & 0 \\ 0 & 1 & 0 & 0 \\ 0 & 0 & 1 & -2 \\ 0 & 0 & 0 & 1 \end{bmatrix}$.

13. Prove: The inverse of the product of two matrices A and B, each of which has an inverse, is the product of the inverses in reverse order, that is,

$$(A \cdot B)^{-1} = B^{-1} \cdot A^{-1}$$

By definition, $(A \cdot B)^{-1} \cdot (A \cdot B) = (A \cdot B)(A \cdot B)^{-1} = I$. Now

$$(B^{-1} \cdot A^{-1}) \cdot (A \cdot B) = B^{-1}(A^{-1} \cdot A)B = B^{-1} \cdot I \cdot B = B^{-1} \cdot B = I$$

and

$$(A \cdot B)(B^{-1} \cdot A^{-1}) = A(B \cdot B^{-1})A^{-1} = A \cdot A^{-1} = I$$

Since $(A \cdot B)^{-1}$ is unique (see Problem 33), we have $(A \cdot B)^{-1} = B^{-1} \cdot A^{-1}$.

14. Compute the inverse of $A = \begin{bmatrix} 1 & 2 & 4 \\ 3 & 1 & 0 \\ 2 & 2 & 1 \end{bmatrix}$ over $I/(5)$.

We have

$$[A \quad I_3] = \begin{bmatrix} 1 & 2 & 4 & 1 & 0 & 0 \\ 3 & 1 & 0 & 0 & 1 & 0 \\ 2 & 2 & 1 & 0 & 0 & 1 \end{bmatrix} \sim \begin{bmatrix} 1 & 2 & 4 & 1 & 0 & 0 \\ 0 & 0 & 3 & 2 & 1 & 0 \\ 0 & 3 & 3 & 3 & 0 & 1 \end{bmatrix} \sim \begin{bmatrix} 1 & 2 & 4 & 1 & 0 & 0 \\ 0 & 3 & 3 & 3 & 0 & 1 \\ 0 & 0 & 3 & 2 & 1 & 0 \end{bmatrix}$$

$$\sim \begin{bmatrix} 1 & 2 & 4 & 1 & 0 & 0 \\ 0 & 1 & 1 & 1 & 0 & 2 \\ 0 & 0 & 1 & 4 & 2 & 0 \end{bmatrix} \sim \begin{bmatrix} 1 & 0 & 2 & 4 & 0 & 1 \\ 0 & 1 & 1 & 1 & 0 & 2 \\ 0 & 0 & 1 & 4 & 2 & 0 \end{bmatrix} \sim \begin{bmatrix} 1 & 0 & 0 & 1 & 1 & 1 \\ 0 & 1 & 0 & 2 & 3 & 2 \\ 0 & 0 & 1 & 4 & 2 & 0 \end{bmatrix}$$

and

$$A^{-1} = \begin{bmatrix} 1 & 1 & 1 \\ 2 & 3 & 2 \\ 4 & 2 & 0 \end{bmatrix}$$

15. Find the minimum polynomial of $A = \begin{bmatrix} 1 & 1 & 1 \\ 0 & 2 & 0 \\ 1 & 1 & 1 \end{bmatrix}$ over R.

Clearly, $A \neq a_0 I$ for all $a_0 \in R$. Set

$$A^2 = \begin{bmatrix} 2 & 4 & 2 \\ 0 & 4 & 0 \\ 2 & 4 & 2 \end{bmatrix} = a_1 \begin{bmatrix} 1 & 1 & 1 \\ 0 & 2 & 0 \\ 1 & 1 & 1 \end{bmatrix} + a_0 \begin{bmatrix} 1 & 0 & 0 \\ 0 & 1 & 0 \\ 0 & 0 & 1 \end{bmatrix} = \begin{bmatrix} a_1 + a_0 & a_1 & a_1 \\ 0 & 2a_1 + a_0 & 0 \\ a_1 & a_1 & a_1 + a_0 \end{bmatrix}$$

which is impossible. Next, set

$$A^3 = \begin{bmatrix} 4 & 12 & 4 \\ 0 & 8 & 0 \\ 4 & 12 & 4 \end{bmatrix} = a_2 \begin{bmatrix} 2 & 4 & 2 \\ 0 & 4 & 0 \\ 2 & 4 & 2 \end{bmatrix} + a_1 \begin{bmatrix} 1 & 1 & 1 \\ 0 & 2 & 0 \\ 1 & 1 & 1 \end{bmatrix} + a_0 \begin{bmatrix} 1 & 0 & 0 \\ 0 & 1 & 0 \\ 0 & 0 & 1 \end{bmatrix}$$

$$= \begin{bmatrix} 2a_2 + a_1 + a_0 & 4a_2 + a_1 & 2a_2 + a_1 \\ 0 & 4a_2 + 2a_1 + a_0 & 0 \\ 2a_2 + a_1 & 4a_2 + a_1 & 2a_2 + a_1 + a_0 \end{bmatrix}$$

From $\begin{cases} 2a_2 + a_1 + a_0 = 4 \\ 4a_2 + a_1 = 12, \\ 2a_2 + a_1 = 4 \end{cases}$ we obtain $a_0 = 0,\ a_1 = -4,\ a_2 = 4$. After checking for every element of A^3 and *not before*, we conclude $m(\lambda) = \lambda^3 - 4\lambda^2 + 4\lambda$.

16. Find all solutions, if any, of the system $\begin{cases} 2x_1 + 2x_2 + 3x_3 + x_4 = 1 \\ 3x_1 - x_2 + x_3 + 3x_4 = 2 \\ -2x_1 + 3x_2 - x_3 - 2x_4 = 4 \\ x_1 + 5x_2 + 3x_3 - 3x_4 = 2 \\ 2x_1 + 7x_2 + 3x_3 - 2x_4 = 8 \end{cases}$ over R.

We have

$$[A\ H] = \begin{bmatrix} 2 & 2 & 3 & 1 & 1 \\ 3 & -1 & 1 & 3 & 2 \\ -2 & 3 & -1 & -2 & 4 \\ 1 & 5 & 3 & -3 & 2 \\ 2 & 7 & 3 & -2 & 8 \end{bmatrix} \sim \begin{bmatrix} 1 & 5 & 3 & -3 & 2 \\ 3 & -1 & 1 & 3 & 2 \\ -2 & 3 & -1 & -2 & 4 \\ 2 & 2 & 3 & 1 & 1 \\ 2 & 7 & 3 & -2 & 8 \end{bmatrix} \sim \begin{bmatrix} 1 & 5 & 3 & -3 & 2 \\ 0 & -16 & -8 & 12 & -4 \\ 0 & 13 & 5 & -8 & 8 \\ 0 & -8 & -3 & 7 & -3 \\ 0 & -3 & -3 & 4 & 4 \end{bmatrix}$$

$$\sim \begin{bmatrix} 1 & 5 & 3 & -3 & 2 \\ 0 & 1 & 1/2 & -3/4 & 1/4 \\ 0 & 13 & 5 & -8 & 8 \\ 0 & -8 & -3 & 7 & -3 \\ 0 & -3 & -3 & 4 & 4 \end{bmatrix} \sim \begin{bmatrix} 1 & 0 & 1/2 & 3/4 & 3/4 \\ 0 & 1 & 1/2 & -3/4 & 1/4 \\ 0 & 0 & -3/2 & 7/4 & 19/4 \\ 0 & 0 & 1 & 1 & -1 \\ 0 & 0 & -3/2 & 7/4 & 19/4 \end{bmatrix} \sim \begin{bmatrix} 1 & 0 & 1/2 & 3/4 & 3/4 \\ 0 & 1 & 1/2 & -3/4 & 1/4 \\ 0 & 0 & 1 & 1 & -1 \\ 0 & 0 & -3/2 & 7/4 & 19/4 \\ 0 & 0 & 0 & 0 & 0 \end{bmatrix}$$

$$\sim \begin{bmatrix} 1 & 0 & 0 & 1/4 & 5/4 \\ 0 & 1 & 0 & -5/4 & 3/4 \\ 0 & 0 & 1 & 1 & -1 \\ 0 & 0 & 0 & 13/4 & 13/4 \\ 0 & 0 & 0 & 0 & 0 \end{bmatrix} \sim \begin{bmatrix} 1 & 0 & 0 & 1/4 & 5/4 \\ 0 & 1 & 0 & -5/4 & 3/4 \\ 0 & 0 & 1 & 1 & -1 \\ 0 & 0 & 0 & 1 & 1 \\ 0 & 0 & 0 & 0 & 0 \end{bmatrix} \sim \begin{bmatrix} 1 & 0 & 0 & 0 & 1 \\ 0 & 1 & 0 & 0 & 2 \\ 0 & 0 & 1 & 0 & -2 \\ 0 & 0 & 0 & 1 & 1 \\ 0 & 0 & 0 & 0 & 0 \end{bmatrix}$$

Both A and $[A\ H]$ have rank 4, the number of unknowns. There is one and only one solution: $x_1 = 1,\ x_2 = 2,\ x_3 = -2,\ x_4 = 1$.

Note. The first move in the reduction was H_{14}. Its purpose, to obtain the element 1 in the first row and column, could also be realized by the use of $H_1(\frac{1}{2})$.

17. Reduce $\begin{bmatrix} 3 & 2 & 1 \\ 6 & 5 & 4 \\ 4 & 2 & 5 \end{bmatrix}$ over $I/(7)$ to normal form.

Using $H_1(5)$; $H_{21}(1), H_{31}(3)$; $H_{12}(4), H_{32}(3)$; $H_3(3)$; $H_{13}(1), H_{23}(5)$, we have

$$\begin{bmatrix} 3 & 2 & 1 \\ 6 & 5 & 4 \\ 4 & 2 & 5 \end{bmatrix} \sim \begin{bmatrix} 1 & 3 & 5 \\ 6 & 5 & 4 \\ 4 & 2 & 5 \end{bmatrix} \sim \begin{bmatrix} 1 & 3 & 5 \\ 0 & 1 & 2 \\ 0 & 4 & 6 \end{bmatrix} \sim \begin{bmatrix} 1 & 0 & 6 \\ 0 & 1 & 2 \\ 0 & 0 & 5 \end{bmatrix} \sim \begin{bmatrix} 1 & 0 & 6 \\ 0 & 1 & 2 \\ 0 & 0 & 1 \end{bmatrix} \sim \begin{bmatrix} 1 & 0 & 0 \\ 0 & 1 & 0 \\ 0 & 0 & 1 \end{bmatrix}$$

18. Find all solutions, if any, of the system $\begin{cases} x_1 + 2x_2 + \ x_3 + 3x_4 = 4 \\ 2x_1 + \ x_2 + 3x_3 + 2x_4 = 1 \\ \qquad 2x_2 + \ x_3 + \ x_4 = 3 \\ 3x_1 + \ x_2 + 3x_3 + 4x_4 = 2 \end{cases}$ over $I/(5)$.

We have

$$[A \ H] = \begin{bmatrix} 1 & 2 & 1 & 3 & 4 \\ 2 & 1 & 3 & 2 & 1 \\ 0 & 2 & 1 & 1 & 3 \\ 3 & 1 & 3 & 4 & 2 \end{bmatrix} \sim \begin{bmatrix} 1 & 2 & 1 & 3 & 4 \\ 0 & 2 & 1 & 1 & 3 \\ 0 & 2 & 1 & 1 & 3 \\ 0 & 0 & 0 & 0 & 0 \end{bmatrix} \sim \begin{bmatrix} 1 & 2 & 1 & 3 & 4 \\ 0 & 1 & 3 & 3 & 4 \\ 0 & 2 & 1 & 1 & 3 \\ 0 & 0 & 0 & 0 & 0 \end{bmatrix} \sim \begin{bmatrix} 1 & 0 & 0 & 2 & 1 \\ 0 & 1 & 3 & 3 & 4 \\ 0 & 0 & 0 & 0 & 0 \\ 0 & 0 & 0 & 0 & 0 \end{bmatrix}$$

Here, $r_A = r_{[A \ H]} = 2$; the system is compatible. Setting $x_3 = s$ and $x_4 = t$, with $s, t \in I/(5)$, all solutions are given by

$$x_1 = 1 + 3t, \quad x_2 = 4 + 2s + 2t, \quad x_3 = s, \quad x_4 = t$$

Since $I/(5)$ is a finite field, there is only a finite number (find it) of solutions.

19. Solve the system $\begin{cases} 2x_1 + \ x_2 + x_3 = 0 \\ x_1 \qquad\ \ + x_3 = 0 \\ \qquad 2x_2 + x_3 = 0 \end{cases}$ over $I/(3)$.

We have

$$A = \begin{bmatrix} 2 & 1 & 1 \\ 1 & 0 & 1 \\ 0 & 2 & 1 \end{bmatrix} \sim \begin{bmatrix} 1 & 2 & 2 \\ 1 & 0 & 1 \\ 0 & 2 & 1 \end{bmatrix} \sim \begin{bmatrix} 1 & 2 & 2 \\ 0 & 1 & 2 \\ 0 & 2 & 1 \end{bmatrix} \sim \begin{bmatrix} 1 & 0 & 1 \\ 0 & 1 & 2 \\ 0 & 0 & 0 \end{bmatrix}$$

Then assigning $x_3 = s \in I/(3)$, we obtain $x_1 = 2s$, $x_2 = x_3 = s$ as the required solution.

20. With each matrix over Q, evaluate:

(*a*) $$\begin{vmatrix} 0 & 1 & -3 \\ 2 & 5 & 4 \\ -3 & 2 & -2 \end{vmatrix} = \begin{vmatrix} -1 & 1 & -3 \\ -3 & 5 & 4 \\ -5 & 2 & -2 \end{vmatrix} = \begin{vmatrix} -1 & 0 & 0 \\ -3 & 2 & 13 \\ -5 & -3 & 13 \end{vmatrix} = \begin{vmatrix} -1 & 0 & 0 \\ 2 & 5 & 0 \\ -5 & -3 & 13 \end{vmatrix} = -65$$

$K_{12}(-1)$ is used to replace $a_{11} = 0$ by a non-zero element. The same result can be obtained by using K_{12}; then

$$\begin{vmatrix} 0 & 1 & -3 \\ 2 & 5 & 4 \\ -3 & 2 & -2 \end{vmatrix} = -\begin{vmatrix} 1 & 0 & -3 \\ 5 & 2 & 4 \\ 2 & -3 & -2 \end{vmatrix} = -\begin{vmatrix} 1 & 0 & 0 \\ 5 & 2 & 19 \\ 2 & -3 & 4 \end{vmatrix} = -\begin{vmatrix} 1 & 0 & 0 \\ 5 & 2 & 0 \\ 2 & -3 & 65/2 \end{vmatrix} = -65$$

An alternate evaluation is as follows:

$$\begin{vmatrix} 0 & 1 & -3 \\ 2 & 5 & 4 \\ -3 & 2 & -2 \end{vmatrix} = \begin{vmatrix} 0 & 1 & 0 \\ 2 & 5 & 19 \\ -3 & 2 & 4 \end{vmatrix} = -(1)\begin{vmatrix} 2 & 19 \\ -3 & 4 \end{vmatrix} = -(8 + 57) = -65$$

(b)
$$\begin{vmatrix} 2 & 3 & -2 & 4 \\ 3 & -2 & 1 & 2 \\ 3 & 2 & 3 & 4 \\ -2 & 4 & 0 & 5 \end{vmatrix} = \begin{vmatrix} -1 & 3 & -2 & 4 \\ 5 & -2 & 1 & 2 \\ 1 & 2 & 3 & 4 \\ -6 & 4 & 0 & 5 \end{vmatrix} = \begin{vmatrix} -1 & 0 & 0 & 0 \\ 5 & 13 & -9 & 22 \\ 1 & 5 & 1 & 8 \\ -6 & -14 & 12 & -19 \end{vmatrix}$$

$$= \begin{vmatrix} -1 & 0 & 0 & 0 \\ -1 & -1 & 3 & 3 \\ 1 & 5 & 1 & 8 \\ -6 & -14 & 12 & -19 \end{vmatrix} = \begin{vmatrix} -1 & 0 & 0 & 0 \\ -1 & -1 & 0 & 0 \\ 1 & 5 & 16 & 23 \\ -6 & -14 & -30 & -61 \end{vmatrix} = \begin{vmatrix} -1 & 0 & 0 & 0 \\ -1 & -1 & 0 & 0 \\ 1 & 5 & 16 & 0 \\ -6 & -14 & -30 & -143/8 \end{vmatrix}$$

$$= (-1)(-1)(16)(-143/8) = -286$$

Supplementary Problems

21. Given $A = \begin{bmatrix} 1 & 0 & 2 \\ 0 & 3 & 1 \\ 4 & 2 & 0 \end{bmatrix}$, $B = \begin{bmatrix} 1 & 2 & 3 \\ 1 & 3 & 4 \\ 1 & 4 & 3 \end{bmatrix}$, $C = \begin{bmatrix} -7 & 6 & -1 \\ 1 & 0 & -1 \\ 1 & -2 & 1 \end{bmatrix}$ over Q, compute:

(a) $A+B = \begin{bmatrix} 2 & 2 & 5 \\ 1 & 6 & 5 \\ 5 & 6 & 3 \end{bmatrix}$ (c) $A \cdot B = \begin{bmatrix} 3 & 10 & 9 \\ 4 & 13 & 15 \\ 6 & 14 & 20 \end{bmatrix}$ (e) $A \cdot C = \begin{bmatrix} -5 & 2 & 1 \\ 4 & -2 & -2 \\ -26 & 24 & -6 \end{bmatrix}$

(b) $3A = \begin{bmatrix} 3 & 0 & 6 \\ 0 & 9 & 3 \\ 12 & 6 & 0 \end{bmatrix}$ (d) $B \cdot C = \begin{bmatrix} -2 & 0 & 0 \\ 0 & -2 & 0 \\ 0 & 0 & -2 \end{bmatrix}$ (f) $A^2 = A \cdot A = \begin{bmatrix} 9 & 4 & 2 \\ 4 & 11 & 3 \\ 4 & 6 & 10 \end{bmatrix}$

22. For the arrays of Problem 21, verify: (a) $(A+B)C = AC+BC$, (b) $(A \cdot B)C = A(B \cdot C)$.

23. For $A = [a_{ij}]$, $(i = 1,2,3;\ j = 1,2,3)$, compute $I_3 \cdot A$ and $A \cdot I_3$ (also $0_3 \cdot A$ and $A \cdot 0_3$) to verify: In the set $\mathcal{R}$ of all n-square matrices over $\mathcal{F}$, the zero matrix and the identity matrix commute with all elements of $\mathcal{R}$.

24. Show that the set of all matrices of the form $\begin{bmatrix} a & b & 0 \\ 0 & a+b & 0 \\ 0 & 0 & c \end{bmatrix}$, where $a, b, c \in Q$, is a subalgebra of $\mathcal{M}_3(Q)$.

25. Show that the set of all matrices of the form $\begin{bmatrix} a & b & c \\ 0 & a+c & 0 \\ c & b & a \end{bmatrix}$ where $a, b, c \in R$, is a subalgebra of $\mathcal{M}_3(R)$.

26. Find the dimension of the vector space spanned by each set of vectors over Q. Select a basis for each.

(a) $\{(1,4,2,4), (1,3,1,2), (0,1,1,2), (3,8,2,4)\}$

(b) $\{(1,2,3,4,5), (5,4,3,2,1), (1,0,1,0,1), (3,2,-1,-2,-5)\}$

(c) $\{(1,1,0,-1,1), (1,0,1,1,-1), (0,1,0,1,0), (1,0,0,1,1), (1,-1,0,1,1)\}$

Ans. (a) 2, (b) 3, (c) 4

27. Show that the linear transformation $A = \begin{bmatrix} 1 & 2 & 3 & 0 \\ 2 & 4 & 3 & 1 \\ 3 & 2 & 1 & 4 \\ 2 & 0 & 4 & 2 \end{bmatrix}$ of $V_4(R)$ into itself is singular and find a vector whose image is 0.

28. Prove: The 3-square matrices $I, H_{12}, H_{13}, H_{23}, H_{12}\cdot H_{13}, H_{12}\cdot H_{23}$ under multiplication form a group isomorphic to the symmetric group on 3 letters.

29. Prove: Under multiplication the set of all non-singular n-square diagonal matrices over $\mathcal{F}$ is a commutative group.

30. Reduce each of the following matrices over R to its row equivalent canonical matrix:

$$(a)\ \begin{bmatrix} 1 & 2 & -3 \\ 2 & 5 & -4 \end{bmatrix} \qquad (c)\ \begin{bmatrix} 1 & 2 & 1 & 2 \\ 1 & 3 & 2 & 2 \\ 2 & 4 & 3 & 4 \\ 3 & 7 & 4 & 6 \end{bmatrix} \qquad (e)\ \begin{bmatrix} 3 & 4 & 5 & 6 & 7 & 8 \\ 4 & 5 & 6 & 7 & 8 & 9 \\ 5 & 6 & 7 & 8 & 9 & 8 \\ 10 & 11 & 12 & 13 & 14 & 15 \end{bmatrix}$$

$$(b)\ \begin{bmatrix} 1 & 1 & 1 & 2 \\ 2 & 1 & -3 & -6 \\ 3 & 3 & 1 & 2 \end{bmatrix} \qquad (d)\ \begin{bmatrix} 1 & 2 & -2 & 3 \\ 2 & 5 & -4 & 6 \\ -1 & -3 & 2 & -2 \\ 2 & 4 & -1 & 6 \end{bmatrix}$$

Ans. $(a)\ \begin{bmatrix} 1 & 0 & -7 \\ 0 & 1 & 2 \end{bmatrix} \quad (b)\ \begin{bmatrix} 1 & 0 & 0 & 0 \\ 0 & 1 & 0 & 0 \\ 0 & 0 & 1 & 2 \end{bmatrix} \quad (c)\ \begin{bmatrix} 1 & 0 & 0 & 2 \\ 0 & 1 & 0 & 0 \\ 0 & 0 & 1 & 0 \\ 0 & 0 & 0 & 0 \end{bmatrix} \quad (d)\ I_4 \quad (e)\ \begin{bmatrix} 1 & 0 & -1 & -2 & -3 & 0 \\ 0 & 1 & 2 & 3 & 4 & 0 \\ 0 & 0 & 0 & 0 & 0 & 1 \\ 0 & 0 & 0 & 0 & 0 & 0 \end{bmatrix}$

31. In Example 11, Page 175, use $H_2(\frac{1}{2}), H_{32}(-1), H_{23}(-5), K_3(2)$ on

$$\begin{matrix} 1 & -2 & 1 & & & \\ 0 & 1 & 0 & & & \\ 0 & 0 & 1 & & & \\ 1 & 0 & 0 & 1 & 0 & 0 \\ 0 & 2 & 5 & -3 & 1 & 0 \\ 0 & 1 & 3 & -4 & 0 & 1 \end{matrix}$$

and obtain $S = \begin{bmatrix} 1 & 0 & 0 \\ 11 & 3 & -5 \\ -5/2 & -1/2 & 1 \end{bmatrix}$ and $T = \begin{bmatrix} 1 & -2 & 2 \\ 0 & 1 & 0 \\ 0 & 0 & 2 \end{bmatrix}$

to show that the non-singular matrices S and T such that $S\cdot A\cdot T = I$ are not unique.

32. Reduce $A = \begin{bmatrix} 1 & 2 & 3 & -2 \\ 2 & -2 & 1 & 3 \\ 3 & 0 & 4 & 1 \end{bmatrix}$ over R to normal form N and compute matrices S and T such that $S\cdot A\cdot T = N$.

33. Prove that if A is non-singular, its inverse A^{-1} is unique.
Hint. Assume $A\cdot B = C\cdot A = I$ and consider $(C\cdot A)B = C(A\cdot B)$.

34. Prove: If A is non-singular, then $A\cdot B = A\cdot C$ implies $B = C$.

35. Show that if the non-singular matrices A and B commute, so also do (a) A^{-1} and B, (b) A and B^{-1}, (c) A^{-1} and B^{-1}.
Hint. (a) $A^{-1}(A\cdot B)A^{-1} = A^{-1}(B\cdot A)A^{-1}$.

36. Find the inverse of:

$$(a)\ \begin{bmatrix} 1 & 3 & 3 \\ 1 & 4 & 3 \\ 1 & 3 & 4 \end{bmatrix} \qquad (c)\ \begin{bmatrix} 1 & 2 & 3 \\ 1 & 3 & 3 \\ 2 & 4 & 3 \end{bmatrix} \qquad (e)\ \begin{bmatrix} 3 & 4 & 2 & 7 \\ 2 & 3 & 3 & 2 \\ 5 & 7 & 3 & 9 \\ 2 & 3 & 2 & 3 \end{bmatrix}$$

$$(b)\ \begin{bmatrix} 1 & 2 & 3 \\ 2 & 4 & 5 \\ 3 & 5 & 6 \end{bmatrix} \qquad (d)\ \begin{bmatrix} 2 & 1 & -1 \\ 1 & 3 & 2 \\ -1 & 2 & 1 \end{bmatrix} \qquad (f)\ \begin{bmatrix} 1 & 1 & 1 & 1 \\ 1 & 2 & 3 & -4 \\ 2 & 3 & 5 & -5 \\ 3 & -4 & -5 & 8 \end{bmatrix} \text{ over } Q.$$

Ans. (a) $\begin{bmatrix} 7 & -3 & -3 \\ -1 & 1 & 0 \\ -1 & 0 & 1 \end{bmatrix}$, (b) $\begin{bmatrix} 1 & -3 & 2 \\ -3 & 3 & -1 \\ 2 & -1 & 0 \end{bmatrix}$, (c) $\frac{1}{3}\begin{bmatrix} 3 & -6 & 3 \\ -3 & 3 & 0 \\ 2 & 0 & -1 \end{bmatrix}$

(d) $\frac{1}{10}\begin{bmatrix} 1 & 3 & -5 \\ 3 & -1 & 5 \\ -5 & 5 & -5 \end{bmatrix}$, (e) $\frac{1}{2}\begin{bmatrix} -1 & 11 & 7 & -26 \\ -1 & -7 & -3 & 16 \\ 1 & 1 & -1 & 0 \\ 1 & -1 & -1 & 2 \end{bmatrix}$, (f) $\frac{1}{18}\begin{bmatrix} 2 & 16 & -6 & 4 \\ 22 & 41 & -30 & -1 \\ -10 & -44 & 30 & -2 \\ 4 & -13 & 6 & -1 \end{bmatrix}$

37. Find the inverse of $A = \begin{bmatrix} 1 & 0 & 1 \\ 1 & 1 & 1 \\ 2 & 1 & 1 \end{bmatrix}$ over $I/(3)$. Does A have an inverse over $I/(5)$?

Ans. $A^{-1} = \begin{bmatrix} 0 & 2 & 1 \\ 2 & 1 & 0 \\ 1 & 1 & 2 \end{bmatrix}$

38. Find the minimum polynomial of (a) $\begin{bmatrix} 0 & 1 & 0 \\ 0 & 0 & 1 \\ 1 & 2 & -1 \end{bmatrix}$, (b) $\begin{bmatrix} 2 & 0 & 0 \\ 0 & 1 & 0 \\ 0 & 0 & 1 \end{bmatrix}$, (c) $\begin{bmatrix} 1 & 1 & 2 \\ 1 & 1 & 2 \\ 1 & 1 & 2 \end{bmatrix}$, (d) $\begin{bmatrix} 2 & 1 & 1 \\ 1 & 2 & 1 \\ 1 & 1 & 2 \end{bmatrix}$.

Ans. (a) $\lambda^3 + \lambda^2 - 2\lambda - 1$, (b) $\lambda^2 - 3\lambda + 2$, (c) $\lambda^2 - 4\lambda$, (d) $\lambda^2 - 5\lambda + 4$

39. Find the inverse of each of the matrices (a), (b), (d) of Problem 36, using its minimum polynomial.

40. Suppose $\lambda^3 + a\lambda^2 + b\lambda$ is the minimum polynomial of a non-singular matrix A and obtain a contradiction.

41. Prove: Theorems XIX, XX and XXI, Page 183.

42. Prove Theorem XXIV (*Hint.* If the ith and jth rows of A are identical, $|A| = |H_{ij}| \cdot |A|$) and Theorem XXVIII.

43. Evaluate:

(a) $\begin{vmatrix} 1 & 2 & 3 \\ 1 & 3 & 4 \\ 1 & 4 & 3 \end{vmatrix}$ (c) $\begin{vmatrix} -7 & 6 & -1 \\ 1 & 0 & -1 \\ 1 & -2 & 1 \end{vmatrix}$ (e) $\begin{vmatrix} 1 & 1 & 1 & 6 \\ 2 & 4 & 1 & 6 \\ 4 & 1 & 2 & 9 \\ 2 & 4 & 2 & 7 \end{vmatrix}$

(b) $\begin{vmatrix} 1 & 0 & 2 \\ 0 & 3 & 1 \\ 4 & 2 & 0 \end{vmatrix}$ (d) $\begin{vmatrix} 2 & -1 & 1 \\ 3 & 2 & 4 \\ -1 & 0 & 3 \end{vmatrix}$ (f) $\begin{vmatrix} 3 & 5 & 7 & 2 \\ 2 & 4 & 1 & 1 \\ -2 & 0 & 0 & 0 \\ 1 & 1 & 3 & 4 \end{vmatrix}$

Ans. (a) -2, (b) -26, (c) 4, (d) 27, (e) 41, (f) 156

44. Evaluate: (a) $\begin{vmatrix} \lambda-1 & 2 & 3 \\ 1 & \lambda-3 & 4 \\ 1 & 4 & \lambda-3 \end{vmatrix}$, (b) $\begin{vmatrix} \lambda-2 & -1 & -4 \\ -1 & \lambda-3 & -5 \\ -4 & -5 & \lambda-6 \end{vmatrix}$

Hint. Expand along the first row or first column.

Ans. (a) $\lambda^3 - 7\lambda^2 - 6\lambda + 42$, (b) $\lambda^3 - 11\lambda^2 - 6\lambda + 28$

45. Denote the row vectors of $A = [a_{ij}]$, $(i, j = 1, 2, 3)$, by ρ_1, ρ_2, ρ_3. Show that

(a) $\rho_1 \times \rho_2$ (see Problem 13, Chapter 13, Page 157) can be found as follows: Write the array

$$\begin{matrix} a_{11} & a_{12} & a_{13} & a_{11} & a_{12} \\ a_{21} & a_{22} & a_{23} & a_{21} & a_{22} \end{matrix}$$

and strike out the first column. Then

$$\rho_1 \times \rho_2 = \left(\begin{vmatrix} a_{12} & a_{13} \\ a_{22} & a_{23} \end{vmatrix}, \begin{vmatrix} a_{13} & a_{11} \\ a_{23} & a_{21} \end{vmatrix}, \begin{vmatrix} a_{11} & a_{12} \\ a_{21} & a_{22} \end{vmatrix} \right)$$

(b) $|A| = \rho_1 \cdot (\rho_2 \times \rho_3) = -\rho_2 \cdot (\rho_1 \times \rho_3) = \rho_3 \cdot (\rho_1 \times \rho_2)$.

46. Show that the set of linear forms

$$(a)\quad \begin{cases} f_1 = a_{11}x_1 + a_{12}x_2 + \cdots + a_{1n}x_n \\ f_2 = a_{21}x_1 + a_{22}x_2 + \cdots + a_{2n}x_n \\ \cdots\cdots\cdots\cdots\cdots\cdots\cdots\cdots \\ f_m = a_{m1}x_1 + a_{m2}x_2 + \cdots + a_{mn}x_n \end{cases} \quad \text{over } \mathcal{F}$$

is linearly dependent if and only if the coefficient matrix

$$A = [a_{ij}], \quad (i = 1, 2, \ldots, m;\ j = 1, 2, \ldots, n)$$

is of rank $r < m$. Thus (a) is necessarily linearly dependent if $m > n$.

47. Find all solutions of:

(a) $x_1 - 2x_2 + 3x_3 - 5x_4 = 1$

(b) $\begin{cases} x_1 + x_2 + x_3 = 4 \\ 2x_1 + 5x_2 - 2x_3 = 3 \end{cases}$

(c) $\begin{cases} x_1 + x_2 + x_3 = 4 \\ 2x_1 + 5x_2 - 2x_3 = 3 \\ x_1 + 7x_2 - 7x_3 = 5 \end{cases}$

(d) $\begin{cases} 2x_1 + x_2 + 5x_3 + x_4 = 5 \\ x_1 + x_2 - 3x_3 - 4x_4 = -1 \\ 3x_1 + 6x_2 - 2x_3 + x_4 = 8 \\ 2x_1 + 2x_2 + 2x_3 - 3x_4 = 2 \end{cases}$

(e) $\begin{cases} x_1 + x_2 + 2x_3 + x_4 = 5 \\ 2x_1 + 3x_2 - x_3 - 2x_4 = 2 \\ 4x_1 + 5x_2 + 3x_3 = 7 \end{cases}$

(f) $\begin{cases} x_1 + x_2 - 2x_3 + x_4 + 3x_5 = 1 \\ 2x_1 - x_2 + 2x_3 + 2x_4 + 6x_5 = 2 \\ 3x_1 + 2x_2 - 4x_3 - 3x_4 - 9x_5 = 3 \end{cases}$

(g) $\begin{cases} x_1 + 3x_2 + x_3 + x_4 + 2x_5 = 0 \\ 2x_1 + 5x_2 - 3x_3 + 2x_4 - x_5 = 3 \\ -x_1 + x_2 + 2x_3 - x_4 + x_5 = 5 \\ 3x_1 + x_2 + x_3 - 2x_4 + 3x_5 = 0 \end{cases}$ over Q.

Ans. (a) $x_1 = 1 + 2r - 3s + 5t,\ x_2 = r,\ x_3 = s,\ x_4 = t$

(b) $x_1 = 17/3 - 7r/3,\ x_2 = -5/3 + 4r/3,\ x_3 = r$

(d) $x_1 = 2,\ x_2 = 1/5,\ x_3 = 0,\ x_4 = 4/5$

(f) $x_1 = 1,\ x_2 = 2r,\ x_3 = r,\ x_4 = -3b,\ x_5 = b$

(g) $x_1 = -11/5 - 4r/5,\ x_2 = 2,\ x_3 = -1 - r,\ x_4 = -14/5 - r/5,\ x_5 = r$

48. (a) Show that the set $M_2 = \{A, B, \ldots\}$ of all matrices over Q of order 2 is isomorphic to the vector space $V_4(Q)$. *Hint.* Use $A = \begin{bmatrix} a_{11} & a_{12} \\ a_{21} & a_{22} \end{bmatrix} \to (a_{11}, a_{12}, a_{21}, a_{22})$. See Prob. 3, Chap. 10, Page 108.

(b) Show that $I_{11} = \begin{bmatrix} 1 & 0 \\ 0 & 0 \end{bmatrix},\ I_{12} = \begin{bmatrix} 0 & 1 \\ 0 & 0 \end{bmatrix},\ I_{21} = \begin{bmatrix} 0 & 0 \\ 1 & 0 \end{bmatrix},\ I_{22} = \begin{bmatrix} 0 & 0 \\ 0 & 1 \end{bmatrix}$ is a basis for the vector space.

(c) Prove: A commutes with $B = \begin{bmatrix} b_{11} & b_{12} \\ b_{21} & b_{22} \end{bmatrix}$ if and only if A commutes with each I_{ij} of (b).

Hint. $B = b_{11}I_{11} + b_{12}I_{12} + b_{21}I_{21} + b_{22}I_{22}$.

49. Define $S_2 = \left\{\begin{bmatrix} x & y \\ -y & x \end{bmatrix} : x, y \in R\right\}$. Show (a) S_2 is a vector space over R, (b) S_2 is a field.

Hint. In (b) show that the mapping $S_2 \to C : \begin{bmatrix} x & y \\ -y & x \end{bmatrix} \to x + yi$ is an isomorphism.

50. Show that the set $\mathcal{Q} = \{(q_1 + q_2 i + q_3 j + q_4 k) : q_1, q_2, q_3, q_4 \in R\}$ with addition and multiplication defined in Problem 27, Chapter 11, Page 123, is isomorphic to the set

$$S_4 = \left\{\begin{bmatrix} q_1 & q_2 & q_3 & q_4 \\ -q_2 & q_1 & -q_4 & q_3 \\ -q_3 & q_4 & q_1 & -q_2 \\ -q_4 & -q_3 & q_2 & q_1 \end{bmatrix} : q_1, q_2, q_3, q_4 \in R\right\}$$

Is S_4 a field?

51. Prove: If $\xi_1, \xi_2, \ldots, \xi_m$ are $m < n$ linearly independent vectors of $V_n(\mathcal{F})$, then the p vectors

$$\eta_j = s_{j1}\xi_1 + s_{j2}\xi_2 + \cdots + s_{jm}\xi_m, \qquad (j = 1, 2, \ldots, p)$$

are linearly dependent if $p > m$ or, when $p \leq m$, if $[s_{ij}]$ is of rank $r < p$.

52. Prove: If $\xi_1, \xi_2, \ldots, \xi_n$ are linearly independent vectors of $V_n(\mathcal{F})$, then the n vectors

$$\eta_j = a_{j1}\xi_1 + a_{j2}\xi_2 + \cdots + a_{jn}\xi_n, \qquad (j = 1, 2, \ldots, n)$$

are linearly independent if and only if $|a_{ij}| \neq 0$.

53. Verify: The ring $T_2 = \left\{\begin{bmatrix} a & b \\ 0 & c \end{bmatrix} : a, b, c \in R\right\}$ has the subrings

$$\left\{\begin{bmatrix} 0 & x \\ 0 & 0 \end{bmatrix} : x \in R\right\}, \quad \left\{\begin{bmatrix} x & y \\ 0 & 0 \end{bmatrix} : x, y \in R\right\}, \quad \text{and} \quad \left\{\begin{bmatrix} 0 & x \\ 0 & y \end{bmatrix} : x, y \in R\right\}$$

as its proper ideals. Write the homomorphism which determines each as an ideal. (See Theorem VI, Chapter 10, Page 105.)

54. Prove: $(A + B)^T = A^T + B^T$ and $(A \cdot B)^T = B^T \cdot A^T$ when A and B are n-square matrices over $\mathcal{F}$.

55. Consider the n-vectors X and Y as $1 \times n$ matrices and verify:

$$X \cdot Y = X \cdot Y^T = Y \cdot X^T$$

56. (*a*) Show that the set of 4-square matrices

$$\mathcal{M} = \{I, H_{12}, H_{13}, H_{14}, H_{23}, H_{24}, H_{34}, H_{12}\cdot H_{13}, H_{12}\cdot H_{23}, H_{12}\cdot H_{14}, H_{12}\cdot H_{24}, H_{13}\cdot H_{14},$$
$$H_{14}\cdot H_{13}, H_{23}\cdot H_{24}, H_{24}\cdot H_{23}, H_{12}\cdot H_{34}, H_{13}\cdot H_{24}, H_{14}\cdot H_{23}, H_{12}\cdot H_{13}\cdot H_{14},$$
$$H_{12}\cdot H_{14}\cdot H_{13}, H_{13}\cdot H_{12}\cdot H_{14}, H_{13}\cdot H_{14}\cdot H_{12}, H_{14}\cdot H_{12}\cdot H_{13}, H_{14}\cdot H_{13}\cdot H_{12}\}$$

is a multiplicative group. *Hint.* Show that the mapping:

$$H_{ij} \to (ij), \quad H_{ij}\cdot H_{ik} \to (ijk), \quad H_{ij}\cdot H_{kl} \to (ij)(kl), \quad H_{ij}\cdot H_{ik}\cdot H_{il} \to (ijkl)$$

of $\mathcal{M}$ into S_n is an isomorphism.

(*b*) Show that the subset $\{I, H_{13}, H_{24}, H_{12}\cdot H_{34}, H_{13}\cdot H_{24}, H_{14}\cdot H_{23}, H_{12}\cdot H_{13}\cdot H_{14}, H_{14}\cdot H_{13}\cdot H_{12}\}$ of $\mathcal{M}$ is a group isomorphic to the octic group of a square. (In Fig. 9-1, Page 92) replace the designations 1, 2, 3, 4 of the vertices by $(1,0,0,0)$, $(0,1,0,0)$, $(0,0,1,0)$, $(0,0,0,1)$ respectively.)

57. Show that the set of 2-square matrices

$$\left\{I, \begin{bmatrix} 0 & 1 \\ -1 & 0 \end{bmatrix}, \begin{bmatrix} -1 & 0 \\ 0 & -1 \end{bmatrix}, \begin{bmatrix} 0 & -1 \\ 1 & 0 \end{bmatrix}, \begin{bmatrix} 1 & 0 \\ 0 & -1 \end{bmatrix}, \begin{bmatrix} -1 & 0 \\ 0 & 1 \end{bmatrix}, \begin{bmatrix} 0 & 1 \\ 1 & 0 \end{bmatrix}, \begin{bmatrix} 0 & -1 \\ -1 & 0 \end{bmatrix}\right\}$$

is a multiplicative group isomorphic to the octic group of a square.

Hint. Place the square of Fig. 9-1, Page 92, in a rectangular coordinate system so that the vertices 1, 2, 3, 4 have coordinates $(1,-1)$, $(1,1)$, $(-1,1)$, $(-1,-1)$ respectively.

58. Let S spanned by $(1,0,1,-1)$, $(1,0,2,3)$, $(3,0,2,-1)$, $(1,0,-2,-7)$ and T spanned by $(2,1,3,2)$, $(0,4,-1,0)$, $(2,3,-4,2)$, $(2,4,-1,2)$ be subspaces of $V_4(Q)$. Find bases for S, T, $S \cap T$, and $S + T$.

Chapter 15

Matrix Polynomials

MATRICES WITH POLYNOMIAL ELEMENTS

Let $\mathcal{F}[\lambda]$ be the polynomial domain consisting of all polynomials in λ with coefficients in $\mathcal{F}$. An $m \times n$ matrix over $\mathcal{F}[\lambda]$, that is, one whose elements are polynomials of $\mathcal{F}[\lambda]$,

$$A(\lambda) = [a_{ij}(\lambda)] = \begin{bmatrix} a_{11}(\lambda) & a_{12}(\lambda) & \dots & a_{1n}(\lambda) \\ a_{21}(\lambda) & a_{22}(\lambda) & \dots & a_{2n}(\lambda) \\ \dots & \dots & \dots & \dots \\ a_{m1}(\lambda) & a_{m2}(\lambda) & \dots & a_{mn}(\lambda) \end{bmatrix}$$

is called a λ-matrix (read, lambda matrix).

Since $\mathcal{F} \subset \mathcal{F}[\lambda]$, the set of all $m \times n$ matrices over $\mathcal{F}$ is a subset of the set of all $m \times n$ λ-matrices over $\mathcal{F}[\lambda]$. It is to be expected then that much of Chapter 14 holds here with, at most, minor changes. For example, with addition and multiplication defined on the set of all n-square λ-matrices over $\mathcal{F}[\lambda]$ precisely as on the set of all n-square matrices over $\mathcal{F}$, we find readily that the former set is also a non-commutative ring with unity I_n. On the other hand, although $A(\lambda) = \begin{bmatrix} \lambda & 0 \\ 0 & \lambda+1 \end{bmatrix}$ is non-singular, i.e. $|A(\lambda)| = \lambda(\lambda+1) \neq 0$, $A(\lambda)$ does not have an inverse over $\mathcal{F}[\lambda]$. The reason, of course, is that generally $\alpha(\lambda)$ does not have a multiplicative inverse in $\mathcal{F}[\lambda]$. Thus it is impossible to extend the notion of elementary transformations on λ-matrices so that, for instance, $A(\lambda) = \begin{bmatrix} \lambda & 0 \\ 0 & \lambda+1 \end{bmatrix} \sim \begin{bmatrix} 1 & 0 \\ 0 & 1 \end{bmatrix}$.

ELEMENTARY TRANSFORMATIONS

The elementary transformations on λ-matrices are defined as follows:

The interchange of the ith and jth rows, denoted by H_{ij}; the interchange of the ith and jth columns, denoted by K_{ij}.

The multiplication of the ith row by a non-zero element $k \in \mathcal{F}$, denoted by $H_i(k)$; the multiplication of the ith column by a non-zero element $k \in \mathcal{F}$, denoted by $K_i(k)$.

The addition to the ith row of the product of $f(\lambda) \in \mathcal{F}[\lambda]$ and the jth row, denoted by $H_{ij}(f(\lambda))$; the addition to the ith column of the product of $f(\lambda) \in \mathcal{F}[\lambda]$ and the jth column, denoted by $K_{ij}(f(\lambda))$.

(Note that the first two transformations are identical with those of Chapter 14 while the third permits all elements of $\mathcal{F}[\lambda]$ as multipliers.)

An elementary transformation and the elementary matrix obtained by performing that transformation on I will again be denoted by the same symbol. Also, a row transformation on $A(\lambda)$ is effected by multiplying it on the left by the appropriate H, and a column transformation is effected by multiplying it on the right by the appropriate K.

Paralleling the results of Chapter 14, we state:

Every elementary matrix is non-singular.

The determinant of every elementary matrix is an element of $\mathcal{F}$.

Every elementary matrix has an inverse which, in turn, is an elementary matrix.

Two $m \times n$ λ-matrices $A(\lambda)$ and $B(\lambda)$ are called equivalent if one can be obtained from the other by a sequence of elementary row and column transformations, i.e. if there exist matrices $S(\lambda) = H_s \ldots H_2 \cdot H_1$ and $T(\lambda) = K_1 \cdot K_2 \ldots K_t$ such that

$$S(\lambda) \cdot A(\lambda) \cdot T(\lambda) \;=\; B(\lambda)$$

The row (column) rank of a λ-matrix is the number of linearly independent rows (columns) of the matrix. The rank of a λ-matrix is its row (column) rank.

Equivalent λ-matrices have the same rank. The converse is not true.

NORMAL FORM OF A λ-MATRIX

Corresponding to Theorem IX′, Chapter 14, Page 174, there is

Theorem I. Every $m \times n$ λ-matrix $A(\lambda)$ over $\mathcal{F}[\lambda]$ of rank r can be reduced by elementary transformations to a canonical form (*normal form*)

$$N(\lambda) \;=\; \begin{bmatrix} f_1(\lambda) & 0 & 0 & \ldots & 0 & 0 & 0 & \ldots & 0 \\ 0 & f_2(\lambda) & 0 & \ldots & 0 & 0 & 0 & \ldots & 0 \\ \cdots & \cdots & \cdots & \cdots & \cdots & \cdots & \cdots & \cdots & \cdots \\ 0 & 0 & 0 & \ldots & 0 & f_r(\lambda) & 0 & \ldots & 0 \\ 0 & 0 & 0 & \ldots & 0 & 0 & 0 & \ldots & 0 \\ \cdots & \cdots & \cdots & \cdots & \cdots & \cdots & \cdots & \cdots & \cdots \\ 0 & 0 & 0 & \ldots & 0 & 0 & 0 & \ldots & 0 \end{bmatrix}$$

in which $f_1(\lambda), f_2(\lambda), \ldots, f_r(\lambda)$ are monic polynomials in $\mathcal{F}[\lambda]$ and $f_i(\lambda)$ divides $f_{i+1}(\lambda)$ for $i = 1, 2, \ldots, r-1$.

We shall not prove this theorem nor that the normal form of a given $A(\lambda)$ is unique. (The proof of the theorem consists in showing how to reach $N(\lambda)$ for any given $A(\lambda)$; uniqueness requires further study of determinants.) A simple procedure for obtaining the normal form is illustrated in the example and problems below.

Example 1:

Reduce $A(\lambda) \;=\; \begin{bmatrix} \lambda+3 & \lambda+1 & \lambda+2 \\ 2\lambda^2+\lambda-3 & \lambda^2+\lambda-1 & 2\lambda^2-2 \\ \lambda^3+\lambda^2+6\lambda+3 & 2\lambda^2+2\lambda+1 & \lambda^3+\lambda^2+5\lambda+2 \end{bmatrix}$ over $R(\lambda)$ to normal form.

The greatest common divisor of the elements of $A(\lambda)$ is 1; take $f_1(\lambda) = 1$. Now use $K_{13}(-1)$ to replace $a_{11}(\lambda)$ by $f_1(\lambda)$, and then by appropriate row and column transformation obtain an equivalent matrix whose first row and first column have zero elements except for the common element $f_1(\lambda)$; thus,

$$A(\lambda) \;\sim\; \begin{bmatrix} 1 & \lambda+1 & \lambda+2 \\ \lambda-1 & \lambda^2+\lambda-1 & 2\lambda^2-2 \\ \lambda+1 & 2\lambda^2+2\lambda+1 & \lambda^3+\lambda^2+5\lambda+2 \end{bmatrix}$$

$$\sim \begin{bmatrix} 1 & 0 & 0 \\ \lambda-1 & \lambda & \lambda^2-\lambda \\ \lambda+1 & \lambda^2 & \lambda^3+2\lambda \end{bmatrix} \;\sim\; \begin{bmatrix} 1 & 0 & 0 \\ 0 & \lambda & \lambda^2-\lambda \\ 0 & \lambda^2 & \lambda^3+2\lambda \end{bmatrix} \;=\; B(\lambda)$$

Consider now the submatrix $\begin{bmatrix} \lambda & \lambda^2-\lambda \\ \lambda^2 & \lambda^3+2\lambda \end{bmatrix}$. The greatest common divisor of its elements is λ; set $f_2(\lambda) = \lambda$. Since $f_2(\lambda)$ occupies the position of $b_{22}(\lambda)$ in $B(\lambda)$, we proceed to clear the second row and second column of non-zero elements, except of course for the common element $f_2(\lambda)$, and obtain

$$A(\lambda) \sim \begin{bmatrix} 1 & 0 & 0 \\ 0 & \lambda & \lambda^2-\lambda \\ 0 & \lambda^2 & \lambda^3+2\lambda \end{bmatrix} \sim \begin{bmatrix} 1 & 0 & 0 \\ 0 & \lambda & \lambda^2-\lambda \\ 0 & 0 & \lambda^2+2\lambda \end{bmatrix} \sim \begin{bmatrix} 1 & 0 & 0 \\ 0 & \lambda & 0 \\ 0 & 0 & \lambda^2+2\lambda \end{bmatrix} = N(\lambda)$$

since $\lambda^2 + 2\lambda$ is monic.

See also Problems 1-3.

The non-zero elements of $N(\lambda)$, the normal form of $A(\lambda)$, are called *invariant factors* of $A(\lambda)$. Under the assumption that the normal form of a λ-matrix is unique, we have

Theorem II. Two $m \times n$ λ-matrices over $\mathcal{F}[\lambda]$ are equivalent if and only if they have the same invariant factors.

POLYNOMIALS WITH MATRIX COEFFICIENTS

In the remainder of this chapter we shall restrict our attention to n-square λ-matrices over $\mathcal{F}[\lambda]$. Let $A(\lambda)$ be such a matrix and suppose the maximum degree of all polynomial elements $a_{ij}(\lambda)$ of $A(\lambda)$ is p. By the addition, when necessary of terms with zero coefficients, $A(\lambda)$ can be written so each of its elements has $p+1$ terms. Then $A(\lambda)$ can be written as a polynomial of degree p in λ with n-square matrices A_i over $\mathcal{F}$ as coefficients, called a *matrix polynomial of degree p in* λ.

Example 2:

For the λ-matrix $A(\lambda)$ of Example 1, we have

$$A(\lambda) = \begin{bmatrix} \lambda+3 & \lambda+1 & \lambda+2 \\ 2\lambda^2+\lambda-3 & \lambda^2+\lambda-1 & 2\lambda^2-2 \\ \lambda^3+\lambda^2+6\lambda+3 & 2\lambda^2+2\lambda+1 & \lambda^3+\lambda^2+5\lambda+2 \end{bmatrix}$$

$$= \begin{bmatrix} 0\lambda^3+0\lambda^2+\lambda+3 & 0\lambda^3+0\lambda^2+\lambda+1 & 0\lambda^3+0\lambda^2+\lambda+2 \\ 0\lambda^3+2\lambda^2+\lambda-3 & 0\lambda^3+\lambda^2+\lambda-1 & 0\lambda^3+2\lambda^2+0\lambda-2 \\ \lambda^3+\lambda^2+6\lambda+3 & 0\lambda^3+2\lambda^2+2\lambda+1 & \lambda^3+\lambda^2+5\lambda+2 \end{bmatrix}$$

$$= \begin{bmatrix} 0 & 0 & 0 \\ 0 & 0 & 0 \\ 1 & 0 & 1 \end{bmatrix}\lambda^3 + \begin{bmatrix} 0 & 0 & 0 \\ 2 & 1 & 2 \\ 1 & 2 & 1 \end{bmatrix}\lambda^2 + \begin{bmatrix} 1 & 1 & 1 \\ 1 & 1 & 0 \\ 6 & 2 & 5 \end{bmatrix}\lambda + \begin{bmatrix} 3 & 1 & 2 \\ -3 & -1 & -2 \\ 3 & 1 & 2 \end{bmatrix}$$

Consider now the n-square λ-matrices or matrix polynomials

$$A(\lambda) = A_p\lambda^p + A_{p-1}\lambda^{p-1} + \cdots + A_1\lambda + A_0 \tag{1}$$

and

$$B(\lambda) = B_q\lambda^q + B_{q-1}\lambda^{q-1} + \cdots + B_1\lambda + B_0 \tag{2}$$

The two λ-matrices (matrix polynomials) are said to be *equal* when $p=q$ and $A_i=B_i$ for $i = 0,1,2,\ldots,p$.

The sum $A(\lambda)+B(\lambda)$ is a λ-matrix (matrix polynomial) obtained by adding corresponding elements (terms) of the λ-matrices (matrix polynomials). If $p>q$, its degree is p; if $p=q$, its degree is at most p.

The product $A(\lambda)\cdot B(\lambda)$ is a λ-matrix (matrix polynomial) of degree at most $p+q$. If either $A(\lambda)$ or $B(\lambda)$ is non-singular (i.e., either $|A(\lambda)| \neq 0$ or $|B(\lambda)| \neq 0$) then both $A(\lambda)\cdot B(\lambda)$ and $B(\lambda)\cdot A(\lambda)$ are of degree $p+q$. Since, in general, matrices do not commute, we shall expect $A(\lambda)\cdot B(\lambda) \neq B(\lambda)\cdot A(\lambda)$.

The equality in (1) is not disturbed if λ is replaced throughout by any $k \in \mathcal{F}$. For example,

$$A(k) \quad = \quad A_p k^p + A_{p-1} k^{p-1} + \cdots + A_1 k + A_0$$

When, however, λ is replaced by an n-square matrix C over $\mathcal{F}$, we obtain two results which are usually different

$$A_R(C) \quad = \quad A_p C^p + A_{p-1} C^{p-1} + \cdots + A_1 C + A_0 \tag{3}$$

and

$$A_L(C) \quad = \quad C^p A_p + C^{p-1} A_{p-1} + \cdots + C A_1 + A_0 \tag{3'}$$

called, respectively, the *right* and *left functional values* of $A(\lambda)$ when $\lambda = C$.

Example 3:

When $A(\lambda) = \begin{bmatrix} \lambda^2 & \lambda - 1 \\ \lambda + 3 & \lambda^2 + 2 \end{bmatrix} = \begin{bmatrix} 1 & 0 \\ 0 & 1 \end{bmatrix} \lambda^2 + \begin{bmatrix} 0 & 1 \\ 1 & 0 \end{bmatrix} \lambda + \begin{bmatrix} 0 & -1 \\ 3 & 2 \end{bmatrix}$ and $C = \begin{bmatrix} 1 & 2 \\ 2 & 3 \end{bmatrix}$,

then $$A_R(C) = \begin{bmatrix} 1 & 0 \\ 0 & 1 \end{bmatrix} \cdot \begin{bmatrix} 1 & 2 \\ 2 & 3 \end{bmatrix}^2 + \begin{bmatrix} 0 & 1 \\ 1 & 0 \end{bmatrix} \cdot \begin{bmatrix} 1 & 2 \\ 2 & 3 \end{bmatrix} + \begin{bmatrix} 0 & -1 \\ 3 & 2 \end{bmatrix} = \begin{bmatrix} 7 & 10 \\ 12 & 17 \end{bmatrix}$$

and $$A_L(C) = \begin{bmatrix} 1 & 2 \\ 2 & 3 \end{bmatrix}^2 \cdot \begin{bmatrix} 1 & 0 \\ 0 & 1 \end{bmatrix} + \begin{bmatrix} 1 & 2 \\ 2 & 3 \end{bmatrix} \cdot \begin{bmatrix} 0 & 1 \\ 1 & 0 \end{bmatrix} + \begin{bmatrix} 0 & -1 \\ 3 & 2 \end{bmatrix} = \begin{bmatrix} 7 & 8 \\ 14 & 17 \end{bmatrix}$$

See also Problem 4.

DIVISION ALGORITHM

The division algorithm for polynomials $\alpha(x), \beta(x)$ in x over a non-commutative ring $\mathcal{R}$ with unity was given in Theorem II, Chapter 12, Page 126. It was assumed there that the divisor $\beta(x)$ was monic. For a non-monic divisor, that is, a divisor $\beta(x)$ whose leading coefficient is $b_n \neq 1$, the theorem holds only if $b_n^{-1} \in \mathcal{R}$.

For the coefficient ring considered here, every non-singular matrix A has an inverse over $\mathcal{F}$; thus the algorithm may be stated as

If $A(\lambda)$ and $B(\lambda)$ are matrix polynomials (1) and (2) and if B_q is non-singular, then there exist unique matrix polynomials $Q_1(\lambda), R_1(\lambda); Q_2(\lambda), R_2(\lambda) \in \mathcal{F}[\lambda]$, where $R_1(\lambda)$ and $R_2(\lambda)$ are either zero or of degree less than that of $B(\lambda)$, such that

$$A(\lambda) \quad = \quad Q_1(\lambda) \cdot B(\lambda) + R_1(\lambda) \tag{4}$$

and

$$A(\lambda) \quad = \quad B(\lambda) \cdot Q_2(\lambda) + R_2(\lambda) \tag{4'}$$

If in (4) $R_1(\lambda) = 0$, $B(\lambda)$ is called a *right divisor* of $A(\lambda)$; if in (4′) $R_2(\lambda) = 0$, $B(\lambda)$ is called a *left divisor* of $A(\lambda)$.

Example 4:

Given

$$A(\lambda) = \begin{bmatrix} \lambda^3 + 3\lambda^2 + 3\lambda & 2\lambda^2 + 5\lambda + 4 \\ \lambda^2 + \lambda - 1 & \lambda^2 + 1 \end{bmatrix} = \begin{bmatrix} 1 & 0 \\ 0 & 0 \end{bmatrix} \lambda^3 + \begin{bmatrix} 3 & 2 \\ 1 & 1 \end{bmatrix} \lambda^2 + \begin{bmatrix} 3 & 5 \\ 1 & 0 \end{bmatrix} \lambda + \begin{bmatrix} 0 & 4 \\ -1 & 1 \end{bmatrix}$$

and

$$B(\lambda) = \begin{bmatrix} \lambda + 1 & 1 \\ \lambda & \lambda + 2 \end{bmatrix} = \begin{bmatrix} 1 & 0 \\ 1 & 1 \end{bmatrix} \lambda + \begin{bmatrix} 1 & 1 \\ 0 & 2 \end{bmatrix}$$

find $Q_1(\lambda), R_1(\lambda); Q_2(\lambda), R_2(\lambda)$ such that

(a) $A(\lambda) = Q_1(\lambda) \cdot B(\lambda) + R_1(\lambda)$, (b) $A(\lambda) = B(\lambda) \cdot Q_2(\lambda) + R_2(\lambda)$

Here $B_1 = \begin{bmatrix} 1 & 0 \\ 1 & 1 \end{bmatrix} \neq 0$ and $B_1^{-1} = \begin{bmatrix} 1 & 0 \\ -1 & 1 \end{bmatrix}$.

(a) We compute

$$A(\lambda) - A_3B_1^{-1}\lambda^2 B(\lambda) = \begin{bmatrix} 2 & 1 \\ 1 & 1 \end{bmatrix}\lambda^2 + \begin{bmatrix} 3 & 5 \\ 1 & 0 \end{bmatrix}\lambda + \begin{bmatrix} 0 & 4 \\ -1 & 1 \end{bmatrix} = C(\lambda)$$

$$C(\lambda) - C_2B_1^{-1}\lambda B(\lambda) = \begin{bmatrix} 2 & 2 \\ 1 & -2 \end{bmatrix}\lambda + \begin{bmatrix} 0 & 4 \\ -1 & 1 \end{bmatrix} = D(\lambda)$$

$$D(\lambda) - D_1B_1^{-1}B(\lambda) = \begin{bmatrix} 0 & 0 \\ -4 & 2 \end{bmatrix} = R_1(\lambda)$$

Then $$Q_1(\lambda) = (A_3\lambda^2 + C_2\lambda + D_1)B_1^{-1} = \begin{bmatrix} 1 & 0 \\ 0 & 0 \end{bmatrix}\lambda^2 + \begin{bmatrix} 1 & 1 \\ 0 & 1 \end{bmatrix}\lambda + \begin{bmatrix} 0 & 2 \\ 3 & -2 \end{bmatrix} = \begin{bmatrix} \lambda^2+\lambda & \lambda+2 \\ 3 & \lambda-2 \end{bmatrix}$$

(b) We compute

$$A(\lambda) - B(\lambda)B_1^{-1}A_3\lambda^2 = \begin{bmatrix} 3 & 2 \\ 3 & 1 \end{bmatrix}\lambda^2 + \begin{bmatrix} 3 & 5 \\ 1 & 0 \end{bmatrix}\lambda + \begin{bmatrix} 0 & 4 \\ -1 & 1 \end{bmatrix} = E(\lambda)$$

$$E(\lambda) - B(\lambda)B_1^{-1}E_2\lambda = \begin{bmatrix} 0 & 4 \\ 1 & 2 \end{bmatrix}\lambda + \begin{bmatrix} 0 & 4 \\ -1 & 1 \end{bmatrix} = F(\lambda)$$

$$F(\lambda) - B(\lambda)B_1^{-1}F_1 = \begin{bmatrix} -1 & 2 \\ -3 & 5 \end{bmatrix} = R_2(\lambda)$$

Then $$Q_2(\lambda) = B_1^{-1}(A_3\lambda^2 + E_2\lambda + F_1) = \begin{bmatrix} 1 & 0 \\ -1 & 0 \end{bmatrix}\lambda^2 + \begin{bmatrix} 3 & 2 \\ 0 & -1 \end{bmatrix}\lambda + \begin{bmatrix} 0 & 4 \\ 1 & -2 \end{bmatrix} = \begin{bmatrix} \lambda^2+3\lambda & 2\lambda+4 \\ -\lambda^2+1 & -\lambda-2 \end{bmatrix}$$

See Problem 5.

For the n-square matrix $B = [b_{ij}]$ over $\mathcal{F}$, define its *characteristic matrix* as

$$\lambda I - B = \begin{bmatrix} \lambda - b_{11} & -b_{12} & -b_{13} & \dots & -b_{1n} \\ -b_{21} & \lambda - b_{22} & -b_{23} & \dots & -b_{2n} \\ -b_{31} & -b_{32} & \lambda - b_{33} & \dots & -b_{3n} \\ \dots & \dots & \dots & \dots & \dots \\ -b_{n1} & -b_{n2} & -b_{n3} & \dots & \lambda - b_{nn} \end{bmatrix}$$

With $A(\lambda)$ as in (*1*) and $B(\lambda) = \lambda I - B$, (*4*) and (*4′*) yield

$$A(\lambda) = Q_1(\lambda)\cdot(\lambda I - B) + R_1 \tag{5}$$

and

$$A(\lambda) = (\lambda I - B)\cdot Q_2(\lambda) + R_2 \tag{5'}$$

in which the remainders R_1 and R_2 are free of λ. It can be shown, moreover, that

$$R_1 = A_R(B) \qquad \text{and} \qquad R_2 = A_L(B)$$

Example 5:

With $A(\lambda) = \begin{bmatrix} \lambda^2 & \lambda-1 \\ \lambda+3 & \lambda^2+2 \end{bmatrix}$ and $B = \begin{bmatrix} 1 & 2 \\ 2 & 3 \end{bmatrix}$, we have $\lambda I - B = \begin{bmatrix} \lambda-1 & -2 \\ -2 & \lambda-3 \end{bmatrix}$

and

$$A(\lambda) = \begin{bmatrix} \lambda+1 & 3 \\ 3 & \lambda+3 \end{bmatrix}(\lambda I - B) + \begin{bmatrix} 7 & 10 \\ 12 & 17 \end{bmatrix} = (\lambda I - B)\begin{bmatrix} \lambda+1 & 3 \\ 3 & \lambda+3 \end{bmatrix} + \begin{bmatrix} 7 & 8 \\ 14 & 17 \end{bmatrix}$$

From Example 3, the remainders are $R_1 = A_R(B) = \begin{bmatrix} 7 & 10 \\ 12 & 17 \end{bmatrix}$ and $R_2 = A_L(B) = \begin{bmatrix} 7 & 8 \\ 14 & 17 \end{bmatrix}$.

THE CHARACTERISTIC ROOTS AND VECTORS OF A MATRIX

We return now to further study of a given linear transformation of $V_n(\mathcal{F})$ into itself. Consider, for example, the transformation of $V = V_3(R)$ given by

$$A = \begin{bmatrix} 2 & 2 & 1 \\ 1 & 3 & 1 \\ 1 & 2 & 2 \end{bmatrix} \quad \text{or} \quad \begin{aligned} \epsilon_1 &\to (2,2,1) \\ \epsilon_2 &\to (1,3,1) \\ \epsilon_3 &\to (1,2,2) \end{aligned}$$

(It is necessary to remind the reader that in our notation the images of the unit vectors $\epsilon_1, \epsilon_2, \epsilon_3$ of the space are the row vectors of A and that the linear transformation is given by

$$V \to V: \ \xi \to \xi A$$

since he may find elsewhere the image vectors written as the column vectors of A. In this case, the transformation is given by

$$V \to V: \ \xi \to A\xi$$

For the same matrix A, the two transformations are generally different.)

The image of $\xi = (1,2,3) \in V$ is

$$\eta = (1,2,3)\begin{bmatrix} 2 & 2 & 1 \\ 1 & 3 & 1 \\ 1 & 2 & 2 \end{bmatrix} = (7,14,9) \in V$$

whose only connection with ξ is through the transformation A. On the other hand, the image of $\xi_1 = (r, 2r, r) \in V$ is $5\xi_1$, that is, the image of any vector of the subspace $V^1 \subset V$, spanned by $(1,2,1)$, is a vector of V^1. Similarly, it is easily verified that the image of any vector of the subspace $V^2 \subset V$, spanned by $(1,-1,0)$, is a vector of V^2; and the image of any vector of $V^3 \subset V$, spanned by $(1,0,-1)$, is a vector of V^3. Moreover, the image of any vector $(s+t, -s, -t)$ of the subspace $V^4 \subset V$, spanned by $(1,-1,0)$ and $(1,0,-1)$, is a vector of the subspace generated by itself. We leave for the reader to show that the same is not true for either the subspace V^5, spanned by $(1,2,1)$ and $(1,-1,0)$, or of V^6, spanned by $(1,2,1)$ and $(1,0,-1)$.

We summarize: The linear transformation A of $V_3(R)$ carries any vector of the subspace V^1, spanned by $(1,2,1)$, into a vector of V^1 and any vector of the subspace V^4, spanned by $(1,-1,0)$ and $(1,0,-1)$, into a vector of the subspace generated by itself. We shall call any non-zero vector of V^1, also of V^4, a *characteristic vector* (*invariant vector* or *eigenvector*) of the transformation.

In general, let a linear transformation of $V = V_n(\mathcal{F})$ relative to the basis $\epsilon_1, \epsilon_2, \ldots, \epsilon_n$ be given by the n-square matrix $A = [a_{ij}]$ over $\mathcal{F}$. Any given non-zero vector $\xi = (x_1, x_2, x_3, \ldots, x_n) \in V$ is a characteristic vector of A provided $\xi A = \lambda\xi$, i.e.,

$$\begin{aligned}(a_{11}x_1 + a_{21}x_2 + \cdots + a_{n1}x_n,\ a_{12}x_1 + a_{22}x_2 + \cdots + a_{n2}x_n,\ \ldots,& \\ a_{1n}x_1 + a_{2n}x_2 + \cdots + a_{nn}x_n) \ = \ (\lambda x_1, \lambda x_2, \ldots, \lambda x_n)&\end{aligned} \tag{6}$$

for some $\lambda \in \mathcal{F}$.

We shall now use (6) to solve the following problem: Given A, find all non-zero vectors ξ such that $\xi A = \lambda\xi$ with $\lambda \in \mathcal{F}$. After equating corresponding components in (6), the resulting system of equations may be written as follows

$$\begin{cases} (\lambda - a_{11})x_1 - a_{21}x_2 - a_{31}x_3 - \cdots - a_{n1}x_n = 0 \\ -a_{12}x_1 + (\lambda - a_{22})x_2 - a_{32}x_3 - \cdots - a_{n2}x_n = 0 \\ -a_{13}x_1 - a_{23}x_2 + (\lambda - a_{33})x_3 - \cdots - a_{n3}x_n = 0 \\ \cdots\cdots\cdots\cdots\cdots\cdots\cdots\cdots\cdots\cdots \\ -a_{1n}x_1 - a_{2n}x_2 - a_{3n}x_3 - \cdots + (\lambda - a_{nn})x_n = 0 \end{cases} \tag{7}$$

which by Theorem XVIII, Chapter 14, Page 181, has a non-trivial solution if and only if the determinant of the coefficient matrix

$$\begin{vmatrix} \lambda - a_{11} & -a_{21} & -a_{31} & \ldots & -a_{n1} \\ -a_{12} & \lambda - a_{22} & -a_{32} & \ldots & -a_{n2} \\ -a_{13} & -a_{23} & \lambda - a_{33} & \ldots & -a_{n3} \\ \ldots & \ldots & \ldots & \ldots & \ldots \\ -a_{1n} & -a_{2n} & -a_{3n} & \ldots & \lambda - a_{nn} \end{vmatrix} = |\lambda I - A^T| = 0$$

where A^T is the transpose of A. Now $\lambda I - A^T = (\lambda I - A)^T$ (check this); hence, by Theorem XXII, Chapter 14, $|\lambda I - A^T| = |\lambda I - A|$, the determinant of the characteristic matrix of A.

For any n-square matrix A over $\mathcal{F}$, $|\lambda I - A^T|$ is called the *characteristic determinant* of A and its expansion, a polynomial $\phi(\lambda)$ of degree n, is called the *characteristic polynomial* of A. The n zeros $\lambda_1, \lambda_2, \lambda_3, \ldots, \lambda_n$ of $\phi(\lambda)$ are called the *characteristic roots* (*latent roots* or *eigenvalues*) of A.

Now $\phi(\lambda) \in \mathcal{F}[\lambda]$ and may or may not have all of its zeros in $\mathcal{F}$. (For example, the characteristic polynomial of a 2-square matrix over R will have either both or neither of its zeros in R; that of a 3-square matrix over R will have either one or three zeros in R. One may then restrict attention solely to the subspaces of $V_3(R)$ associated with the real zeros, if any, or one may enlarge the space to $V_3(C)$ and find the subspaces associated with all of the zeros.) For any characteristic root λ_i, the matrix $\lambda_i I - A^T$ is singular so that the system of linear equations (7) is linearly dependent and a characteristic vector ξ always exists. Also, $k\xi$ is a characteristic vector associated with λ_i for every scalar k. Moreover, by Theorem XVIII, Chapter 14, Page 181, when $\lambda_i I - A^T$ has rank r, then (7) has $n - r$ linearly independent solutions which span a subspace of dimension $n - r$. Every non-zero vector of this subspace is a characteristic vector of A associated with the characteristic root λ_i.

Example 6: Determine the characteristic roots and associated characteristic vectors of $V_3(R)$, given $A = \begin{bmatrix} 1 & 1 & 2 \\ 0 & 2 & 2 \\ -1 & 1 & 3 \end{bmatrix}$.

The characteristic polynomial of A is

$$|\lambda I - A^T| = \begin{vmatrix} \lambda - 1 & 0 & 1 \\ -1 & \lambda - 2 & -1 \\ -2 & -2 & \lambda - 3 \end{vmatrix} = \lambda^3 - 6\lambda^2 + 11\lambda - 6;$$

the characteristic roots are $\lambda_1 = 1$, $\lambda_2 = 2$, $\lambda_3 = 3$; the system of linear equations (7) is

$$(a) \qquad \begin{cases} (\lambda - 1)x_1 & & + \quad x_3 & = 0 \\ -x_1 & + (\lambda - 2)x_2 & - \quad x_3 & = 0 \\ -2x_1 & - \quad 2x_2 & + (\lambda - 3)x_3 & = 0 \end{cases}$$

When $\lambda = \lambda_1 = 1$, the system (a) reduces to $\begin{cases} x_1 + x_2 = 0 \\ x_3 = 0 \end{cases}$, having $x_1 = 1$, $x_2 = -1$, $x_3 = 0$ as a solution. Thus, associated with the characteristic root $\lambda_1 = 1$ is the one-dimensional vector space spanned by $\xi_1 = (1, -1, 0)$. Every vector $(k, -k, 0)$, $k \neq 0$, of this subspace is a characteristic vector of A.

When $\lambda = \lambda_2 = 2$, system (a) reduces to $\begin{cases} x_1 + x_3 = 0 \\ x_1 + 2x_2 = 0 \end{cases}$, having $x_1 = 2$, $x_2 = -1$, $x_3 = -2$ as a solution. Thus, associated with the characteristic root $\lambda_2 = 2$ is the one-dimensional vector space spanned by $\xi_2 = (2, -1, -2)$, and every vector $(2k, -k, -2k)$, $k \neq 0$, is a characteristic vector of A.

When $\lambda = \lambda_3 = 3$, system (a) reduces to $\begin{cases} x_1 + x_2 = 0 \\ 2x_1 + x_3 = 0 \end{cases}$, having $x_1 = 1$, $x_2 = -1$, $x_3 = -2$ as a solution. Thus, associated with the characteristic root $\lambda_3 = 3$ is the one-dimensional vector space spanned by $\xi_3 = (1, -1, -2)$, and every vector $(k, -k, -2k)$, $k \neq 0$, is a characteristic vector of A.

Example 7: Determine the characteristic roots and associated characteristic vectors of $V_3(R)$, given $A = \begin{bmatrix} 2 & 2 & 1 \\ 1 & 3 & 1 \\ 1 & 2 & 2 \end{bmatrix}$.

The characteristic polynomial is

$$|\lambda I - A^T| = \begin{vmatrix} \lambda-2 & -1 & -1 \\ -2 & \lambda-3 & -2 \\ -1 & -1 & \lambda-2 \end{vmatrix} = \lambda^3 - 7\lambda^2 + 11\lambda - 5;$$

the characteristic roots are $\lambda_1 = 5$, $\lambda_2 = 1$, $\lambda_3 = 1$; and the system of linear equations (7) is

$$(a) \qquad \begin{cases} (\lambda-2)x_1 - x_2 - x_3 = 0 \\ -2x_1 + (\lambda-3)x_2 - 2x_3 = 0 \\ -x_1 - x_2 + (\lambda-2)x_3 = 0 \end{cases}$$

When $\lambda = \lambda_1 = 5$, the system (*a*) reduces to $\begin{cases} x_1 + x_2 - 3x_3 = 0 \\ x_1 - x_3 = 0 \end{cases}$ having $x_1 = 1$, $x_2 = 2$, $x_3 = 1$ as a solution. Thus, associated with $\lambda_1 = 5$ is the one-dimensional vector space spanned by $\xi_1 = (1, 2, 1)$. When $\lambda = \lambda_2 = 1$, the system (*a*) reduces to $x_1 + x_2 + x_3 = 0$ having $x_1 = 1$, $x_2 = 0$, $x_3 = -1$ and $x_1 = 1$, $x_2 = -1$, $x_3 = 0$ as linearly independent solutions. Thus, associated with $\lambda_2 = 1$ is the two-dimensional vector space spanned by $\xi_2 = (1, 0, -1)$ and $\xi_3 = (1, -1, 0)$.

The matrix of Example 7 was considered at the beginning of this section. Examples 6 and 7, also Problem 6, suggest that associated with each simple characteristic root is a one-dimensional vector space and associated with each characteristic root of multiplicity $m > 1$ is an m-dimensional vector space. The first is true but (see Problem 7) the second is not. We shall not investigate this matter here (the interested reader may consult any book on matrices); we simply state

If λ is a characteristic root of multiplicity $m \geq 1$ of A, then associated with λ is a vector space whose dimension is *at least* 1 and *at most* m.

In Problem 8 we prove

Theorem III. If $\lambda_1, \xi_1; \lambda_2, \xi_2$ are distinct characteristic roots and associated characteristic vectors of an n-square matrix, then ξ_1 and ξ_2 are linearly independent.

We leave for the reader to prove

Theorem IV. The diagonal matrix $D = \text{diag}(\lambda_1, \lambda_2, \ldots, \lambda_n)$ has $\lambda_1, \lambda_2, \ldots, \lambda_n$ as characteristic roots and $\epsilon_1, \epsilon_2, \ldots, \epsilon_n$ as respective associated characteristic vectors.

SIMILAR MATRICES

Two n-square matrices A and B over $\mathcal{F}$ are called *similar* over $\mathcal{F}$ provided there exists a non-singular matrix P over $\mathcal{F}$ such that $B = PAP^{-1}$.

In Problems 9 and 10, Page 213, we prove

Theorem V. Two similar matrices have the same characteristic roots.

and

Theorem VI. If ζ_i is a characteristic vector associated with the characteristic root λ_i of $B = PAP^{-1}$, then $\xi_i = \zeta_i P$ is a characteristic vector associated with the same characteristic root λ_i of A.

Let A, an n-square matrix over $\mathcal{F}$ having $\lambda_1, \lambda_2, \ldots, \lambda_n$ as characteristic roots, be similar to $D = \text{diag}(\lambda_1, \lambda_2, \ldots, \lambda_n)$ and let P be a non-singular matrix such that $PAP^{-1} = D$. By Theorem IV, ϵ_i is a characteristic vector associated with the characteristic root λ_i of D and, by Theorem VI, $\xi_i = \epsilon_i P$ is a characteristic vector associated with the same characteristic root λ_i of A. Now $\epsilon_i P$ is the ith row vector of P; hence, A has n linearly independent characteristic vectors $\epsilon_i P$ which constitute a basis of $V_n(\mathcal{F})$.

Conversely, suppose that the set S of all characteristic vectors of an n-square matrix A spans $V_n(\mathcal{F})$. Then we can select a subset $\{\xi_1, \xi_2, \ldots, \xi_n\}$ of S which is a basis of $V_n(\mathcal{F})$. Since each ξ_i is a characteristic vector,

$$\xi_1 A = \lambda_1 \xi_1, \; \xi_2 A = \lambda_2 \xi_2, \; \ldots, \; \xi_n A = \lambda_n \xi_n$$

where $\lambda_1, \lambda_2, \ldots, \lambda_n$ are the characteristic roots of A. With $P = \begin{bmatrix} \xi_1 \\ \xi_2 \\ \cdot \\ \xi_n \end{bmatrix}$, we find

$$PA = \begin{bmatrix} \lambda_1 & 0 & \ldots & 0 \\ 0 & \lambda_2 & \ldots & 0 \\ \ldots & \ldots & \ldots & \ldots \\ 0 & 0 & \ldots & \lambda_n \end{bmatrix} P \quad \text{or}$$

$$PAP^{-1} = \text{diag}(\lambda_1, \lambda_2, \ldots, \lambda_n) = D$$

and A is similar to D. We have proved

Theorem VII. An n-square matrix A over $\mathcal{F}$, having $\lambda_1, \lambda_2, \ldots, \lambda_n$ as characteristic roots is similar to $D = \text{diag}(\lambda_1, \lambda_2, \ldots, \lambda_n)$ if and only if the set S of all characteristic vectors of A spans $V_n(\mathcal{F})$.

Example 8: For the matrix $A = \begin{bmatrix} 2 & 2 & 1 \\ 1 & 3 & 1 \\ 1 & 2 & 2 \end{bmatrix}$ of Example 7, take $P = \begin{bmatrix} \xi_1 \\ \xi_2 \\ \xi_3 \end{bmatrix} = \begin{bmatrix} 1 & 2 & 1 \\ 1 & 0 & -1 \\ 1 & -1 & 0 \end{bmatrix}$.

Then $P^{-1} = \begin{bmatrix} \frac{1}{4} & \frac{1}{4} & \frac{1}{2} \\ \frac{1}{4} & \frac{1}{4} & -\frac{1}{2} \\ \frac{1}{4} & -\frac{3}{4} & \frac{1}{2} \end{bmatrix}$ and

$$PAP^{-1} = \begin{bmatrix} 1 & 2 & 1 \\ 1 & 0 & -1 \\ 1 & -1 & 0 \end{bmatrix} \cdot \begin{bmatrix} 2 & 2 & 1 \\ 1 & 3 & 1 \\ 1 & 2 & 2 \end{bmatrix} \cdot \begin{bmatrix} \frac{1}{4} & \frac{1}{4} & \frac{1}{2} \\ \frac{1}{4} & \frac{1}{4} & -\frac{1}{2} \\ \frac{1}{4} & -\frac{3}{4} & \frac{1}{2} \end{bmatrix} = \begin{bmatrix} 5 & 0 & 0 \\ 0 & 1 & 0 \\ 0 & 0 & 1 \end{bmatrix} = \text{diag}(\lambda_1, \lambda_2, \lambda_3)$$

Not every n-square matrix is similar to a diagonal matrix. In Problem 7, Page 213, for instance, the condition in Theorem VII is not met since there the set of all characteristic vectors spans only a two-dimensional subspace of $V_3(R)$.

REAL SYMMETRIC MATRICES

An n-square matrix $A = [a_{ij}]$ over R is called *symmetric* provided $A^T = A$, i.e., $a_{ij} = a_{ji}$ for all i and j. The matrix A of Problem 6, Page 212, is symmetric; the matrices of Examples 6 and 7 are not.

In Problem 11, Page 214, we prove

Theorem VIII. The characteristic roots of a real symmetric matrix are real.

In Problem 12, Page 214, we prove

Theorem IX. If λ_1, ξ_1; λ_2, ξ_2 are distinct characteristic roots and associated characteristic vectors of an n-square real symmetric matrix, then ξ_1 and ξ_2 are mutually orthogonal.

Although the proof will not be given here, it can be shown that every real symmetric matrix A is similar to a diagonal matrix whose diagonal elements are the characteristic roots of A. Then A has n real characteristic roots and n real, mutually orthogonal associated characteristic vectors, say,

$$\lambda_1, \xi_1;\ \lambda_2, \xi_2;\ \ldots;\ \lambda_n, \xi_n$$

If, now, we define $\quad \eta_i = \xi_i/|\xi_i|, \quad (i = 1, 2, \ldots, n),$

A has n real characteristic roots and n real, mutually orthogonal associated characteristic unit vectors

$$\lambda_1, \eta_1;\ \lambda_2, \eta_2;\ \ldots;\ \lambda_n, \eta_n$$

Finally, with $S = \begin{bmatrix} \eta_1 \\ \eta_2 \\ \cdot \\ \cdot \\ \cdot \\ \eta_n \end{bmatrix}$, we have $SAS^{-1} = \text{diag}(\lambda_1, \lambda_2, \ldots, \lambda_n)$.

The vectors $\eta_1, \eta_2, \ldots, \eta_n$ constitute a basis of $V_n(R)$. Such bases, consisting of mutually orthogonal unit vectors, are called *normal orthogonal* or *orthonormal* bases.

ORTHOGONAL MATRICES

The matrix S, defined in the preceding section, is called an *orthogonal matrix*. We develop below a number of its unique properties.

1. Since the row vectors η_i of S are mutually orthogonal unit vectors, i.e., $\eta_i \cdot \eta_j = \begin{cases} 1, & \text{when } i = j \\ 0, & \text{when } i \neq j \end{cases}$, it follows readily that

$$S \cdot S^T = \begin{bmatrix} \eta_1 \\ \eta_2 \\ \cdot \\ \cdot \\ \cdot \\ \eta_n \end{bmatrix} \cdot [\eta_1, \eta_2, \ldots, \eta_n] = \begin{bmatrix} \eta_1 \cdot \eta_1 & \eta_1 \cdot \eta_2 & \cdots & \eta_1 \cdot \eta_n \\ \eta_2 \cdot \eta_1 & \eta_2 \cdot \eta_2 & \cdots & \eta_2 \cdot \eta_n \\ \cdots & \cdots & \cdots & \cdots \\ \eta_n \cdot \eta_1 & \eta_n \cdot \eta_2 & \cdots & \eta_n \cdot \eta_n \end{bmatrix} = I$$

   and $S^T = S^{-1}$.

2. Since $S \cdot S^T = S^T \cdot S = I$, the column vectors of S are also mutually orthogonal unit vectors. Thus,

   A real matrix H is *orthogonal* provided $H \cdot H^T = H^T \cdot H = I$.

3. Consider the *orthogonal transformation* $Y = XH$ of $V_n(R)$, whose matrix H is orthogonal, and denote by Y_1, Y_2 respectively the images of arbitrary $X_1, X_2 \in V_n(R)$. Since

$$Y_1 \cdot Y_2 = Y_1 Y_2^T = (X_1 H)(X_2 H)^T = X_1(H \cdot H^T)X_2^T = X_1 X_2^T = X_1 \cdot X_2,$$

   an orthogonal transformation preserves *inner* or *dot products* of vectors.

4. Since $|Y_1| = (Y_1 \cdot Y_1)^{1/2} = (X_1 \cdot X_1)^{1/2} = |X_1|$, an orthogonal transformation preserves *length* of vectors.

5. Since $\cos\theta' = \dfrac{Y_1 \cdot Y_2}{|Y_1| \cdot |Y_2|} = \dfrac{X_1 \cdot X_2}{|X_1| \cdot |X_2|} = \cos\theta$, where $0 \leq \theta, \theta' < \pi$, we have $\theta' = \theta$.

   In particular, if $X_1 \cdot X_2 = 0$ then $Y_1 \cdot Y_2 = 0$, that is, under an orthogonal transformation the image vectors of mutually orthogonal vectors are mutually orthogonal.

An orthogonal transformation $Y = XH$ (also, the orthogonal matrix H) is called *proper* or *improper* according as $|H| = 1$ or $|H| = -1$.

Example 9:

For the matrix A of Problem 6, we obtain

$$\eta_1 = \xi_1/|\xi_1| = (2/\sqrt{6}, -1/\sqrt{6}, -1/\sqrt{6}), \quad \eta_2 = (1/\sqrt{3}, 1/\sqrt{3}, 1/\sqrt{3}), \quad \eta_3 = (0, 1/\sqrt{2}, -1/\sqrt{2})$$

Then, with

$$S = \begin{bmatrix} \eta_1 \\ \eta_2 \\ \eta_3 \end{bmatrix} = \begin{bmatrix} 2/\sqrt{6} & -1/\sqrt{6} & -1/\sqrt{6} \\ 1/\sqrt{3} & 1/\sqrt{3} & 1/\sqrt{3} \\ 0 & 1/\sqrt{2} & -1/\sqrt{2} \end{bmatrix}, \qquad S^{-1} = S^T = \begin{bmatrix} 2/\sqrt{6} & 1/\sqrt{3} & 0 \\ -1/\sqrt{6} & 1/\sqrt{3} & 1/\sqrt{2} \\ -1/\sqrt{6} & 1/\sqrt{3} & -1/\sqrt{2} \end{bmatrix}$$

and we have $S \cdot A \cdot S^{-1} = \text{diag}(9, 3, -3)$.

The matrix S of Example 9 is improper, i.e., $|S| = -1$. It can be verified easily that had the negative of any one of the vectors η_1, η_2, η_3 been used in forming S, the matrix then would have been proper. Thus, for any real symmetric matrix A, a proper orthogonal matrix S can always be found such that $S \cdot A \cdot S^{-1}$ is a diagonal matrix whose diagonal elements are the characteristic roots of A.

CONICS AND QUADRIC SURFACES

One of the problems of analytic geometry of the plane and of ordinary space is the reduction of the equations of conics and quadric surfaces to standard forms which make apparent the nature of these curves and surfaces.

Relative to rectangular coordinate axes OX and OY, let the equation of a conic be given as

$$ax^2 + by^2 + 2cxy + 2dx + 2ey + f = 0 \tag{8}$$

and, relative to rectangular coordinate axes OX, OY and OZ, let the equation of a quadric surface be given as

$$ax^2 + by^2 + cz^2 + 2dxy + 2exz + 2fyz + 2gx + 2hy + 2kz + m = 0 \tag{9}$$

It will be recalled that the necessary reductions are effected by a rotation of the axes to remove all cross-product terms and a translation of the axes to remove, whenever possible, terms of degree less than two. It will be our purpose here to outline a standard procedure for handling both conics and quadric surfaces.

Consider the general conic of equation (8). Its terms of degree two, $ax^2 + by^2 + 2cxy$, may be written in matrix notation as

$$ax^2 + by^2 + 2cxy = (x, y) \cdot \begin{bmatrix} a & c \\ c & b \end{bmatrix} \cdot \begin{pmatrix} x \\ y \end{pmatrix} = X \cdot E \cdot X^T$$

where $X = (x, y)$. Now E is real and symmetric; hence there exists a proper orthogonal matrix $S = \begin{bmatrix} \eta_1 \\ \eta_2 \end{bmatrix}$ such that $S \cdot E \cdot S^{-1} = \text{diag}(\lambda_1, \lambda_2)$ where $\lambda_1, \eta_1; \lambda_2, \eta_2$ are the characteristic roots and associated characteristic unit vectors of E. Thus there exists a proper orthogonal transformation $X = (x', y')S = X'S$ such that

$$X'S \cdot E \cdot S^{-1}X'^T = X' \begin{bmatrix} \lambda_1 & 0 \\ 0 & \lambda_2 \end{bmatrix} X'^T = \lambda_1 x'^2 + \lambda_2 y'^2$$

in which the cross-product term has 0 as coefficient.

Let $S = \begin{bmatrix} \eta_1 \\ \eta_2 \end{bmatrix} = \begin{bmatrix} \eta_{11} & \eta_{12} \\ \eta_{21} & \eta_{22} \end{bmatrix}$; then

$$(x,y) = X = X'S = (x',y')\begin{bmatrix} \eta_{11} & \eta_{12} \\ \eta_{21} & \eta_{22} \end{bmatrix} = [\eta_{11}x' + \eta_{21}y',\ \eta_{12}x' + \eta_{22}y']$$

and we have
$$\begin{cases} x = \eta_{11}x' + \eta_{21}y' \\ y = \eta_{12}x' + \eta_{22}y' \end{cases}$$

This transformation reduces (8) to

$$\lambda_1 x'^2 + \lambda_2 y'^2 + 2(d\eta_{11} + e\eta_{12})x' + 2(d\eta_{21} + e\eta_{22})y' + f = 0 \tag{8'}$$

which is then to be reduced to standard form by a translation.

An alternate procedure for obtaining (8′) is as follows:

(i) Obtain the proper orthogonal matrix S.

(ii) Form the associate of (8)

$$ax^2 + by^2 + 2cxy + 2dxu + 2eyu + fu^2 = (x,y,u)\cdot\begin{bmatrix} a & c & d \\ c & b & e \\ d & e & f \end{bmatrix}\cdot\begin{pmatrix} x \\ y \\ u \end{pmatrix} = \bar{X}\cdot F\cdot \bar{X}^T = 0$$

where $\bar{X} = (x,y,u)$.

(iii) Use the transformation $\bar{X} = \bar{X}'\begin{bmatrix} S & 0 \\ 0 & 1 \end{bmatrix}$, where $\bar{X}' = (x',y',u')$, to obtain

$$\bar{X}'\begin{bmatrix} S & 0 \\ 0 & 1 \end{bmatrix}\cdot F\cdot\begin{bmatrix} S^T & 0 \\ 0 & 1 \end{bmatrix}\bar{X}'^T = 0$$

the associate of (8′).

Example 10: Identify the conic $5x^2 - 2\sqrt{3}\,xy + 7y^2 + 20\sqrt{3}\,x - 44y + 75 = 0$.

For the matrix $E = \begin{bmatrix} 5 & -\sqrt{3} \\ -\sqrt{3} & 7 \end{bmatrix}$ of terms of degree two, we find $4, (\frac{1}{2}\sqrt{3}, \frac{1}{2});\ 8, (-\frac{1}{2}, \frac{1}{2}\sqrt{3})$ as the characteristic roots and associated characteristic unit vectors and form $S = \begin{bmatrix} \frac{1}{2}\sqrt{3} & \frac{1}{2} \\ -\frac{1}{2} & \frac{1}{2}\sqrt{3} \end{bmatrix}$.

Then $\bar{X} = \bar{X}'\begin{bmatrix} S & 0 \\ 0 & 1 \end{bmatrix}$ reduces $\bar{X}\cdot F\cdot\bar{X}^T = \bar{X}\begin{bmatrix} 5 & -\sqrt{3} & 10\sqrt{3} \\ -\sqrt{3} & 7 & -22 \\ 10\sqrt{3} & -22 & 75 \end{bmatrix}\bar{X}^T = 0$ to

$$\bar{X}'\begin{bmatrix} \frac{1}{2}\sqrt{3} & \frac{1}{2} & 0 \\ -\frac{1}{2} & \frac{1}{2}\sqrt{3} & 0 \\ 0 & 0 & 1 \end{bmatrix}\begin{bmatrix} 5 & -\sqrt{3} & 10\sqrt{3} \\ -\sqrt{3} & 7 & -22 \\ 10\sqrt{3} & -22 & 75 \end{bmatrix}\begin{bmatrix} \frac{1}{2}\sqrt{3} & -\frac{1}{2} & 0 \\ \frac{1}{2} & \frac{1}{2}\sqrt{3} & 0 \\ 0 & 0 & 1 \end{bmatrix}\bar{X}'^T$$

$$= (x',y',u')\begin{bmatrix} 4 & 0 & 4 \\ 0 & 8 & -16\sqrt{3} \\ 4 & -16\sqrt{3} & 75 \end{bmatrix}\cdot\begin{pmatrix} x' \\ y' \\ u' \end{pmatrix} = 4x'^2 + 8y'^2 + 8x'u' - 32\sqrt{3}\,y'u' + 75u'^2 = 0$$

the associate of $4x'^2 + 8y'^2 + 8x' - 32\sqrt{3}\,y' + 75 = 4(x'+1)^2 + 8(y' - 2\sqrt{3})^2 - 25 = 0$.

Under the translation $\begin{cases} x'' = x' + 1 \\ y'' = y' - 2\sqrt{3} \end{cases}$ this becomes $4x''^2 + 8y''^2 = 25$. The conic is an ellipse.

Using $\begin{cases} x = \frac{1}{2}\sqrt{3}\,x' - \frac{1}{2}y' \\ y = \frac{1}{2}x' + \frac{1}{2}\sqrt{3}\,y' \end{cases}$ and $\begin{cases} x' = x'' - 1 \\ y' = y'' + 2\sqrt{3} \end{cases}$ it follows readily that, in terms of the original coordinate system, the new origin is at $O''(-3\sqrt{3}/2, 5/2)$ and the new axes $O''X''$ and $O''Y''$ have respectively the directions of the characteristic unit vectors $(\frac{1}{2}\sqrt{3}, \frac{1}{2})$ and $(-\frac{1}{2}, \frac{1}{2}\sqrt{3})$.

See Problem 14.

Solved Problems

1. Reduce $A(\lambda) = \begin{bmatrix} \lambda & 2\lambda+1 & \lambda+2 \\ \lambda^2+\lambda & 2\lambda^2+2\lambda & \lambda^2+2\lambda \\ \lambda^2-2\lambda & 2\lambda^2-2\lambda-1 & \lambda^2+\lambda-3 \end{bmatrix}$ to normal form.

The greatest common divisor of the elements of $A(\lambda)$ is 1; set $f_1(\lambda) = 1$. Now use $K_{21}(-2)$ followed by K_{12} and then proceed to clear the first row and first column to obtain

$$A(\lambda) \sim \begin{bmatrix} 1 & \lambda & \lambda+2 \\ 0 & \lambda^2+\lambda & \lambda^2+2\lambda \\ 2\lambda-1 & \lambda^2-2\lambda & \lambda^2+\lambda-3 \end{bmatrix} \sim \begin{bmatrix} 1 & 0 & 0 \\ 0 & \lambda^2+\lambda & \lambda^2+2\lambda \\ 2\lambda-1 & -\lambda^2-\lambda & -\lambda^2-2\lambda-1 \end{bmatrix}$$

$$\sim \begin{bmatrix} 1 & 0 & 0 \\ 0 & \lambda^2+\lambda & \lambda^2+2\lambda \\ 0 & -\lambda^2-\lambda & -\lambda^2-2\lambda-1 \end{bmatrix} = B(\lambda)$$

The greatest common divisor of the elements of the submatrix $\begin{bmatrix} \lambda^2+\lambda & \lambda^2+2\lambda \\ -\lambda^2-\lambda & -\lambda^2-2\lambda-1 \end{bmatrix}$ is 1; set $f_2(\lambda) = 1$. On $B(\lambda)$ use $H_{23}(1)$ and $K_{23}(-1)$ and then proceed to clear the second row and second column to obtain

$$A(\lambda) \sim \begin{bmatrix} 1 & 0 & 0 \\ 0 & 1 & -1 \\ 0 & \lambda+1 & -\lambda^2-2\lambda-1 \end{bmatrix} \sim \begin{bmatrix} 1 & 0 & 0 \\ 0 & 1 & 0 \\ 0 & \lambda+1 & -\lambda^2-\lambda \end{bmatrix}$$

$$\sim \begin{bmatrix} 1 & 0 & 0 \\ 0 & 1 & 0 \\ 0 & 0 & -\lambda^2-\lambda \end{bmatrix} \sim \begin{bmatrix} 1 & 0 & 0 \\ 0 & 1 & 0 \\ 0 & 0 & \lambda^2+\lambda \end{bmatrix} = N(\lambda)$$

the final step being necessary in order that $f_3(\lambda) = \lambda^2+\lambda$ be monic.

2. Reduce (a) $A(\lambda) = \begin{bmatrix} \lambda & 0 \\ 0 & \lambda+1 \end{bmatrix}$ and (b) $B(\lambda) = \begin{bmatrix} \lambda^3 & 0 & 0 \\ 0 & \lambda^2-\lambda & 0 \\ 0 & 0 & \lambda^2 \end{bmatrix}$ to normal form.

(a) The greatest common divisor of the elements of $A(\lambda)$ is 1. We obtain

$$A(\lambda) = \begin{bmatrix} \lambda & 0 \\ 0 & \lambda+1 \end{bmatrix} \sim \begin{bmatrix} \lambda & \lambda+1 \\ 0 & \lambda+1 \end{bmatrix} \sim \begin{bmatrix} -1 & \lambda+1 \\ -\lambda-1 & \lambda+1 \end{bmatrix}$$

$$\sim \begin{bmatrix} -1 & 0 \\ -\lambda-1 & -\lambda^2-\lambda \end{bmatrix} \sim \begin{bmatrix} -1 & 0 \\ 0 & -\lambda^2-\lambda \end{bmatrix} \sim \begin{bmatrix} 1 & 0 \\ 0 & \lambda^2+\lambda \end{bmatrix} = N(\lambda)$$

(b) The greatest common divisor of $B(\lambda)$ is λ. We obtain

$$B(\lambda) = \begin{bmatrix} \lambda^3 & 0 & 0 \\ 0 & \lambda^2-\lambda & 0 \\ 0 & 0 & \lambda^2 \end{bmatrix} \sim \begin{bmatrix} \lambda^3 & \lambda^2-\lambda & 0 \\ 0 & \lambda^2-\lambda & 0 \\ 0 & 0 & \lambda^2 \end{bmatrix} \sim \begin{bmatrix} \lambda^2 & \lambda^2-\lambda & 0 \\ -\lambda^3+\lambda^2 & \lambda^2-\lambda & 0 \\ 0 & 0 & \lambda^2 \end{bmatrix} \sim \begin{bmatrix} \lambda & \lambda^2-\lambda & 0 \\ -\lambda^3+\lambda & \lambda^2-\lambda & 0 \\ 0 & 0 & \lambda^2 \end{bmatrix}$$

$$\sim \begin{bmatrix} \lambda & \lambda^2-\lambda & 0 \\ -\lambda^3 & 0 & 0 \\ 0 & 0 & \lambda^2 \end{bmatrix} \sim \begin{bmatrix} \lambda & 0 & 0 \\ 0 & \lambda^4-\lambda^3 & 0 \\ 0 & 0 & \lambda^2 \end{bmatrix} \sim \begin{bmatrix} \lambda & 0 & 0 \\ 0 & \lambda^2 & 0 \\ 0 & 0 & \lambda^4-\lambda^3 \end{bmatrix} = N(\lambda)$$

3. Reduce $A(\lambda) = \begin{bmatrix} \lambda-2 & -1 & -1 \\ -2 & \lambda-3 & -2 \\ -1 & -1 & \lambda-2 \end{bmatrix}$ to normal form.

The greatest common divisor of the elements of $A(\lambda)$ is 1. We use K_{13} followed by $K_1(-1)$ and then proceed to clear the first row and column to obtain

$$A(\lambda) \sim \begin{bmatrix} 1 & -1 & \lambda-2 \\ 2 & \lambda-3 & -2 \\ 2-\lambda & -1 & -1 \end{bmatrix} \sim \begin{bmatrix} 1 & 0 & 0 \\ 0 & \lambda-1 & 2-2\lambda \\ 0 & 1-\lambda & \lambda^2-4\lambda+3 \end{bmatrix} = \begin{bmatrix} 1 & 0 \\ 0 & B(\lambda) \end{bmatrix}$$

The greatest common divisor of the elements of $B(\lambda)$ is $\lambda - 1$; then

$$A(\lambda) \sim \begin{bmatrix} 1 & 0 & 0 \\ 0 & \lambda-1 & 2-2\lambda \\ 0 & 1-\lambda & \lambda^2-4\lambda+3 \end{bmatrix} \sim \begin{bmatrix} 1 & 0 & 0 \\ 0 & \lambda-1 & 0 \\ 0 & 0 & \lambda^2-6\lambda+5 \end{bmatrix} = N(\lambda)$$

4. Write $A(\lambda) = \begin{bmatrix} \lambda+2 & \lambda+1 & \lambda+3 \\ \lambda & \lambda & -3\lambda^2+\lambda \\ \lambda^2+2\lambda & \lambda^2+\lambda & 3\lambda^2+5\lambda \end{bmatrix}$ as a polynomial in λ and compute $A(-2)$,

$A_R(C)$ and $A_L(C)$ when $C = \begin{bmatrix} 1 & 0 & 2 \\ -1 & -1 & -4 \\ -1 & 0 & -2 \end{bmatrix}$.

We obtain
$$A(\lambda) = \begin{bmatrix} 0 & 0 & 0 \\ 0 & 0 & -3 \\ 1 & 1 & 3 \end{bmatrix}\lambda^2 + \begin{bmatrix} 1 & 1 & 1 \\ 1 & 1 & 1 \\ 2 & 1 & 5 \end{bmatrix}\lambda + \begin{bmatrix} 2 & 1 & 3 \\ 0 & 0 & 0 \\ 0 & 0 & 0 \end{bmatrix}$$

and
$$A(-2) = 4\begin{bmatrix} 0 & 0 & 0 \\ 0 & 0 & -3 \\ 1 & 1 & 3 \end{bmatrix} - 2\begin{bmatrix} 1 & 1 & 1 \\ 1 & 1 & 1 \\ 2 & 1 & 5 \end{bmatrix} + \begin{bmatrix} 2 & 1 & 3 \\ 0 & 0 & 0 \\ 0 & 0 & 0 \end{bmatrix} = \begin{bmatrix} 0 & -1 & 1 \\ -2 & -2 & -14 \\ 0 & 2 & 2 \end{bmatrix}$$

Since $C^2 = \begin{bmatrix} -1 & 0 & -2 \\ 4 & 1 & 10 \\ 1 & 0 & 2 \end{bmatrix}$, we have

$$A_R(C) = \begin{bmatrix} 0 & 0 & 0 \\ 0 & 0 & -3 \\ 1 & 1 & 3 \end{bmatrix}\begin{bmatrix} -1 & 0 & -2 \\ 4 & 1 & 10 \\ 1 & 0 & 2 \end{bmatrix} + \begin{bmatrix} 1 & 1 & 1 \\ 1 & 1 & 1 \\ 2 & 1 & 5 \end{bmatrix}\begin{bmatrix} 1 & 0 & 2 \\ -1 & -1 & -4 \\ -1 & 0 & -2 \end{bmatrix} + \begin{bmatrix} 2 & 1 & 3 \\ 0 & 0 & 0 \\ 0 & 0 & 0 \end{bmatrix} = \begin{bmatrix} 1 & 0 & -1 \\ -4 & -1 & -10 \\ 2 & 0 & 4 \end{bmatrix}$$

$$A_L(C) = \begin{bmatrix} -1 & 0 & -2 \\ 4 & 1 & 10 \\ 1 & 0 & 2 \end{bmatrix}\begin{bmatrix} 0 & 0 & 0 \\ 0 & 0 & -3 \\ 1 & 1 & 3 \end{bmatrix} + \begin{bmatrix} 1 & 0 & 2 \\ -1 & -1 & -4 \\ -1 & 0 & -2 \end{bmatrix}\begin{bmatrix} 1 & 1 & 1 \\ 1 & 1 & 1 \\ 2 & 1 & 5 \end{bmatrix} + \begin{bmatrix} 2 & 1 & 3 \\ 0 & 0 & 0 \\ 0 & 0 & 0 \end{bmatrix} = \begin{bmatrix} 5 & 2 & 8 \\ 0 & 4 & 5 \\ -3 & -1 & -5 \end{bmatrix}$$

5. Given $A(\lambda) = \begin{bmatrix} \lambda^4+\lambda^3+3\lambda^2+\lambda & \lambda^4+\lambda^3+2\lambda^2+\lambda+1 \\ \lambda^3-2\lambda+1 & 2\lambda^3-3\lambda^2-2 \end{bmatrix}$ and $B(\lambda) = \begin{bmatrix} \lambda^2+1 & \lambda^2-\lambda \\ \lambda^2+\lambda & 2\lambda^2+1 \end{bmatrix}$,

find $Q_1(\lambda), R_1(\lambda); Q_2(\lambda), R_2(\lambda)$ such that

(a) $A(\lambda) = Q_1(\lambda)\cdot B(\lambda) + R_1(\lambda)$ and (b) $A(\lambda) = B(\lambda)\cdot Q_2(\lambda) + R_2(\lambda)$.

We have
$$A(\lambda) = \begin{bmatrix} 1 & 1 \\ 0 & 0 \end{bmatrix}\lambda^4 + \begin{bmatrix} 1 & 1 \\ 1 & 2 \end{bmatrix}\lambda^3 + \begin{bmatrix} 3 & 2 \\ 0 & -3 \end{bmatrix}\lambda^2 + \begin{bmatrix} 1 & 1 \\ -2 & 0 \end{bmatrix}\lambda + \begin{bmatrix} 0 & 1 \\ 1 & -2 \end{bmatrix}$$

$$B(\lambda) \;=\; \begin{bmatrix} 1 & 1 \\ 1 & 2 \end{bmatrix}\lambda^2 + \begin{bmatrix} 0 & -1 \\ 1 & 0 \end{bmatrix}\lambda + \begin{bmatrix} 1 & 0 \\ 0 & 1 \end{bmatrix}$$

$$B_2 \;=\; \begin{bmatrix} 1 & 1 \\ 1 & 2 \end{bmatrix} \quad \text{and} \quad B_2^{-1} \;=\; \begin{bmatrix} 2 & -1 \\ -1 & 1 \end{bmatrix}$$

(*a*) $$A(\lambda) - A_4B_2^{-1}\cdot\lambda^2\cdot B(\lambda) \;=\; \begin{bmatrix} 1 & 2 \\ 1 & 2 \end{bmatrix}\lambda^3 + \begin{bmatrix} 2 & 2 \\ 0 & -3 \end{bmatrix}\lambda^2 + \begin{bmatrix} 1 & 1 \\ -2 & 0 \end{bmatrix}\lambda + \begin{bmatrix} 0 & 1 \\ 1 & -2 \end{bmatrix} \;=\; C(\lambda)$$

$$C(\lambda) - C_3B_2^{-1}\cdot\lambda\cdot B(\lambda) \;=\; \begin{bmatrix} 1 & 2 \\ -1 & -3 \end{bmatrix}\lambda^2 + \begin{bmatrix} 1 & 0 \\ -2 & -1 \end{bmatrix}\lambda + \begin{bmatrix} 0 & 1 \\ 1 & -2 \end{bmatrix} \;=\; D(\lambda)$$

$$D(\lambda) - D_2B_2^{-1}\cdot B(\lambda) \;=\; 0 \;=\; R_1(\lambda)$$

and

$$Q_1(\lambda) \;=\; (A_4\lambda^2 + C_3\lambda + D_2)B_2^{-1} \;=\; \begin{bmatrix} \lambda^2 & \lambda+1 \\ 1 & \lambda-2 \end{bmatrix}$$

Here, $B(\lambda)$ is a right divisor of $A(\lambda)$.

(*b*) $$A(\lambda) - B(\lambda)\cdot B_2^{-1}A_4\lambda^2 \;=\; \begin{bmatrix} 0 & 0 \\ -1 & 0 \end{bmatrix}\lambda^3 + \begin{bmatrix} 1 & 0 \\ 1 & -2 \end{bmatrix}\lambda^2 + \begin{bmatrix} 1 & 1 \\ -2 & 0 \end{bmatrix}\lambda + \begin{bmatrix} 0 & 1 \\ 1 & -2 \end{bmatrix} \;=\; E(\lambda)$$

$$E(\lambda) - B(\lambda)\cdot B_2^{-1}E_3\lambda \;=\; \begin{bmatrix} 0 & 0 \\ 0 & -2 \end{bmatrix}\lambda^2 + \begin{bmatrix} 0 & 1 \\ -1 & 0 \end{bmatrix}\lambda + \begin{bmatrix} 0 & 1 \\ 1 & -2 \end{bmatrix} \;=\; F(\lambda)$$

$$F(\lambda) - B(\lambda)\cdot B_2^{-1}F_2 \;=\; \begin{bmatrix} 0 & -1 \\ -1 & -2 \end{bmatrix}\lambda + \begin{bmatrix} 0 & -1 \\ 1 & 0 \end{bmatrix} \;=\; \begin{bmatrix} 0 & -\lambda-1 \\ -\lambda+1 & -2\lambda \end{bmatrix} \;=\; R_2(\lambda)$$

and

$$Q_2(\lambda) \;=\; B_2^{-1}(A_4\lambda^2 + E_3\lambda + F_2) \;=\; \begin{bmatrix} 2\lambda^2+\lambda & 2\lambda^2+2 \\ -\lambda^2-\lambda & -\lambda^2-2 \end{bmatrix}$$

6. Find the characteristic roots and associated characteristic vectors of $A = \begin{bmatrix} 7 & -2 & -2 \\ -2 & 1 & 4 \\ -2 & 4 & 1 \end{bmatrix}$ over R.

The characteristic polynomial of A is $|\lambda I - A^T| \;=\; \begin{vmatrix} \lambda-7 & 2 & 2 \\ 2 & \lambda-1 & -4 \\ 2 & -4 & \lambda-1 \end{vmatrix} \;=\; \lambda^3 - 9\lambda^2 - 9\lambda + 81;$

the characteristic roots are $\lambda_1 = 9$, $\lambda_2 = 3$, $\lambda_3 = -3$; and the system of linear equations (*7*) is

$$(a) \qquad \begin{cases} (\lambda-7)x_1 + 2x_2 + 2x_3 = 0 \\ 2x_1 + (\lambda-1)x_2 - 4x_3 = 0 \\ 2x_1 - 4x_2 + (\lambda-1)x_3 = 0 \end{cases}$$

When $\lambda = \lambda_1 = 9$, (*a*) reduces to $\begin{cases} x_1 + 2x_2 = 0 \\ x_1 + 2x_3 = 0 \end{cases}$ having $x_1 = 2$, $x_2 = -1$, $x_3 = -1$ as a solution.
Thus, associated with $\lambda_1 = 9$ is the one-dimensional vector space spanned by $\xi_1 = (2, -1, -1)$.

When $\lambda = \lambda_2 = 3$, (*a*) reduces to $\begin{cases} x_1 - x_3 = 0 \\ x_2 - x_3 = 0 \end{cases}$ having $x_1 = 1$, $x_2 = 1$, $x_3 = 1$ as a solution.
Thus, associated with $\lambda_2 = 3$ is the one-dimensional vector space spanned by $\xi_2 = (1, 1, 1)$.

When $\lambda = \lambda_3 = -3$, (*a*) reduces to $\begin{cases} x_1 = 0 \\ x_2 + x_3 = 0 \end{cases}$ having $x_1 = 0$, $x_2 = 1$, $x_3 = -1$ as a solution.
Thus, associated with $\lambda_3 = -3$ is the one-dimensional vector space spanned by $\xi_3 = (0, 1, -1)$.

7. Find the characteristic roots and associated characteristic vectors of $A = \begin{bmatrix} 0 & -2 & -2 \\ -1 & 1 & 2 \\ -1 & -1 & 2 \end{bmatrix}$ over R.

The characteristic polynomial of A is $|\lambda I - A^T| = \begin{vmatrix} \lambda & 1 & 1 \\ 2 & \lambda-1 & 1 \\ 2 & -2 & \lambda-2 \end{vmatrix} = \lambda^3 - 3\lambda^2 + 4$; the characteristic roots are $\lambda_1 = -1$, $\lambda_2 = 2$, $\lambda_3 = 2$; and the system of linear equations (7) is

$$(a) \quad \begin{cases} \lambda x_1 + x_2 + x_3 = 0 \\ 2x_1 + (\lambda - 1)x_2 + x_3 = 0 \\ 2x_1 - 2x_2 + (\lambda - 2)x_3 = 0 \end{cases}$$

When $\lambda = \lambda_1 = -1$, the system (a) reduces to $\begin{cases} x_1 - x_2 = 0 \\ x_3 = 0 \end{cases}$ having $x_1 = 1$, $x_2 = 1$, $x_3 = 0$ as a solution. Thus, associated with $\lambda_1 = -1$ is the one-dimensional vector space spanned by $\xi_1 = (1, 1, 0)$.

When $\lambda = \lambda_2 = 2$, the system (a) reduces to $\begin{cases} 3x_1 + x_3 = 0 \\ x_1 - x_2 = 0 \end{cases}$ having $x_1 = 1$, $x_2 = 1$, $x_3 = -3$ as a solution. Thus, associated with $\lambda_2 = 2$ is the one-dimensional vector space spanned by $\xi_2 = (1, 1, -3)$.

Note that here a vector space of dimension one is associated with the double root $\lambda_2 = 2$ whereas in Example 7 a vector space of dimension two was associated with the double root.

8. Prove: If λ_1, ξ_1; λ_2, ξ_2 are distinct characteristic roots and associated characteristic vectors of A, then ξ_1 and ξ_2 are linearly independent.

Suppose, on the contrary, that ξ_1 and ξ_2 are linearly dependent; then there exist scalars a_1 and a_2, not both 0, such that

$$\text{(i)} \qquad a_1\xi_1 + a_2\xi_2 = 0$$

Multiplying (i) on the right by A and using $\xi_i A = \lambda_i \xi_i$, we have

$$\text{(ii)} \qquad a_1\xi_1 A + a_2\xi_2 A = a_1\lambda_1\xi_1 + a_2\lambda_2\xi_2 = 0$$

Now (i) and (ii) hold if and only if $\begin{vmatrix} 1 & 1 \\ \lambda_1 & \lambda_2 \end{vmatrix} = 0$. But then $\lambda_1 = \lambda_2$, a contradiction; hence, ξ_1 and ξ_2 are linearly independent.

9. Prove: Two similar matrices have the same characteristic roots.

Let A and $B = PAP^{-1}$ be the similar matrices; then

$$\lambda I - B = \lambda I - PAP^{-1} = P\lambda IP^{-1} - PAP^{-1} = P(\lambda I - A)P^{-1}$$

and

$$|\lambda I - B| = |P(\lambda I - A)P^{-1}| = |P| \cdot |\lambda I - A| \cdot |P^{-1}| = |\lambda I - A|$$

Now A and B, having the same characteristic polynomial, must have the same characteristic roots.

10. Prove: If ζ_i is a characteristic vector associated with the characteristic root λ_i of $B = PAP^{-1}$, then $\xi_i = \zeta_i P$ is a characteristic vector associated with the same characteristic root λ_i of A.

By hypothesis, $\zeta_i B = \lambda_i \zeta_i$ and $BP = (PAP^{-1})P = PA$. Then $\xi_i A = \zeta_i PA = \zeta_i BP = \lambda_i \zeta_i P = \lambda_i \xi_i$ and ξ_i is a characteristic vector associated with the characteristic root λ_i of A.

11. Prove: The characteristic roots of a real symmetric n-square matrix are real.

Let A be any real symmetric matrix and suppose $h+ik$ is a complex characteristic root. Now $(h+ik)I - A$ is singular as also is

$$B = [(h+ik)I-A]\cdot[(h-ik)I-A] = (h^2+k^2)I - 2hA + A^2 = (hI-A)^2 + k^2I$$

Since B is real and singular, there exists a real non-zero vector ξ such that $\xi B = 0$ and, hence,

$$\xi B\,\xi^T = \xi\{(hI-A)^2 + k^2I\}\xi^T = \{\xi(hI-A)\}\{(hI-A)^T\xi^T\} + k^2\xi\xi^T$$
$$= \eta\cdot\eta + k^2\xi\cdot\xi = 0$$

where $\eta = \xi(hI-A)$. Now $\eta\cdot\eta \geqq 0$ while, since ξ is real and non-zero, $\xi\cdot\xi > 0$. Hence, $k=0$ and A has only real characteristic roots.

12. Prove: If $\lambda_1, \xi_1;\ \lambda_2, \xi_2$ are distinct characteristic roots and associated characteristic vectors of an n-square real symmetric matrix A, then ξ_1 and ξ_2 are mutually orthogonal.

By hypothesis, $\xi_1A = \lambda_1\xi_1$ and $\xi_2A = \lambda_2\xi_2$. Then

$$\xi_1A\,\xi_2^T = \lambda_1\xi_1\,\xi_2^T \quad\text{and}\quad \xi_2A\,\xi_1^T = \lambda_2\xi_2\,\xi_1^T$$

and, taking transposes,

$$\xi_2A\,\xi_1^T = \lambda_1\xi_2\,\xi_1^T \quad\text{and}\quad \xi_1A\,\xi_2^T = \lambda_2\xi_1\,\xi_2^T$$

Now $\xi_1A\,\xi_2^T = \lambda_1\xi_1\xi_2^T = \lambda_2\xi_1\xi_2^T$ and $(\lambda_1-\lambda_2)\xi_1\xi_2^T = 0$. Since $\lambda_1-\lambda_2 \neq 0$, it follows that $\xi_1\xi_2^T = \xi_1\cdot\xi_2 = 0$ and ξ_1 and ξ_2 are mutually orthogonal.

13. (a) Show that $\alpha = (2,1,3)$ and $\beta = (1,1,-1)$ are mutually orthogonal.
(b) Find a vector γ which is orthogonal to each.
(c) Use α, β, γ to form an orthogonal matrix S such that $|S| = 1$.

(a) $\alpha\cdot\beta = 0$; α and β are mutually orthogonal.

(b) $\gamma = \alpha\times\beta = (-4,5,1)$.

(c) Take $\rho_1 = \alpha/|\alpha| = (2/\sqrt{14}, 1/\sqrt{14}, 3/\sqrt{14})$, $\rho_2 = \beta/|\beta| = (1/\sqrt{3}, 1/\sqrt{3}, -1/\sqrt{3})$ and $\rho_3 = \gamma/|\gamma| = (-4/\sqrt{42}, 5/\sqrt{42}, 1/\sqrt{42})$. Then

$$\begin{vmatrix}\rho_1\\ \rho_2\\ \rho_3\end{vmatrix} = 1 \quad\text{and}\quad S = \begin{bmatrix}\rho_1\\ \rho_2\\ \rho_3\end{bmatrix} = \begin{bmatrix}2/\sqrt{14} & 1/\sqrt{14} & 3/\sqrt{14}\\ 1/\sqrt{3} & 1/\sqrt{3} & -1/\sqrt{3}\\ -4/\sqrt{42} & 5/\sqrt{42} & 1/\sqrt{42}\end{bmatrix}$$

14. Identify the quadric surface

$$3x^2 - 2y^2 - z^2 - 4xy - 8xz - 12yz - 8x - 16y - 34z - 31 = 0$$

For the matrix $E = \begin{bmatrix}3 & -2 & -4\\ -2 & -2 & -6\\ -4 & -6 & -1\end{bmatrix}$ of terms of degree two, take

$$3, (2/3, -2/3, 1/3);\quad 6, (2/3, 1/3, -2/3);\quad -9, (1/3, 2/3, 2/3)$$

as characteristic roots and associated characteristic unit vectors. Then, with $S = \begin{bmatrix}2/3 & -2/3 & 1/3\\ 2/3 & 1/3 & -2/3\\ 1/3 & 2/3 & 2/3\end{bmatrix}$,

$$\bar{X} = (x,y,z,u) = (x',y',z',u')\begin{bmatrix}S & 0\\ 0 & 1\end{bmatrix} = \bar{X}'\begin{bmatrix}S & 0\\ 0 & 1\end{bmatrix}$$

reduces
$$\bar{X}\cdot F\cdot \bar{X}^T \;=\; \bar{X}\begin{bmatrix} 3 & -2 & -4 & -4 \\ -2 & -2 & -6 & -8 \\ -4 & -6 & -1 & -17 \\ -4 & -8 & -17 & -31 \end{bmatrix}\bar{X}^T$$

to
$$\bar{X}'\begin{bmatrix} 2/3 & -2/3 & 1/3 & 0 \\ 2/3 & 1/3 & -2/3 & 0 \\ 1/3 & 2/3 & 2/3 & 0 \\ 0 & 0 & 0 & 1 \end{bmatrix}\cdot\begin{bmatrix} 3 & -2 & -4 & -4 \\ -2 & -2 & -6 & -8 \\ -4 & -6 & -1 & -17 \\ -4 & -8 & -17 & -31 \end{bmatrix}\cdot\begin{bmatrix} 2/3 & 2/3 & 1/3 & 0 \\ -2/3 & 1/3 & 2/3 & 0 \\ 1/3 & -2/3 & 2/3 & 0 \\ 0 & 0 & 0 & 1 \end{bmatrix}\bar{X}'^T$$

$$= \quad (x',y',z',u')\begin{bmatrix} 3 & 0 & 0 & -3 \\ 0 & 6 & 0 & 6 \\ 0 & 0 & -9 & -18 \\ -3 & 6 & -18 & -31 \end{bmatrix}\begin{pmatrix} x' \\ y' \\ z' \\ u' \end{pmatrix}$$

$$= \quad 3x'^2 + 6y'^2 - 9z'^2 - 6x'u' + 12y'u' - 36z'u' - 31u'^2 \;=\; 0$$

the associate of
$$3x'^2 + 6y'^2 - 9z'^2 - 6x' + 12y' - 36z' - 31 \;=\; 3(x'-1)^2 + 6(y'+1)^2 - 9(z'+2)^2 - 4 \;=\; 0$$

Under the translation $\begin{cases} x'' = x' - 1 \\ y'' = y' + 1 \\ z'' = z' + 2 \end{cases}$, this becomes $3x''^2 + 6y''^2 - 9z''^2 = 4$; the surface is a hyperboloid of one sheet.

Using $(x,y,z) = (x',y',z')S$ and the equations of the translation, it follows readily that, in terms of the original coordinate system, the new origin has coordinates $(-2/3, -7/3, -1/3)$ and the new axes have the directions of the characteristic unit vectors $(2/3, -2/3, 1/3)$, $(2/3, 1/3, -2/3)$, $(1/3, 2/3, 2/3)$.

Supplementary Problems

15. Given $A(\lambda) = \begin{bmatrix} \lambda^2+\lambda & \lambda+1 \\ \lambda^2+2 & \lambda \end{bmatrix}$ and $B(\lambda) = \begin{bmatrix} \lambda^2 & \lambda^2+\lambda \\ \lambda-1 & \lambda \end{bmatrix}$, find

(a) $A(\lambda) + B(\lambda) = \begin{bmatrix} 2\lambda^2+\lambda & \lambda^2+2\lambda+1 \\ \lambda^2+\lambda+1 & 2\lambda \end{bmatrix}$

(b) $A(\lambda) - B(\lambda) = \begin{bmatrix} \lambda & -\lambda^2+1 \\ \lambda^2-\lambda+3 & 0 \end{bmatrix}$

(c) $A(\lambda)\cdot B(\lambda) = \begin{bmatrix} \lambda^4+\lambda^3+\lambda^2-1 & \lambda^4+2\lambda^3+2\lambda^2+\lambda \\ \lambda^4+3\lambda^2-\lambda & \lambda^4+\lambda^3+3\lambda^2+2\lambda \end{bmatrix}$

(d) $B(\lambda)\cdot A(\lambda) = \begin{bmatrix} 2\lambda^4+2\lambda^3+2\lambda^2+2\lambda & 2\lambda^3+2\lambda^2 \\ 2\lambda^3+\lambda & 2\lambda^2-1 \end{bmatrix}$

16. In each of the following find $Q_1(\lambda), R_1(\lambda)$; $Q_2(\lambda), R_2(\lambda)$, where $R_1(\lambda)$ and $R_2(\lambda)$ are either 0 or of degree less than that of $B(\lambda)$, such that $A(\lambda) = Q_1(\lambda)\cdot B(\lambda) + R_1(\lambda)$ and $A(\lambda) = B(\lambda)\cdot Q_2(\lambda) + R_2(\lambda)$.

(a) $A(\lambda) = \begin{bmatrix} \lambda^3-2\lambda^2+2\lambda-2 & \lambda^4+\lambda-1 \\ \lambda^4+\lambda^3+\lambda-2 & 2\lambda^2+\lambda-1 \end{bmatrix}$; $B(\lambda) = \begin{bmatrix} \lambda^2+1 & \lambda \\ 1 & \lambda^2+\lambda \end{bmatrix}$

(b) $A(\lambda) = \begin{bmatrix} 2\lambda^2+2\lambda & 2\lambda^2 \\ \lambda^2+\lambda+2 & \lambda^2+2\lambda-1 \end{bmatrix}$; $B(\lambda) = \begin{bmatrix} \lambda & \lambda \\ 1 & \lambda \end{bmatrix}$

(c) $A(\lambda) = \begin{bmatrix} \lambda^3-2\lambda^2+\lambda-3 & 4\lambda^2+2\lambda+3 \\ -2\lambda-2 & \lambda^3+4\lambda^2+6\lambda+3 \end{bmatrix}$; $B(\lambda) = \begin{bmatrix} \lambda-2 & 3 \\ -1 & \lambda+3 \end{bmatrix}$

(d) $A(\lambda) = \begin{bmatrix} \lambda^3-\lambda^2+\lambda+4 & 2\lambda^2+\lambda & \lambda^2+4\lambda-2 \\ 2\lambda & \lambda^3+\lambda^2+2\lambda-1 & 4\lambda^2-2\lambda+1 \\ 3\lambda^2-4\lambda-1 & \lambda^2+\lambda & \lambda^3-2\lambda^2+2\lambda+2 \end{bmatrix}$; $B(\lambda) = \begin{bmatrix} \lambda-1 & 1 & 0 \\ -1 & \lambda+1 & 3 \\ 2 & 0 & \lambda-2 \end{bmatrix}$

Ans. (a) $Q_1(\lambda) = \begin{bmatrix} \lambda-3 & \lambda^2-\lambda \\ \lambda^2+\lambda-1 & -\lambda+2 \end{bmatrix}$; $R_1(\lambda) = \begin{bmatrix} 2\lambda+1 & 4\lambda-1 \\ \lambda-3 & -1 \end{bmatrix}$

$Q_2(\lambda) = \begin{bmatrix} -2 & \lambda^2-1 \\ \lambda^2 & 1 \end{bmatrix}$; $R_2(\lambda) = \begin{bmatrix} 2\lambda & 0 \\ \lambda & 0 \end{bmatrix}$

(b) $Q_1(\lambda) = \begin{bmatrix} 2\lambda+2 & -2 \\ \lambda+1 & 1 \end{bmatrix}$; $R_1(\lambda) = \begin{bmatrix} 2 & 0 \\ 1 & -1 \end{bmatrix}$

$Q_2(\lambda) = \begin{bmatrix} \lambda+2 & \lambda-1 \\ \lambda & \lambda+1 \end{bmatrix}$; $R_2(\lambda) = 0$

(c) $Q_1(\lambda) = \begin{bmatrix} \lambda^2+2 & \lambda-1 \\ \lambda+1 & \lambda^2+\lambda \end{bmatrix}$; $R_1(\lambda) = 0$

$Q_2(\lambda) = \begin{bmatrix} \lambda^2-2 & \lambda+1 \\ \lambda-5 & \lambda^2+\lambda+4 \end{bmatrix}$; $R_2(\lambda) = \begin{bmatrix} 8 & -7 \\ 11 & -8 \end{bmatrix}$

(d) $Q_1(\lambda) = \begin{bmatrix} \lambda^2 & \lambda & \lambda+3 \\ \lambda+1 & \lambda^2+1 & \lambda \\ \lambda-2 & \lambda-1 & \lambda^2-1 \end{bmatrix}$; $R_1(\lambda) = \begin{bmatrix} -2 & 0 & 4 \\ 2 & -3 & -2 \\ -2 & 3 & 3 \end{bmatrix}$

$Q_2(\lambda) = \begin{bmatrix} \lambda^2 & \lambda+2 & \lambda+4 \\ \lambda-2 & \lambda^2 & \lambda-2 \\ \lambda-2 & \lambda+1 & \lambda^2 \end{bmatrix}$; $R_2(\lambda) = \begin{bmatrix} 6 & 2 & 4 \\ 8 & -2 & 7 \\ -5 & -2 & -6 \end{bmatrix}$

17. Reduce each of the following to its normal form:

(a) $\begin{bmatrix} \lambda & 2\lambda & 2\lambda-1 \\ \lambda^2+2\lambda & 2\lambda^2+3\lambda & 2\lambda^2+\lambda-1 \\ \lambda^2-2\lambda & 3\lambda^2-4\lambda & 4\lambda^2-5\lambda+2 \end{bmatrix}$

(b) $\begin{bmatrix} \lambda^2+1 & \lambda^3+\lambda & \lambda^3-\lambda^2 \\ \lambda-1 & \lambda^2+1 & -2\lambda \\ \lambda^2 & \lambda^3 & \lambda^3-\lambda^2+1 \end{bmatrix}$

(c) $\begin{bmatrix} -\lambda & \lambda+1 & \lambda+2 \\ -\lambda^2 & \lambda^2+\lambda-1 & \lambda^2+2\lambda-1 \\ \lambda^2+\lambda+1 & -\lambda^2-2\lambda-1 & -\lambda^2-3\lambda-2 \end{bmatrix}$

(d) $\begin{bmatrix} \lambda^2 & 0 & 0 \\ 0 & \lambda+1 & 0 \\ 0 & 0 & \lambda+1 \end{bmatrix}$

(e) $\begin{bmatrix} \lambda-1 & 3 & -2 \\ -2 & \lambda+1 & 0 \\ 3 & 1 & \lambda+2 \end{bmatrix}$

(f) $\begin{bmatrix} \lambda-2 & 2 & 3 \\ -3 & \lambda+3 & 4 \\ 2 & 0 & \lambda+1 \end{bmatrix}$

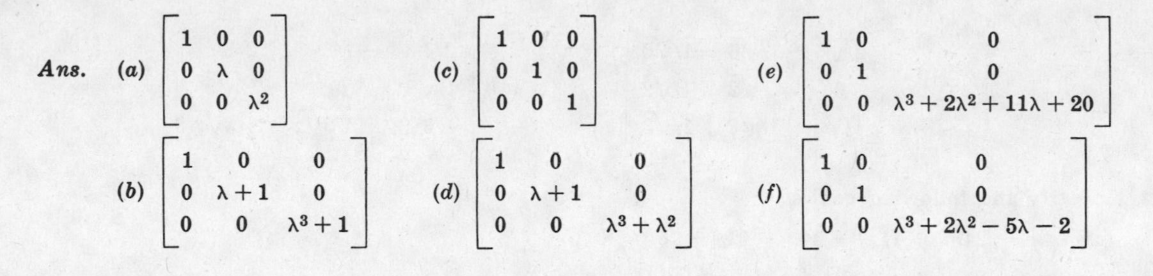

Ans. (a) $\begin{bmatrix} 1 & 0 & 0 \\ 0 & \lambda & 0 \\ 0 & 0 & \lambda^2 \end{bmatrix}$ (c) $\begin{bmatrix} 1 & 0 & 0 \\ 0 & 1 & 0 \\ 0 & 0 & 1 \end{bmatrix}$ (e) $\begin{bmatrix} 1 & 0 & 0 \\ 0 & 1 & 0 \\ 0 & 0 & \lambda^3+2\lambda^2+11\lambda+20 \end{bmatrix}$

(b) $\begin{bmatrix} 1 & 0 & 0 \\ 0 & \lambda+1 & 0 \\ 0 & 0 & \lambda^3+1 \end{bmatrix}$ (d) $\begin{bmatrix} 1 & 0 & 0 \\ 0 & \lambda+1 & 0 \\ 0 & 0 & \lambda^3+\lambda^2 \end{bmatrix}$ (f) $\begin{bmatrix} 1 & 0 & 0 \\ 0 & 1 & 0 \\ 0 & 0 & \lambda^3+2\lambda^2-5\lambda-2 \end{bmatrix}$

18. Find the characteristic roots and associated characteristic vectors of each of the following matrices A over R.

(a) $\begin{bmatrix} 1 & -2 \\ -5 & 4 \end{bmatrix}$ (c) $\begin{bmatrix} 3 & 1 \\ -1 & 1 \end{bmatrix}$ (e) $\begin{bmatrix} 2 & 0 & -1 \\ 0 & 2 & -2 \\ 1 & -1 & 2 \end{bmatrix}$ (g) $\begin{bmatrix} 1 & -1 & 0 \\ 1 & 2 & 1 \\ -2 & 1 & -1 \end{bmatrix}$

(b) $\begin{bmatrix} 2 & -1 \\ -8 & 4 \end{bmatrix}$ (d) $\begin{bmatrix} 1 & -2 \\ -2 & 4 \end{bmatrix}$ (f) $\begin{bmatrix} 3 & 2 & 4 \\ 2 & 0 & 2 \\ 4 & 2 & 3 \end{bmatrix}$

Ans. (a) $6, (k, -k);\ -1, (5k, 2k)$

(b) $0, (4k, k);\ 6, (2k, -k)$

(c) $2, (k, k)$

(d) $0, (2k, k);\ 5, (k, -2k)$

(e) $1, (k, -k, -k);\ 2, (2k, -k, 0);\ 3, (k, -k, k)$

(f) $-1, (k, 2l, -k-l);\ 8, (2k, k, 2k)$

(g) $1, (3k, 2k, k);\ 2, (k, 3k, k);\ -1, (k, 0, k)$

where $k \neq 0$ and $l \neq 0$.

19. For an n-square matrix A, show

(a) the constant term of its characteristic polynomial is $(-1)^n |A|$,

(b) the product of its characteristic roots is $|A|$,

(c) one or more of its characteristic roots is 0 if and only if $|A| = 0$.

20. Prove: The characteristic polynomial of an n-square matrix A is the product of the invariant factors of $\lambda I - A$.

Hint. From $P(\lambda) \cdot (\lambda I - A) \cdot S(\lambda) = \text{diag}\,(f_1(\lambda), f_2(\lambda), \ldots, f_n(\lambda))$ obtain

$$|P(\lambda)| \cdot |S(\lambda)| \cdot \phi(\lambda) = f_1(\lambda) \cdot f_2(\lambda) \ldots f_n(\lambda)$$

with $|P(\lambda)| \cdot |S(\lambda)| = 1$.

21. For each of the following real symmetric matrices A, find a proper orthogonal matrix S such that SAS^{-1} is diagonal.

(a) $\begin{bmatrix} 2 & 2 \\ 2 & -1 \end{bmatrix}$ (c) $\begin{bmatrix} 1 & -6 \\ -6 & -4 \end{bmatrix}$ (e) $\begin{bmatrix} 3 & -2 & -2 \\ -2 & 8 & -2 \\ -2 & -2 & 3 \end{bmatrix}$ (g) $\begin{bmatrix} 3 & -1 & -1 \\ -1 & 2 & 0 \\ -1 & 0 & 2 \end{bmatrix}$

(b) $\begin{bmatrix} 4 & -3 \\ -3 & -4 \end{bmatrix}$ (d) $\begin{bmatrix} 1 & -2 \\ -2 & 4 \end{bmatrix}$ (f) $\begin{bmatrix} 2 & -5 & 0 \\ -5 & -1 & 3 \\ 0 & 3 & -6 \end{bmatrix}$ (h) $\begin{bmatrix} 4 & -2 & 4 \\ -2 & 1 & -2 \\ 4 & -2 & 4 \end{bmatrix}$

Ans. (a) $\begin{bmatrix} 2/\sqrt{5} & 1/\sqrt{5} \\ -1/\sqrt{5} & 2/\sqrt{5} \end{bmatrix}$ (c) $\begin{bmatrix} 3/\sqrt{13} & -2/\sqrt{13} \\ 2/\sqrt{13} & 3/\sqrt{13} \end{bmatrix}$ (e) $\begin{bmatrix} 1/3\sqrt{2} & -4/3\sqrt{2} & 1/3\sqrt{2} \\ 1/\sqrt{2} & 0 & -1/\sqrt{2} \\ 2/3 & 1/3 & 2/3 \end{bmatrix}$

(b) $\begin{bmatrix} 3/\sqrt{10} & -1/\sqrt{10} \\ 1/\sqrt{10} & 3/\sqrt{10} \end{bmatrix}$ (d) $\begin{bmatrix} 1/\sqrt{5} & -2/\sqrt{5} \\ 2/\sqrt{5} & 1/\sqrt{5} \end{bmatrix}$ (f) $\begin{bmatrix} 5/\sqrt{42} & -4/\sqrt{42} & -1/\sqrt{42} \\ 1/\sqrt{14} & 2/\sqrt{14} & -3/\sqrt{14} \\ 1/\sqrt{3} & 1/\sqrt{3} & 1/\sqrt{3} \end{bmatrix}$

$$(g)\ \begin{bmatrix} 2/\sqrt{6} & -1/\sqrt{6} & -1/\sqrt{6} \\ 0 & 1/\sqrt{2} & -1/\sqrt{2} \\ 1/\sqrt{3} & 1/\sqrt{3} & 1/\sqrt{3} \end{bmatrix} \qquad (h)\ \begin{bmatrix} 2/3 & -1/3 & 2/3 \\ 1/\sqrt{2} & 0 & -1/\sqrt{2} \\ 1/3\sqrt{2} & 4/3\sqrt{2} & 1/3\sqrt{2} \end{bmatrix}$$

22. Identify the following conics:

(a) $4x^2 + 24xy + 11y^2 + 16x + 42y + 15 = 0$

(b) $9x^2 - 12xy + 4y^2 + 8\sqrt{13}\,x + 12\sqrt{13}\,y + 52 = 0$

(c) $3x^2 + 2xy + 3y^2 + 4\sqrt{2}\,x + 12\sqrt{2}\,y - 4 = 0$

Ans. (a) Hyperbola, (b) Parabola, (c) Ellipse

23. Identify the following quadric surfaces:

(a) $3x^2 + 8y^2 + 3z^2 - 4xy - 4xz - 4yz - 4x - 2y - 4z + 12 = 0$

(b) $2x^2 - y^2 - 6z^2 - 10xy + 6yz + 50x - 74y + 42z + 107 = 0$

(c) $4x^2 + y^2 + z^2 - 4xy - 4xz + 2yz - 6y + 6z + 2 = 0$

(d) $2xy + 2xz + 2yz + 1 = 0$

Ans. (a) Elliptic paraboloid, (b) Hyperboloid of two sheets, (c) Parabolic cylinder

24. Let A have $\lambda_1, \lambda_2, \ldots, \lambda_n$ as characteristic roots and let S be such that

$$S \cdot A \cdot S^{-1} = \operatorname{diag}(\lambda_1, \lambda_2, \ldots, \lambda_n) = D$$

Show that $\bar{S} A^T \bar{S}^{-1} = D$ when $\bar{S} = S^{-1}$. Thus any matrix A similar to a diagonal matrix is similar to its transpose A^T.

25. Prove: If Q is orthonormal, then $Q^T = Q^{-1}$.

26. Prove: Every real 2-square matrix A for which $|A| < 0$ is similar to a diagonal matrix.

27. Prove by direct substitution that $A = \begin{bmatrix} a & b \\ c & d \end{bmatrix}$ is a zero of its characteristic polynomial.

28. Under what conditions will the real matrix $A = \begin{bmatrix} a & b \\ c & d \end{bmatrix}$ have

(a) equal characteristic roots,

(b) the characteristic roots ± 1.

Chapter 16

Linear Algebras

LINEAR ALGEBRA

A set $\mathcal{L}$ having binary operations of addition and multiplication, together with scalar multiplication by elements of a field $\mathcal{F}$, is called a *linear algebra* over $\mathcal{F}$ provided

(i) Under addition and scalar multiplication, $\mathcal{L}$ is a vector space $\mathcal{L}(\mathcal{F})$ over $\mathcal{F}$.

(ii) Multiplication is associative.

(iii) Multiplication is both left and right distributive relative to addition.

(iv) $\mathcal{L}$ has a multiplicative identity (unity) element.

(v) $(k\alpha)\beta = \alpha(k\beta) = k(\alpha\cdot\beta)$ for all $\alpha, \beta \in \mathcal{L}$ and $k \in \mathcal{F}$.

Example 1: (*a*) The field C of complex numbers is a linear algebra of *dimension* (*order*) 2 over the field R of real numbers since (see Chapter 13) $C(R)$ is a vector space of dimension 2 and satisfies the postulates (ii)-(v).

(*b*) In general, if $\mathcal{L}$ is any field having $\mathcal{F}$ as a subfield, then $\mathcal{L}(\mathcal{F})$ is a linear algebra over $\mathcal{F}$.

Example 2: Clearly, the algebra of all linear transformations of the vector space $V_n(\mathcal{F})$ is a linear algebra of order n^2. Hence the isomorphic algebra $M_n(\mathcal{F})$ of all n-square matrices over $\mathcal{F}$ is also a linear algebra.

AN ISOMORPHISM

The linear algebra of Example 2 plays a role here similar to that of the symmetric group S_n in group theory. In Chapter 9 it was shown that every abstract group of order n is isomorphic to a subgroup of S_n. We shall now show that every linear algebra of order n over $\mathcal{F}$ is isomorphic to some subalgebra of $M_n(\mathcal{F})$.

Let $\mathcal{L}$ be a linear algebra of order n over $\mathcal{F}$ having $\{x_1, x_2, x_3, \ldots, x_n\}$ as basis. With each $\alpha \in \mathcal{L}$, associate the mapping

$$T_\alpha: \quad x\,T_\alpha = x\cdot\alpha, \quad x \in \mathcal{L}$$

By (iii), $$x\,T_\alpha + y\,T_\alpha = x\cdot\alpha + y\cdot\alpha = (x+y)\alpha = (x+y)\,T_\alpha$$

and by (v), $$(kx)\,T_\alpha = (kx)\alpha = k(x\cdot\alpha) = k(x\,T_\alpha)$$

for any $x, y \in \mathcal{L}$ and $k \in \mathcal{F}$. Hence T_α is a linear transformation of the vector space $\mathcal{L}(\mathcal{F})$. Moreover, the linear transformations T_α and T_β associated with the distinct elements α and β of $\mathcal{L}$ are distinct. For, if $\alpha \neq \beta$, then $u\cdot\alpha \neq u\cdot\beta$, where u is the unity of $\mathcal{L}$, implies $T_\alpha \neq T_\beta$.

Now, by (iii) and (v),

$$x\,T_\alpha + x\,T_\beta = x\cdot\alpha + x\cdot\beta = x(\alpha+\beta) = x\,T_{\alpha+\beta}$$

$$(x\,T_\alpha)\,T_\beta = (x\cdot\alpha)\beta = x(\alpha\cdot\beta) = x\,T_{\alpha\cdot\beta}$$

and $$(kx)\,T_\alpha = (kx)\alpha = k(x\cdot\alpha) = x\cdot k\alpha = x\,T_{k\alpha}$$

Thus the mapping $\alpha \to T_\alpha$ is an isomorphism of $\mathcal{L}$ onto a subalgebra of the algebra of all linear transformations of the vector space $\mathcal{L}(\mathcal{F})$. Since this, in turn, is isomorphic to a subalgebra of $M_n(\mathcal{F})$, we have proved

Theorem I. Every linear algebra of order n over $\mathcal{F}$ is isomorphic to a subalgebra of $M_n(\mathcal{F})$.

Example 3: Consider the linear algebra $Q[\sqrt[3]{2}]$ of order 3 with basis $\{1, \sqrt[3]{2}, \sqrt[3]{4}\}$. For any element $\alpha = a_1 1 + a_2\sqrt[3]{2} + a_3\sqrt[3]{4}$ of $Q[\sqrt[3]{2}]$, we have

$$\begin{aligned} 1\cdot\alpha &= a_1 1 + a_2\sqrt[3]{2} + a_3\sqrt[3]{4} \\ \sqrt[3]{2}\cdot\alpha &= 2a_3 1 + a_1\sqrt[3]{2} + a_2\sqrt[3]{4} \\ \sqrt[3]{4}\cdot\alpha &= 2a_2 1 + 2a_3\sqrt[3]{2} + a_1\sqrt[3]{4} \end{aligned}$$

Then the mapping

$$a_1 1 + a_2\sqrt[3]{2} + a_3\sqrt[3]{4} \to \begin{bmatrix} a_1 & a_2 & a_3 \\ 2a_3 & a_1 & a_2 \\ 2a_2 & 2a_3 & a_1 \end{bmatrix}$$

is an isomorphism of the linear algebra $Q[\sqrt[3]{2}]$ onto the algebra of all matrices of $M_n(Q)$ of the form $\begin{bmatrix} r & s & t \\ 2t & r & s \\ 2s & 2t & r \end{bmatrix}$.

See also Problem 1.

Solved Problems

1. Show that $\mathcal{L} = \{a_1\cdot 1 + a_2\alpha + a_3\beta:\ a_i \in R\}$ with multiplication defined such that 1 is the unity, $0 = 0\cdot 1 + 0\cdot\alpha + 0\cdot\beta$ is the additive identity, and

(a)

$\cdot$	α	β
α	α	β
β	0	0

and (b)

$\cdot$	α	β
α	α	0
β	0	0

are linear algebras over R.

We may simply verify in each case that the postulates (i)-(v) are satisfied. Instead, we prefer to show that in each case $\mathcal{L}$ is isomorphic to a certain subalgebra of $M_3(R)$.

(a) For any $\alpha = a_1\cdot 1 + a_2\alpha + a_3\beta$, we have

$$\begin{aligned} 1\cdot\alpha &= a_1 1 + a_2\alpha + a_3\beta \\ \alpha\cdot\alpha &= (a_1 + a_2)\alpha + a_3\beta \\ \beta\cdot\alpha &= a_1\beta \end{aligned}$$

Hence $\mathcal{L}$ is isomorphic to the algebra of all matrices $M_3(R)$ of the form $\begin{bmatrix} a_1 & a_2 & a_3 \\ 0 & a_1+a_2 & a_3 \\ 0 & 0 & a_1 \end{bmatrix}$ and is a linear algebra over R.

(b) For any $\alpha = a_1\cdot 1 + a_2\alpha + a_3\beta$, we have

$$\begin{aligned} 1\cdot\alpha &= a_1 1 + a_2\alpha + a_3\beta \\ \alpha\cdot\alpha &= (a_1 + a_2)\alpha \\ \beta\cdot\alpha &= a_1\beta \end{aligned}$$

Hence $\mathcal{L}$ is isomorphic to the algebra of all matrices $M_3(R)$ of the form $\begin{bmatrix} a_1 & a_2 & a_3 \\ 0 & a_1+a_2 & 0 \\ 0 & 0 & a_1 \end{bmatrix}$.

Supplementary Problems

2. Verify that each of the following, with addition and multiplication defined as on R, is a linear algebra over Q.

   (a) $Q[\sqrt{3}] = \{a1 + b\sqrt{3} : a,b \in Q\}$

   (b) $\mathcal{L} = \{a1 + b\sqrt{3} + c\sqrt{5} + d\sqrt{15} : a,b,c,d \in Q\}$

3. Show that the linear algebra $\mathcal{L} = Q[\sqrt{t}]$, where $t \in N$ is not a perfect square, is isomorphic to the algebra of all matrices of $M_2(Q)$ of the form $\begin{bmatrix} a & b \\ tb & a \end{bmatrix}$.

4. Show that the linear algebra C over R is isomorphic to the algebra of all matrices of $M_2(R)$ of the form $\begin{bmatrix} a & b \\ -b & a \end{bmatrix}$.

5. Show that each of the following is a linear algebra over R. Obtain the set of matrices isomorphic to each.

   (a) $\mathcal{L} = \{a1 + b\alpha + c\alpha^2 : a, b, c \in R\}$, where $G = \{\alpha, \alpha^2, \alpha^3 = 1\}$ is the cyclic group of order 3.

   (b) $\mathcal{L} = \{a_1 1 + a_2 x + a_3 y : a_i \in R\}$, with multiplication $\cdot$ defined so that 1 is the unity, $0 = 0 \cdot 1 + 0 \cdot x + 0 \cdot y$ is the additive identity, and

$\cdot$	x	y
x	1	y
y	y	0

   (c) $\mathcal{Q} = \{a_1 + a_2 i + a_3 j + a_4 k : a_i \in R\}$ with multiplication table

$\cdot$	i	j	k
i	-1	k	$-j$
j	$-k$	-1	i
k	j	$-i$	-1

Ans. (a) $\begin{bmatrix} a & b & c \\ c & a & b \\ b & c & a \end{bmatrix}$ (b) $\begin{bmatrix} a_1 & a_2 & a_3 \\ a_2 & a_1 & a_3 \\ 0 & 0 & (a_1 + a_2) \end{bmatrix}$ (c) $\begin{bmatrix} a_1 & a_2 & a_3 & a_4 \\ -a_2 & a_1 & -a_4 & a_3 \\ -a_3 & a_4 & a_1 & -a_2 \\ -a_4 & -a_3 & a_2 & a_1 \end{bmatrix}$

Chapter 17

Boolean Algebras

BOOLEAN ALGEBRA

A set $\mathcal{B}$, on which binary operations $\cup$ and $\cap$ are defined, is called a *Boolean algebra* provided the following postulates hold:

(i) $\cup$ and $\cap$ are commutative.

(ii) $\mathcal{B}$ contains an identity element 0 with respect to $\cup$ and an identity element 1 with respect to $\cap$.

(iii) Each operation is distributive with respect to the other, i.e., for all $a, b, c \in \mathcal{B}$

$$a \cup (b \cap c) = (a \cup b) \cap (a \cup c)$$

and

$$a \cap (b \cup c) = (a \cap b) \cup (a \cap c)$$

(iv) For each $a \in \mathcal{B}$ there exists an $a' \in \mathcal{B}$ such that

$$a \cup a' = 1 \quad \text{and} \quad a \cap a' = 0$$

The more familiar symbols $+$ and $\cdot$ are frequently used here instead of $\cup$ and $\cap$. We use the latter since if the empty set $\emptyset$ be now denoted by 0 and the universal set U be now denoted by 1, it is clear that the identities 1.9-1.9′, 1.4-1.4′, 1.10-1.10′, 1.7-1.7′, proved valid in Chapter I for the algebra of all subsets of a given set, are precisely the postulates (i)-(iv) for a Boolean algebra. Our first task then will be to prove, without recourse to subsets of a given set, that the identities 1.1, 1.2-1.2′, 1.5-1.5′, 1.6-1.6′, 1.8-1.8′, 1.11-1.11′ of Chapter 1 are also valid for any Boolean algebra, that is, that these identities are consequences of the postulates (i)-(iv). It will be noted that there is complete symmetry in the postulates with respect to the operations $\cup$ and $\cap$ and also in the identities of (iv). Hence there follows for any Boolean algebra the

Principle of Duality. Any theorem deducible from the postulates (i)-(iv) of a Boolean algebra remains valid when the operation symbols $\cup$ and $\cap$ and the identity elements 0 and 1 are interchanged throughout.

As a consequence of the duality principle it is necessary to prove only one of each pair of dual statements.

Example 1: Prove: For every $a \in \mathcal{B}$,

$$a \cup a = a \quad \text{and} \quad a \cap a = a \tag{1}$$

(See 1.6-1.6′, Chapter 1, Page 5.)

Using in turn (ii), (iv), (iii), (iv), (ii):

$$a \cup a = (a \cup a) \cap 1 = (a \cup a) \cap (a \cup a') = a \cup (a \cap a') = a \cup 0 = a$$

Example 2: Prove: For every $a \in \mathcal{B}$,

$$a \cup 1 = 1 \quad \text{and} \quad a \cap 0 = 0 \tag{2}$$

(See 1.5-1.5′, Chapter 1, Page 5.)

Using in turn (ii), (iv), (iii), (ii), (iv):

$$a \cap 0 = 0 \cup (a \cap 0) = (a \cap a') \cup (a \cap 0) = a \cap (a' \cup 0) = a \cap a' = 0$$

Example 3: Prove: For every $a, b \in \mathcal{B}$,

$$a \cup (a \cap b) = a \quad \text{and} \quad a \cap (a \cup b) = a \tag{3}$$

Using in turn (ii), (iii), (2), (ii):

$$a \cup (a \cap b) = (a \cap 1) \cup (a \cap b) = a \cap (1 \cup b) = a \cap 1 = a$$

See also Problems 1-4.

BOOLEAN FUNCTIONS

Let $\mathcal{B} = \{a, b, c, \ldots\}$ be a Boolean algebra. By a *constant* we shall mean any symbol, as 0 and 1, which represents a specified element of $\mathcal{B}$; by a *variable* we shall mean a symbol which represents an arbitrary element of $\mathcal{B}$. If in the expression $x' \cup (y \cap z)$ we replace $\cup$ by $+$ and $\cap$ by $\cdot$ to obtain $x' + y \cdot z$, it seems natural to call x' and $y \cap z$ *monomials* and the entire expression $x' \cup (y \cap z)$ a *polynomial*.

Any expression as $x \cup x'$, $a \cap b'$, $[a \cap (b \cup c')] \cup (a' \cap b' \cap c)$ consisting of combinations by $\cup$ and $\cap$ of a finite number of elements of a Boolean algebra $\mathcal{B}$ will be called a *Boolean function*. The number of variables in any function is the number of distinct letters appearing without regard to whether the letter is primed or unprimed. Thus, $x \cup x'$ is a function of one variable x while $a \cap b'$ is a function of two variables a and b.

In ordinary algebra every integral function of several variables can always be expressed as a polynomial (including 0) but cannot always be expressed as a product of linear factors. In Boolean algebra, on the contrary, Boolean functions can generally be expressed in polynomial form (including 0 and 1), i.e. a union of distinct intersections, and in factored form, i.e. an intersection of distinct unions.

Example 4:

Simplify:

(a) $(x \cap y) \cup [(x \cup y') \cap y]'$, (b) $[(x \cup y') \cap (x \cap y' \cap z)']'$, (c) $\{[(x' \cap y')' \cup z] \cap (x \cup z)\}'$

(a)
$$\begin{aligned}(x \cap y) \cup [(x \cup y') \cap y]' &= (x \cap y) \cup [(x \cup y')' \cup y'] = (x \cap y) \cup [(x' \cap y) \cup y'] \\ &= (x \cap y) \cup [(x' \cup y') \cap (y \cup y')] = (x \cap y) \cup [(x' \cup y') \cap 1] \\ &= (x \cap y) \cup (x' \cup y') = (x \cap y) \cup (x \cap y)' = 1\end{aligned}$$

(b)
$$\begin{aligned}[(x \cup y') \cap (x \cap y' \cap z)']' &= (x \cup y')' \cup [(x \cap y' \cap z)']' \\ &= (x' \cap y) \cup (x \cap y' \cap z), \text{ a union of intersections.}\end{aligned}$$

$$\begin{aligned}[(x \cup y') \cap (x \cap y' \cap z)']' &= (x' \cap y) \cup (x \cap y' \cap z) \\ &= (x' \cup x) \cap (x' \cup y') \cap (x' \cup z) \cap (x \cup y) \cap (y \cup y') \cap (y \cup z) \\ &= 1 \cap (x' \cup y') \cap (x' \cup z) \cap (x \cup y) \cap 1 \cap (y \cup z) \\ &= (x \cup y) \cap (y \cup z) \cap (x' \cup z) \cap (x' \cup y'), \text{ an intersection of unions.}\end{aligned}$$

(c)
$$\begin{aligned}\{[(x' \cap y')' \cup z] \cap (x \cup z)\}' &= [(x' \cap y')' \cup z]' \cup (x \cup z)' \\ &= (x' \cap y' \cap z') \cup (x' \cap z') = x' \cap z' \quad \text{(by Example 3)}\end{aligned}$$

See also Problem 5.

Since (see Problem 15, Page 234) there exists a Boolean algebra having only the elements 0 and 1, any identity may be verified by assigning in all possible ways the values 0 and 1 to the variables.

Example 5: To verify the proposed identity (see Example 4(a))

$$(x \cap y) \cup [(x \cup y') \cap y]' = 1$$

we form the following Table 17-1.

x	y	$a = x \cap y$	$x \cup y'$	$b = (x \cup y') \cap y$	$a \cup b'$
1	1	1	1	1	1
1	0	0	1	0	1
0	1	0	0	0	1
0	0	0	1	0	1

Table 17-1

NORMAL FORMS

The Boolean function in three variables of Example 4(*b*) when expressed as a union of intersections $(x' \cap y) \cup (x \cap y' \cap z)$ contains one term in which only two of the variables are present. In the next section we shall see that there is good reason at times to replace this expression by a less simple one in which each term present involves all of the variables. Since the variable z is missing in the first term of the above expression, we obtain the required form, called the *canonical form* or the *disjunctive normal form* of the given function, as follows

$$\begin{aligned}(x' \cap y) \cup (x \cap y' \cap z) &= (x' \cap y \cap 1) \cup (x \cap y' \cap z) \\ &= [(x' \cap y) \cap (z \cup z')] \cup (x \cap y' \cap z) \\ &= (x' \cap y \cap z) \cup (x' \cap y \cap z') \cup (x \cap y' \cap z)\end{aligned}$$

See also Problem 6.

It is easy to show that the canonical form of a Boolean function in three variables can contain at most 2^3 distinct terms. For, if x, y, z are the variables, a term is obtained by selecting x or x', y or y', z or z' and forming their intersection. In general, the canonical form of a Boolean function in n variables can contain at most 2^n distinct terms. The canonical form containing all of these 2^n terms is called the complete *canonical form* or *complete disjunctive normal form in n variables.*

The complete canonical form in n variables is identically 1. This is shown for the case $n = 3$ in Problem 7, Page 231, while the general case can be proved by induction. It follows immediately that the complement F' of a Boolean function F expressed in canonical form is the union of all terms of the complete canonical form which do not appear in the canonical form of F. For example, if $F = (x \cap y) \cup (x' \cap y) \cup (x' \cap y')$, then $F' = (x \cap y')$.

In Problems 8 and 9, we prove

Theorem I. If, in the complete canonical form in n variables, each variable is assigned arbitrarily the value 0 or 1, then just one term will have the value 1 while all other terms will have the value 0.

and

Theorem II. Two Boolean functions are equal if and only if their respective canonical forms are identical, i.e., consist of the same terms.

The Boolean function in three variables of Example 4(*b*), when expressed as an intersection of unions in which each union contains all of the variables, is

$$\begin{aligned}&(x \cup y) \cap (y \cup z) \cap (x' \cup z) \cap (x' \cup y') \\ &= [(x \cup y) \cup (z \cap z')] \cap [(y \cup z) \cup (x \cap x')] \cap [(x' \cup z) \cup (y \cap y')] \cap [(x' \cup y') \cup (z \cap z')] \\ &= (x \cup y \cup z) \cap (x \cup y \cup z') \cap (x' \cup y \cup z) \cap (x' \cup y' \cup z) \cap (x' \cup y' \cup z')\end{aligned}$$

This expression is called the *dual canonical form* or the *conjunctive normal form* of the function. Note that it is *not* the dual of the canonical form of that function.

The dual of each statement concerning the canonical form of a Boolean function is a valid statement concerning the dual canonical form of that function. (Note that the dual of *term* is *factor*.) The dual canonical form of a Boolean function in n variables can contain at most 2^n distinct factors. The dual canonical form containing all of these factors is called the *complete dual canonical form* or the *complete conjunctive normal form in n variables*; its value is identically 0. The complement F' of a Boolean function F expressed in dual canonical form is the intersection of all factors of the complete dual canonical form which do not appear in the dual canonical form of F. Also, we have

Theorem I′. If, in the complete dual canonical form in n variables, each variable is assigned arbitrarily the value 0 or 1, then just one factor will have the value 0 while all other factors will have the value 1.

and

Theorem II′. Two Boolean functions are equal if and only if their respective dual canonical forms are identical, i.e., consist of the same factors.

In the next section we will use these theorems to determine the Boolean function when its values for all possible assignments of the values 0 and 1 to the variables are given.

CHANGING THE FORM OF A BOOLEAN FUNCTION

Denote by $F(x,y,z)$ the Boolean function whose values for all possible assignments of 0 or 1 to the variables is given by Table 17-2.

x	y	z	$F(x,y,z)$
1	1	1	1
1	1	0	0
1	0	1	1
0	1	1	0
1	0	0	0
0	1	0	0
0	0	1	1
0	0	0	1

Table 17-2

The proof of Theorem I suggests that the terms appearing in the canonical form of $F(x,y,z)$ are precisely those of the complete canonical form in three variables which have the value 1 whenever $F(x,y,z)=1$. For example, the first row of the table establishes $x \cap y \cap z$ as a term while the third row yields $x \cap y' \cap z$ as another. Thus,

$$\begin{aligned} F(x,y,z) &= (x \cap y \cap z) \cup (x \cap y' \cap z) \cup (x' \cap y' \cap z) \cup (x' \cap y' \cap z') \\ &= (x \cap z) \cup (x' \cap y') \end{aligned}$$

Similarly, the factors appearing in the dual canonical form of $F(x,y,z)$ are precisely those of the complete dual canonical form which have the value 0 whenever $F(x,y,z)=0$. We have

$$\begin{aligned} F(x,y,z) &= (x' \cup y' \cup z) \cap (x \cup y' \cup z') \cap (x' \cup y \cup z) \cap (x \cup y' \cup z) \\ &= (x' \cup z) \cap (x \cup y') \end{aligned}$$

See Problem 10.

If a Boolean function F is given in either the canonical or dual canonical form, the change to the other form may be readily made by using successively the two rules for finding the complement. The order in their use is immaterial; however, at times one order will require less computing than the other.

Example 6: Find the dual canonical form for

$$F = (x \cap y \cap z') \cup (x \cap y' \cap z) \cup (x \cap y' \cap z') \cup (x' \cap y \cap z) \cup (x' \cap y' \cap z')$$

Here $$F' = (x \cap y \cap z) \cup (x' \cap y' \cap z) \cup (x' \cap y \cap z')$$

(the union of all terms of the complete canonical form not appearing in F) and

$$F = (F')' = (x \cup y \cup z') \cap (x \cup y' \cup z) \cap (x' \cup y' \cup z') \quad \text{(by Problem 4)}$$

See also Problem 11.

The procedure for changing from canonical form to dual canonical form and vice versa may also be used advantageously in simplifying certain Boolean functions.

Example 7: Simplify $F = [(y \cap z') \cup (y' \cap z)] \cap [(x' \cap y) \cup (x' \cap z) \cup (x \cap y' \cap z')]$

Set $F_1 = (y \cap z') \cup (y' \cap z)$ and $F_2 = (x' \cap y) \cup (x' \cap z) \cup (x \cap y' \cap z')$.

Then $F_1 = (x \cap y \cap z') \cup (x' \cap y \cap z') \cup (x \cap y' \cap z) \cup (x' \cap y' \cap z)$

$F_1' = (x' \cap y \cap z) \cup (x' \cap y' \cap z') \cup (x \cap y \cap z) \cup (x \cap y' \cap z')$

and $F_1 = (x \cup y' \cup z') \cap (x \cup y \cup z) \cap (x' \cup y' \cup z') \cap (x' \cup y \cup z)$

Also $F_2 = (x' \cap y \cap z) \cup (x' \cap y \cap z') \cup (x' \cap y' \cap z) \cup (x \cap y' \cap z')$

$F_2' = (x' \cap y' \cap z') \cup (x \cap y \cap z) \cup (x \cap y' \cap z) \cup (x \cap y \cap z')$

and $F_2 = (x \cup y \cup z) \cap (x' \cup y' \cup z') \cap (x' \cup y \cup z') \cap (x' \cup y' \cup z)$

Then
$$\begin{aligned} F &= F_1 \cap F_2 \\ &= (x \cup y \cup z) \cap (x \cup y' \cup z') \cap (x' \cup y \cup z) \cap (x' \cup y \cup z') \\ &\qquad \cap (x' \cup y' \cup z) \cap (x' \cup y' \cup z') \end{aligned}$$

Now $F' = (x \cup y' \cup z) \cap (x \cup y \cup z')$

and $F = (x' \cap y \cap z') \cup (x' \cap y' \cap z) = x' \cap [(y \cap z') \cup (y' \cap z)]$

ORDER RELATION IN A BOOLEAN ALGEBRA

Let $U = \{a, b, c\}$ and $S = \{\emptyset, A, B, C, D, E, F, U\}$ where $A = \{a\}$, $B = \{b\}$, $C = \{c\}$, $D = \{a, b\}$, $E = \{a, c\}$, $F = \{b, c\}$. The relation $\subseteq$, defined in Chapter 1, when applied to S satisfies the following laws:

For any $X, Y, Z \in S$,

(*a*) $X \subseteq X$

(*b*) If $X \subseteq Y$ and $Y \subseteq X$, then $X = Y$.

(*c*) If $X \subseteq Y$ and $Y \subseteq Z$, then $X \subseteq Z$.

(*d*) If $X \subseteq Y$ and $X \subseteq Z$, then $X \subseteq (Y \cap Z)$.

(*e*) If $X \subseteq Y$, then $X \subseteq (Y \cup Z)$.

(*f*) $X \subseteq Y$ if and only if $Y' \subseteq X'$.

(*g*) $X \subseteq Y$ if and only if $X \cup Y = Y$ or the equivalent $X \cap Y' = \emptyset$.

The first three laws assure (see Chapter 2) that $\subseteq$ effects a partial ordering in S illustrated by

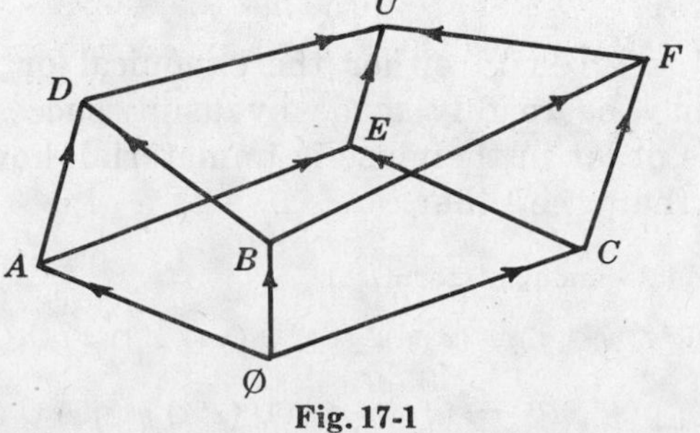

Fig. 17-1

We shall now define the relation $\subseteq$ (read, "under") in any Boolean algebra $\mathcal{B}$ by

$$a \subseteq b \quad \text{if and only if} \quad a \cup b = b \quad \text{or the equivalent} \quad a \cap b' = 0$$

for every $a, b \in \mathcal{B}$. (Note that this is merely a restatement of (g) in terms of the elements of $\mathcal{B}$.) There follows readily

(a_1) $a \subseteq a$

(b_1) If $a \subseteq b$ and $b \subseteq a$, then $a = b$.

(c_1) If $a \subseteq b$ and $b \subseteq c$, then $a \subseteq c$.

so that $\subseteq$ defines a partial order in $\mathcal{B}$. We leave for the reader to prove

(d_1) If $a \subseteq b$ and $a \subseteq c$, then $a \subseteq (b \cap c)$.

(e_1) If $a \subseteq b$, then $a \subseteq (b \cup c)$ for any $c \in \mathcal{B}$.

(f_1) $a \subseteq b$ if and only if $b' \subseteq a'$.

In Problem 12, we prove

Theorem III. For every $a, b \in \mathcal{B}$, $a \cup b$ is the least upper bound and $a \cap b$ is the greatest lower bound of a and b.

There follows easily

Theorem IV. $0 \subseteq b \subseteq 1$, for every $b \in \mathcal{B}$.

ALGEBRA OF ELECTRICAL NETWORKS

The algebra of electrical networks is an interesting and highly important example of the Boolean algebra (see Problem 15) of the two elements 0 and 1. The discussion here will be limited to the simplest kind of networks, that is, a network consisting only of switches. The simplest such network consists of a wire containing a single switch r:

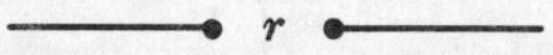

r

When the switch is closed so that current flows through the wire, we assign the value 1 to r; when the switch is open so that no current flows through the wire, we assign the value 0 to r. Also, we shall assign the value 1 or 0 to any network according as current does or does not flow through it. In this simple case, the network has value 1 if and only if r has value 1 and the network has value 0 if and only if r has value 0.

Consider now a network consisting of two switches r and s. When connected in *series*:

r s

Fig. 17-2

it is clear that the network has value 1 if and only if both r and s have value 1, while the network has value 0 in all other assignments of 0 or 1 to r and s. Hence this network can be represented by the function $F = F(r, s)$ which satisfies Table 17-3. We find easily $F = r \cap s$. When connected in *parallel*:

r	s	F
1	1	1
1	0	0
0	1	0
0	0	0

Table 17-3

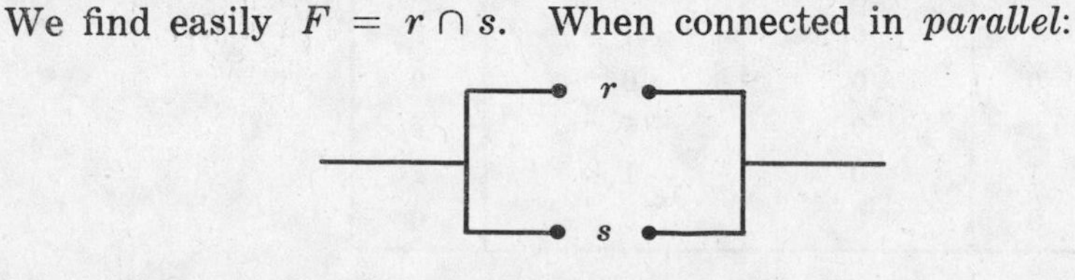

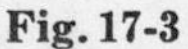

Fig. 17-3

it is clear that the network has value 1 if and only if at least one of r and s has value 1, and the network has value 0 if and only if both r and s have value 0. This network can be represented by the function $F = F(r, s)$ which satisfies Table 17-4. We find easily $F = r \cup s$. For the various networks consisting of three switches, see Problem 13.

r	s	F
1	1	1
1	0	1
0	1	1
0	0	0

Table 17-4

Using more switches, various networks of a more complicated nature may be devised; for example,

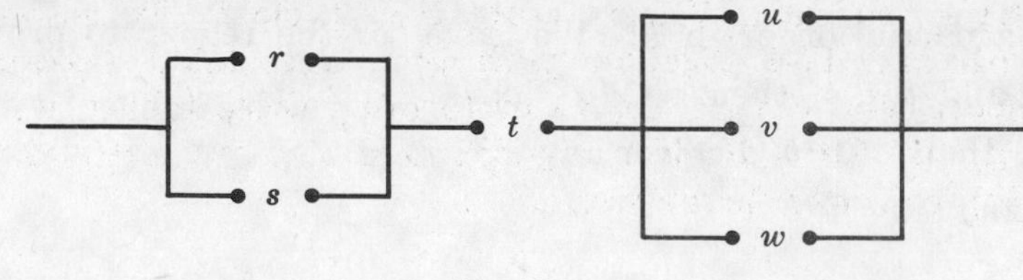

Fig. 17-4

The corresponding function for this network consists of the intersection of three factors:

$$(r \cup s) \cap t \cap (u \cup v \cup w)$$

So far all switches in a network have been tacitly assumed to act independently of one another. Two or more switches may, however, be connected so that (*a*) they open and close simultaneously or (*b*) the closing (opening) of one switch will open (close) all of the others. In case (*a*), we shall denote all switches by the same letter; in case (*b*), we shall denote some one of the switches by, say, r and all others by r'. In this case, any primed letter has value 0 when the unprimed letter has value 1 and vice versa. Thus the network

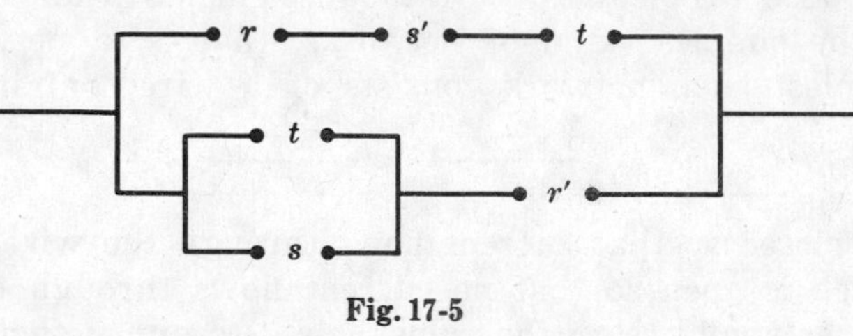

Fig. 17-5

consists of three pairs of switches: one pair, each of which is denoted by t, open and close simultaneously and two other pairs, denoted by r, r' and s, s', such that in each pair the closing of either switch opens the other. The corresponding function is basically a union of two terms each involving the three variables. For the upper wire we have $r \cap s' \cap t$ and for the lower $(t \cup s) \cap r'$. Thus, the function corresponding to the network is

$$F = (r \cap s' \cap t) \cup [(t \cup s) \cap r']$$

and the table giving the values (closure properties) of the function is

r	s	t	$r \cap s' \cap t$	$(t \cup s) \cap r'$	F
1	1	1	0	0	0
1	1	0	0	0	0
1	0	1	1	0	1
0	1	1	0	1	1
1	0	0	0	0	0
0	1	0	0	1	1
0	0	1	0	1	1
0	0	0	0	0	0

Table 17-5

It is clear that current will flow through the network only in the following cases: (1) r and t are closed, s is open; (2) s and t are closed, r is open; (3) s is closed, r and t are open; (4) t is closed, r and s are open.

In further analysis of series-parallel networks it will be helpful to know that the algebra of such networks is truly a Boolean algebra. In terms of a given network, the problem is this: Suppose F is the (switching) function associated with the network and suppose that by means of the laws of Boolean algebra this function is changed in form to G associated with a different network. Are the two networks interchangeable; in other words, will they have the same closure properties (table)? To settle the matter, first consider Tables 17-3 and 17-4 together with their associated networks and functions $r \cap s$ and $r \cup s$ respectively. In the course of forming these tables, we have verified that the postulates (i), (ii), (iv) for a Boolean algebra hold also for network algebra. For the case of postulate (iii), consider the networks

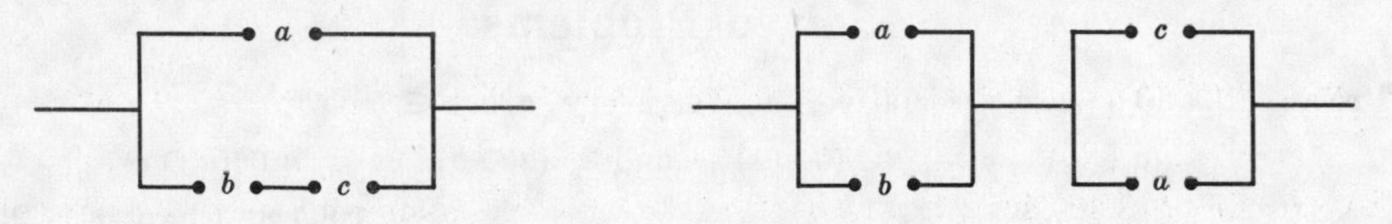

Fig. 17-6

corresponding to the Boolean identity $a \cup (b \cap c) = (a \cup b) \cap (a \cup c)$. It is then clear from the table of closure properties

a	b	c	$a \cup (b \cap c)$	$(a \cup b) \cap (a \cup c)$
1	1	1	1	1
1	1	0	1	1
1	0	1	1	1
0	1	1	1	1
1	0	0	1	1
0	1	0	0	0
0	0	1	0	0
0	0	0	0	0

Table 17-6

that the networks are interchangeable.

We leave for the reader to consider the case of the Boolean identity

$$a \cap (b \cup c) = (a \cap b) \cup (a \cap c)$$

and conclude that network algebra *is* a Boolean algebra.

SIMPLIFICATION OF NETWORKS

Suppose now that the first three and last columns of Table 17-5 are given and we are asked to devise a network having the given closure properties. Using the rows in which $F = 1$, we obtain

$$F = (r \cap s' \cap t) \cup (r' \cap s \cap t) \cup (r' \cap s \cap t') \cup (r' \cap s' \cap t) = (r' \cap s) \cup (s' \cap t)$$

Since this is not the function (network) from which the table was originally computed, the network of Fig. 17-5 is unnecessarily complex and can be replaced by the simpler network, shown in Fig. 17-7.

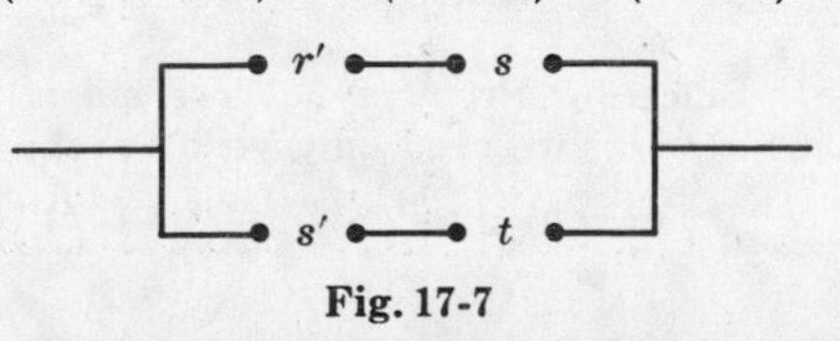

Fig. 17-7

Note. The fact that one of the switches of Fig. 17-7 is denoted by r' when there is no switch denoted by r has no significance here. If the reader has any doubt about this, let him interchange r and r' in Fig. 17-5 and obtain $F = (r \cap s) \cup (s' \cap t)$ with diagram

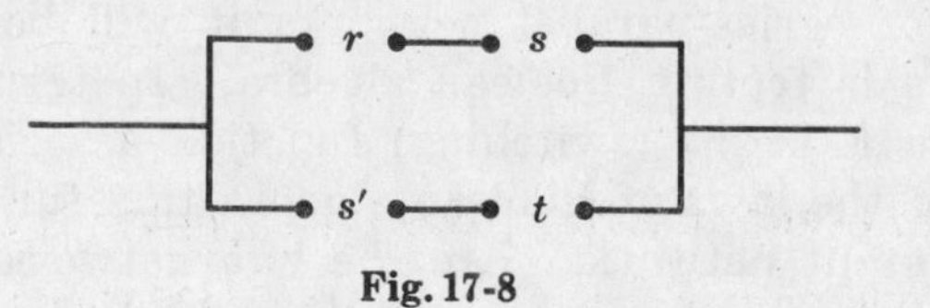

Fig. 17-8

See Problem 14.

Solved Problems

1. Prove: $\cup$ and $\cap$ are associative, i.e., for every $a, b, c \in \mathcal{B}$

$$(a \cup b) \cup c = a \cup (b \cup c) \quad \text{and} \quad (a \cap b) \cap c = a \cap (b \cap c) \tag{4}$$

(See 1.8-1.8′, Chapter 1, Page 5.)

Let $x = (a \cap b) \cap c$ and $y = a \cap (b \cap c)$. We are to prove $x = y$. Now, using (iii) and (3),

$$\begin{aligned} a \cup x &= a \cup [(a \cap b) \cap c] = [a \cup (a \cap b)] \cap (a \cup c) \\ &= a \cap (a \cup c) = a = a \cup [a \cap (b \cap c)] = a \cup y \end{aligned}$$

and

$$\begin{aligned} a' \cup x &= a' \cup [(a \cap b) \cap c] = [a' \cup (a \cap b)] \cap (a' \cup c) = [(a' \cup a) \cap (a' \cup b)] \cap (a' \cup c) \\ &= [1 \cap (a' \cup b)] \cap (a' \cup c) = (a' \cup b) \cap (a' \cup c) = a' \cup (b \cap c) \\ &= (a' \cup a) \cap [a' \cup (b \cap c)] = a' \cup [a \cap (b \cap c)] = a' \cup y \end{aligned}$$

Hence

$$(a \cup x) \cap (a' \cup x) = (a \cup y) \cap (a' \cup y)$$
$$(a \cap a') \cup x = (a \cap a') \cup y$$

and

$$x = y$$

We leave for the reader to show that as a consequence parentheses may be inserted in $a_1 \cup a_2 \cup \cdots \cup a_n$ and $a_1 \cap a_2 \cap \cdots \cap a_n$ at will.

2. Prove: For each $a \in \mathcal{B}$, the element a' defined in (iv) is unique.

Suppose the contrary, i.e., suppose for any $a \in \mathcal{B}$ there are two elements $a', a'' \in \mathcal{B}$ such that

$$a \cup a' = 1 \qquad a \cap a' = 0$$
$$\text{and}$$
$$a \cup a'' = 1 \qquad a \cap a'' = 0$$

Then

$$\begin{aligned} a' &= 1 \cap a' = (a \cup a'') \cap a' = (a \cap a') \cup (a'' \cap a') \\ &= (a \cap a'') \cup (a'' \cap a') = a'' \cap (a \cup a') = a'' \cap 1 = a'' \end{aligned}$$

and a' is unique.

3. Prove: For every $a, b \in \mathcal{B}$

$$(a \cup b)' = a' \cap b' \quad \text{and} \quad (a \cap b)' = a' \cup b' \tag{5}$$

(See 1.11-1.11′, Chapter 1, Page 5.)

Since by Problem 2 there exists for every $x \in \mathcal{B}$ a unique x' such that $x \cup x' = 1$ and $x \cap x' = 0$, we need only verify that

$$\begin{aligned} (a \cup b) \cup (a' \cap b') &= [(a \cup b) \cup a'] \cap [(a \cup b) \cup b'] = [(a \cup a') \cup b] \cap [a \cup (b \cup b')] \\ &= (1 \cup b) \cap (a \cup 1) = 1 \cap 1 = 1 \end{aligned}$$

and (we leave it for the reader) $(a \cup b) \cap (a' \cap b') = 0$

Using the results of Problem 2, it follows readily that

$$(a_1 \cup a_2 \cup \cdots \cup a_n)' = a_1' \cap a_2' \cap \cdots \cap a_n'$$

and

$$(a_1 \cap a_2 \cap \cdots \cap a_n)' = a_1' \cup a_2' \cup \cdots \cup a_n'$$

4. Prove: $(a')' = a$ for every $a \in \mathcal{B}$. (See 1.1, Chapter 1, Page 5.)

$$(a')' = 1 \cap (a')' = (a \cup a') \cap (a')' = [a \cap (a')'] \cup [a' \cap (a')'] = [a \cap (a')'] \cup 0$$
$$= 0 \cup [a \cap (a')'] = (a \cap a') \cup [a \cap (a')'] = a \cap [a' \cup (a')'] = a \cap 1 = a$$

5. Simplify: $[x \cup (x' \cup y)'] \cap [x \cup (y' \cap z')']$.

$$[x \cup (x' \cup y)'] \cap [x \cup (y' \cap z')'] = [x \cup (x \cap y')] \cap [x \cup (y \cup z)] = x \cap [x \cup (y \cup z)] = x$$

6. Obtain the canonical form of $[x \cup (x' \cup y)'] \cap [x \cup (y' \cap z')']$.

Using the identity of Problem 5,

$$[x \cup (x' \cup y)'] \cap [x \cup (y' \cap z')'] = x = x \cap (y \cup y') \cap (z \cup z')$$
$$= (x \cap y \cap z) \cup (x \cap y' \cap z) \cup (x \cap y \cap z') \cup (x \cap y' \cap z')$$

7. Prove: The complete canonical form in 3 variables is identically 1.

First, we show that the complete canonical form in 2 variables

$$(x \cap y) \cup (x \cap y') \cup (x' \cap y) \cup (x' \cap y') = [(x \cap y) \cup (x \cap y')] \cup [(x' \cap y) \cup (x' \cap y')]$$
$$= [x \cap (y \cup y')] \cup [x' \cap (y \cup y')]$$
$$= (x \cup x') \cap (y \cup y') = 1 \cap 1 = 1$$

Then the canonical form in 3 variables

$$[(x \cap y \cap z) \cup (x \cap y \cap z')] \cup [(x \cap y' \cap z) \cup (x \cap y' \cap z')]$$
$$\cup [(x' \cap y \cap z) \cup (x' \cap y \cap z')] \cup [(x' \cap y' \cap z) \cup (x' \cap y' \cap z')]$$
$$= [(x \cap y) \cup (x \cap y') \cup (x' \cap y) \cup (x' \cap y')] \cap (z \cup z') = 1 \cap 1 = 1$$

8. Prove: If, in the complete canonical form in n variables, each variable is assigned arbitrarily the value 0 or 1, then just one term will have the value 1 while all others will have the value 0.

Let the values be assigned to the variables $x_1, x_2, \ldots, x_n$. The term whose value is 1 contains x_1 if x_1 has the value 1 assigned or x_1' if x_1 has the value 0 assigned, x_2 if x_2 has the value 1 or x_2' if x_2 has the value 0, ..., x_n if x_n has the value 1 or x_n' if x_n has the value 0. Every other term of the complete canonical form will then have 0 as at least one factor and, hence, has 0 as value.

9. Prove: Two Boolean functions are equal if and only if their respective canonical forms are identical, i.e. consist of the same terms.

Clearly two functions are equal if their canonical forms consist of the same terms. Conversely, if the two functions are equal, they must have the same value for each of the 2^n possible assignments of 0 or 1 to the variables. Moreover, each of the 2^n assignments for which the function has the value 1 determines a term of the canonical form of that function. Hence the two normal forms contain the same terms.

10. Find the Boolean function F defined by

x	y	z	F
1	1	1	0
1	1	0	1
1	0	1	1
0	1	1	0
1	0	0	1
0	1	0	1
0	0	1	0
0	0	0	1

Table 17-7

It is clear that the canonical form of F will consist of 5 terms while the dual canonical form will consist of 3 factors. We use the latter form. Then

$$F = (x' \cup y' \cup z') \cap (x \cup y' \cup z') \cap (x \cup y \cup z')$$
$$= (y' \cup z') \cap (x \cup y \cup z') = [y' \cap (x \cup y)] \cup z' = (x \cap y') \cup z'$$

11. Find the canonical form for $F = (x \cup y \cup z) \cap (x' \cup y' \cup z)$.

Here $$F' = (x' \cap y' \cap z') \cup (x \cap y \cap z')$$

(by the identity of Problem 3) and

$$F = (F')' = (x \cap y \cap z) \cup (x \cap y' \cap z) \cup (x \cap y' \cap z') \cup (x' \cap y \cap z) \cup (x' \cap y' \cap z) \cup (x' \cap y \cap z')$$

(the union of all terms of the complete canonical form not appearing in F').

12. Prove: For every $a, b \in \mathcal{B}$, $a \cup b$ is the least upper bound and $a \cap b$ is the greatest lower bound of a and b.

That $a \cup b$ is an upper bound of a and b follows from

$$a \cup (a \cup b) = a \cup b = b \cup (a \cup b)$$

Let c be any other upper bound of a and b. Then $a \subseteq c$ and $b \subseteq c$ so that $a \cup c = c$ and $b \cup c = c$. Now

$$(a \cup b) \cup c = a \cup (b \cup c) = a \cup c = c$$

Thus, $(a \cup b) \subseteq c$ and $a \cup b$ is the least upper bound as required.

Similarly, $a \cap b$ is a lower bound of a and b since

$$(a \cap b) \cup a = a \quad \text{and} \quad (a \cap b) \cup b = b$$

Let c be any other lower bound of a and b. Then $c \subseteq a$ and $c \subseteq b$ so that $c \cup a = a$ and $c \cup b = b$. Now

$$c \cup (a \cap b) = (c \cup a) \cap (c \cup b) = a \cap b$$

Thus, $c \subseteq (a \cap b)$ and $a \cap b$ is the greatest lower bound as required.

13. Discuss the possible networks consisting of three switches r, s, t.

There are four cases:

(i) The switches are connected in series. The diagram is

——• r •——• s •——• t •——

and the function is $r \cap s \cap t$.

(ii) The switches are connected in parallel. The diagram is

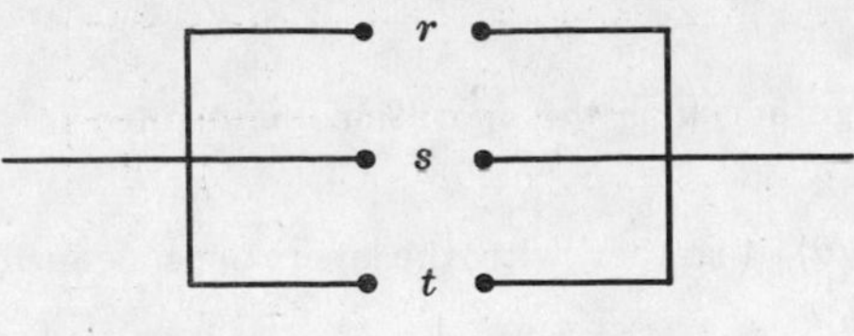

and the function is $r \cup s \cup t$.

(iii) The series-parallel combination

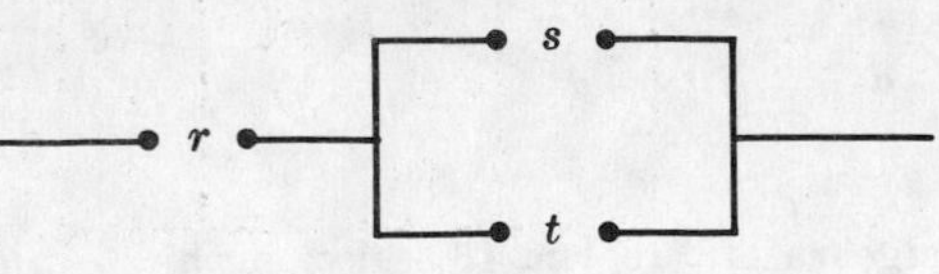

with function $r \cap (s \cup t)$.

(iv) The series-parallel combination

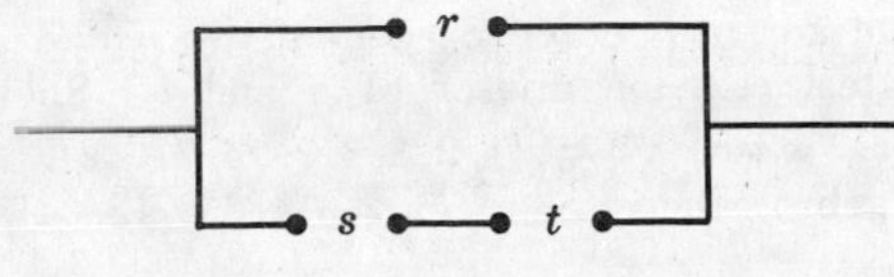

with function $r \cup (s \cap t)$.

14. If possible, replace the network

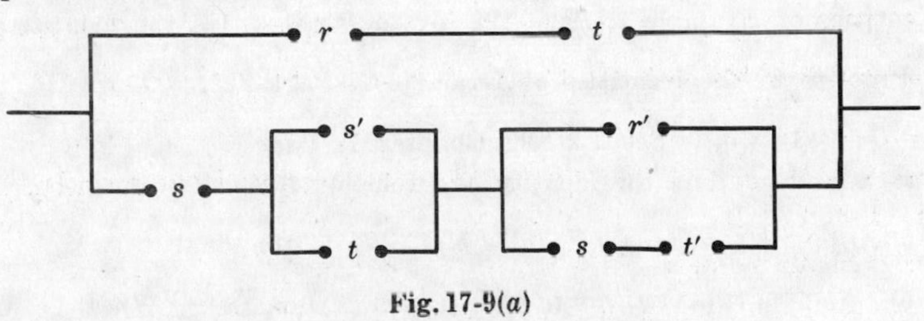

Fig. 17-9(*a*)

by a simpler one.

The Boolean function for the given network is

$$\begin{aligned} F &= (r \cap t) \cup \{s \cap (s' \cup t) \cap [r' \cup (s \cap t')]\} \\ &= (r \cap t) \cup \{s \cap [(s' \cup t) \cap (r' \cup t')]\} \\ &= (r \cap t) \cup [r' \cap (s \cap t)] \;=\; (r \cup s) \cap t \end{aligned}$$

The simpler network is

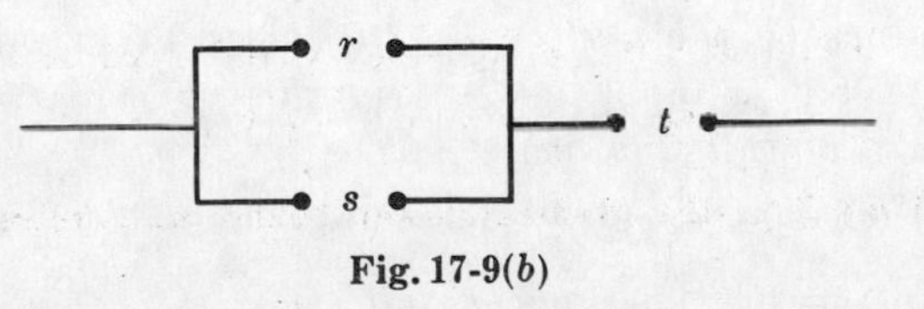

Fig. 17-9(*b*)

Supplementary Problems

15. Show that the set $\{0, 1\}$ together with the operations as defined in

$\cup$	0	1
0	0	1
1	1	1

and

$\cap$	0	1
0	0	0
1	0	1

is a Boolean algebra.

16. Show that the set $\{a, b, c, d\}$ together with the operations defined in

$\cup$	a	b	c	d
a	a	b	c	d
b	b	b	d	d
c	c	d	c	d
d	d	d	d	d

and

$\cap$	a	b	c	d
a	a	a	a	a
b	a	b	a	b
c	a	a	c	c
d	a	b	c	d

is a Boolean algebra.

17. Show that the Boolean algebra of Problem 16 is isomorphic to the algebra of all subsets of a set of two elements.

18. Why is there no Boolean algebra having just three distinct elements?

19. Let S be a subset of N, and for any $a, b \in S$ define $a \cup b$ and $a \cap b$ to be respectively the least common multiple and greatest common divisor of a and b. Show
(*a*) $\mathcal{B}$ is a Boolean algebra when $S = \{1, 2, 3, 6, 7, 14, 21, 42\}$.
(*b*) $\mathcal{B}$ is not a Boolean algebra when $S = \{1, 2, 3, 4, 6, 8, 12, 24\}$.

20. Show that $a \cup (a \cap b) = a \cap (a \cup b)$ without using Example 3, Page 223. State the dual of the identity and prove it.

21. Prove: For every $a, b \in \mathcal{B}$, $a \cup (a' \cap b) = a \cup b$. State the dual and prove it.

22. Obtain the identities of Example 1, Page 222, by taking $b = a$ in the identities of Problem 21.

23. Obtain as in Problem 22 the identities of Example 2, Page 222.

24. Prove: $0' = 1$ and $1' = 0$. (See 1.2-1.2′, Chapter 1, Page 5.)
Hint. Take $a = 0$ and $b = 1$ in the identity of Problem 21.

25. Prove: $(a \cap b') \cup (b \cap a') = (a \cup b) \cap (a' \cup b')$. Write the dual.

26. Prove: $(a \cup b) \cap (b \cup c) \cap (c \cup a) = (a \cap b) \cup (b \cap c) \cup (c \cap a)$. What is its dual?

27. Prove: If $a \cup x = b \cup x$ and $a \cup x' = b \cup x'$, then $a = b$.
Hint. Consider $(a \cup x) \cap (a \cup x') = (b \cup x) \cap (b \cup x')$.

28. State the dual of Problem 27 and prove it.

29. Prove: If $a \cap b = a \cap c$ and $a \cup b = a \cup c$ for any $a, b, c \in \mathcal{B}$, then $b = c$.

30. Simplify:
(*a*) $(a \cup b) \cap a' \cap b'$
(*b*) $(a \cap b \cap c) \cup a' \cup b' \cup c'$
(*c*) $(a \cap b) \cup [c \cap (a' \cup b')]$
(*d*) $[a \cup (a' \cap b)] \cap [b \cup (b \cap c)]$
(*e*) $[(x' \cap y')' \cup z] \cap (x \cup y')'$
(*f*) $(a \cup b') \cap (a' \cup b) \cap (a' \cup b')$
(*g*) $[(a \cup b) \cap (c \cup b')] \cup [b \cap (a' \cup c')]$

Ans. (*a*) 0, (*b*) 1, (*c*) $(a \cap b) \cup c$, (*d*) b, (*e*) $x' \cap y$, (*f*) $a' \cap b'$, (*g*) $a \cup b$

31. Prove: $(a \cup b) \cap (a' \cup c) = (a' \cap b) \cup (a \cap c) \cup (b \cap c)$
$= (a \cup b) \cap (a' \cup c) \cap (b \cup c) = (a \cap c) \cup (a' \cap b)$

32. Find, by inspection, the complement of each of the following in two ways:
(*a*) $(x \cap y) \cup (x \cap y')$
(*b*) $(x \cap y' \cap z) \cup (x' \cap y \cap z')$
(*c*) $(x \cup y' \cup z) \cap (x \cup y \cup z') \cap (x \cup y' \cup z')$
(*d*) $(x \cup y' \cup z) \cap (x' \cup y \cup z)$

33. Express each of the following in both canonical form and dual canonical form in three variables:
(*a*) $x' \cup y'$, (*b*) $(x \cap y') \cup (x' \cap y)$, (*c*) $(x \cup y) \cap (x' \cup z')$, (*d*) $x \cap z$, (*e*) $x \cap (y' \cup z)$

Partial Ans.

(*a*) $(x \cap y' \cap z) \cup (x \cap y' \cap z') \cup (x' \cap y \cap z) \cup (x' \cap y' \cap z) \cup (x' \cap y \cap z') \cup (x' \cap y' \cap z')$

(*b*) $(x \cup y \cup z) \cap (x \cup y \cup z') \cap (x' \cup y' \cup z) \cap (x' \cup y' \cup z')$

(*c*) $(x \cap y \cap z') \cup (x \cap y' \cap z') \cup (x' \cap y \cap z) \cup (x' \cap y \cap z')$

(*d*) $(x \cup y \cup z) \cap (x \cup y' \cup z) \cap (x \cup y \cup z') \cap (x \cup y' \cup z') \cap (x' \cup y \cup z) \cap (x' \cup y' \cup z)$

(*e*) $(x \cup y \cup z) \cap (x \cup y' \cup z) \cap (x \cup y \cup z') \cap (x \cup y' \cup z') \cap (x' \cup y' \cup z)$

34. Express each of the following in both canonical and dual canonical form in the minimum number of variables:

(*a*) $x \cup (x' \cap y)$

(*b*) $[x \cap (y \cup z)] \cup [x \cap (y \cup z')]$

(*c*) $(x \cup y \cup z) \cap [(x \cap y) \cup (x' \cap z)]$

(*d*) $(x \cap y \cap z) \cup [(x \cup y) \cap (x \cup z)]$

(*e*) $(x \cup y) \cap (x \cup z') \cap (x' \cup y') \cap (x' \cup z)$

(*f*) $(x \cap y) \cup (x \cap z') \cup (x' \cap z)$

Partial Ans.

(*a*) $(x \cap y) \cup (x \cap y') \cup (x' \cap y)$

(*b*) $(x \cup y) \cap (x \cup y')$

(*c*) $(x \cap y \cap z) \cup (x \cap y \cap z') \cup (x' \cap y \cap z) \cup (x' \cap y' \cap z)$

(*d*) $(x \cup y \cup z) \cap (x \cup y \cup z') \cap (x \cup y' \cup z)$

(*e*) $(x \cap y' \cap z) \cup (x' \cap y \cap z')$

(*f*) $(x \cup y \cup z) \cap (x' \cup y \cup z') \cap (x \cup y' \cup z)$

35. Write the term of the complete canonical form in x, y, z having the value 1 when:
(*a*) $x = z = 0,\ y = 1$; (*b*) $x = y = 1,\ z = 0$; (*c*) $x = 0,\ y = z = 1$.
Ans. (*a*) $x' \cap y \cap z'$, (*b*) $x \cap y \cap z'$

36. Write the term of the complete canonical form in x, y, z, w having the value 1 when:
(*a*) $x = y = 1,\ z = w = 0$; (*b*) $x = y = w = 0,\ z = 1$; (*c*) $x = 0,\ y = z = w = 1$.
Ans. (*a*) $x \cap y \cap z' \cap w'$, (*c*) $x' \cap y \cap z \cap w$

37. Write the factor of the complete dual canonical form in x, y, z having the value 0 when:
(*a*) $x = z = 0,\ y = 1$; (*b*) $x = y = 1,\ z = 0$; (*c*) $x = 0,\ y = z = 1$.
Ans. (*a*) $x \cup y' \cup z$, (*b*) $x' \cup y' \cup z$

38. Write the factor of the complete dual canonical form in x, y, z, w having the value 0 when:
(*a*) $x = y = 1,\ z = w = 0$; (*b*) $x = y = w = 0,\ z = 1$; (*c*) $x = 0,\ y = z = w = 1$.
Ans. (*a*) $x' \cup y' \cup z \cup w$, (*c*) $x \cup y' \cup z' \cup w'$

39. Write the function in three variables whose value is 1
(*a*) if and only if two of the variables are 1 and the other is 0,
(*b*) if and only if more than one variable is 1.
Ans. (*a*) $(x \cap y \cap z') \cup (x \cap y' \cap z) \cup (x' \cap y \cap z)$, (*b*) $[x \cap (y \cup z)] \cup (y \cap z)$

40. Write the function in three variables whose value is 0
(*a*) if and only if two of the variables are 0 and the other is 1,
(*b*) if and only if more than one variable is 0.
Ans. The duals in Problem 39.

41. Obtain in simplest form the Boolean functions $F_1, F_2, \ldots, F_8$ defined as follows:

x	y	z	F_1	F_2	F_3	F_4	F_5	F_6	F_7	F_8
1	1	1	0	1	0	1	1	1	0	1
1	1	0	1	0	1	0	0	0	0	0
1	0	1	0	1	1	1	1	1	1	1
0	1	1	0	1	1	1	0	0	0	1
1	0	0	1	0	0	0	1	0	1	0
0	1	0	0	1	1	1	0	0	0	0
0	0	1	0	1	1	1	1	1	1	1
0	0	0	1	1	1	0	0	1	1	0

Ans. $F_1 = (x \cup y') \cap z'$, $F_3 = x' \cup (y' \cap z) \cup (y \cap z')$, $F_5 = (x \cup z) \cap [y' \cup (x \cap z)]$, $F_7 = y'$

42. Show that F_7 and F_8 of Problem 41 can be found by inspection.

43. Prove: (d_1) If $a \subseteq b$ and $a \subseteq c$, then $a \subseteq (b \cap c)$.
(e_1) If $a \subseteq b$ then $a \subseteq (b \cup c)$ for any $c \in \mathcal{B}$.
(f_1) $a \subseteq b$ if and only if $b' \subseteq a'$.

44. Prove: If $a, b \in \mathcal{B}$ such that $a \subseteq b$ then, for any $c \in \mathcal{B}$, $a \cup (b \cap c) = b \cap (a \cup c)$.

45. Prove: For every $b \in \mathcal{B}$, $0 \subseteq b \subseteq 1$.

46. Construct a diagram similar to Fig. 17-1 for the Boolean algebra of all subsets of $B = \{a, b, c, d\}$.

47. Diagram the networks represented by $a \cup (a' \cap b)$ and $a \cup b$ and show by tables that they have the same closure properties.

48. Diagram the networks (i) $(a \cup b) \cap a' \cap b'$ and (ii) $(a \cap b \cap c) \cup (a' \cup b' \cup c')$. Construct tables of closure properties for each. What do you conclude?

49. Diagram each of the following networks

(i) $(a \cup b') \cap (a' \cup b) \cap (a' \cup b')$
(ii) $(a \cap b) \cup [c \cap (a' \cup b')]$
(iii) $[(a \cup b) \cap (c \cup b')] \cup [b \cap (a' \cup c')]$
(iv) $(a \cap b \cap c) \cup a' \cup b' \cup c'$

Using the results obtained in Problem 30, diagram the simpler network for each.
Partial Ans.

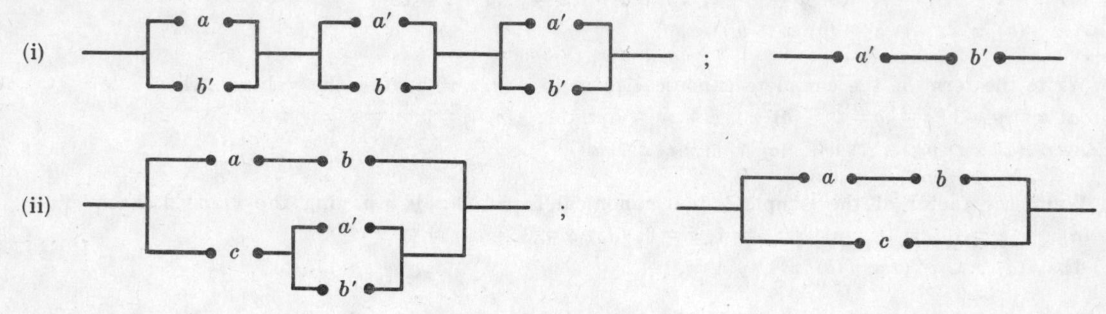

50. Diagram each of the networks $(r \cup s') \cap (r' \cup s)$ and $(r \cap s) \cup (r' \cap s')$ and show that they have the same closure properties.

51. Diagram the simplest network having the closure properties of each of F_1-F_6 in Problem 41.
Partial Ans.

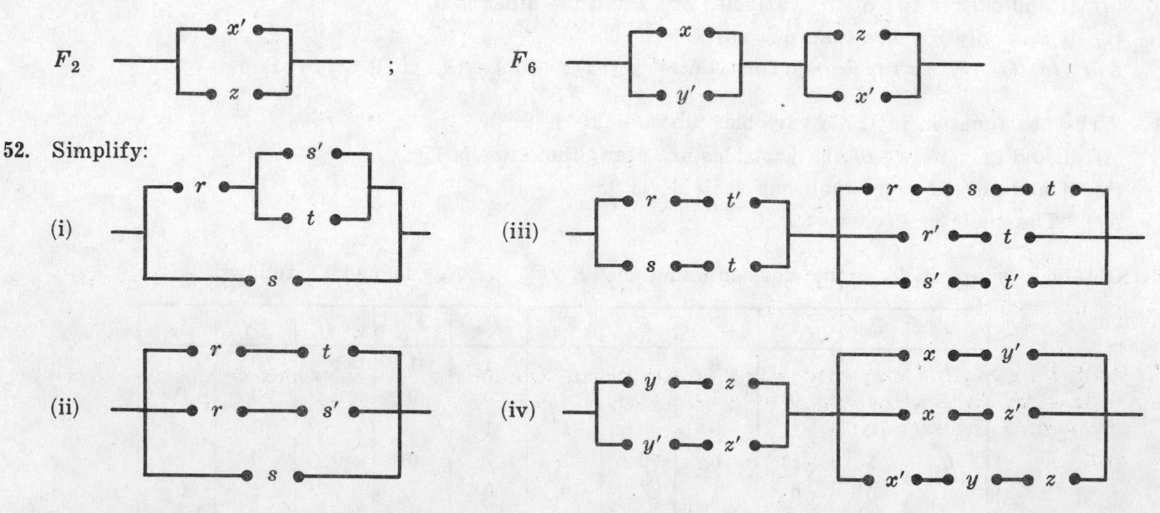

To afford practice and also to check the results, it is suggested that (iii) and (iv) be solved by forming the table of closure properties and also by the procedure of Example 7, Page 226.

Partial Ans. (i) and (ii) — r in parallel with s

53. Simplify:

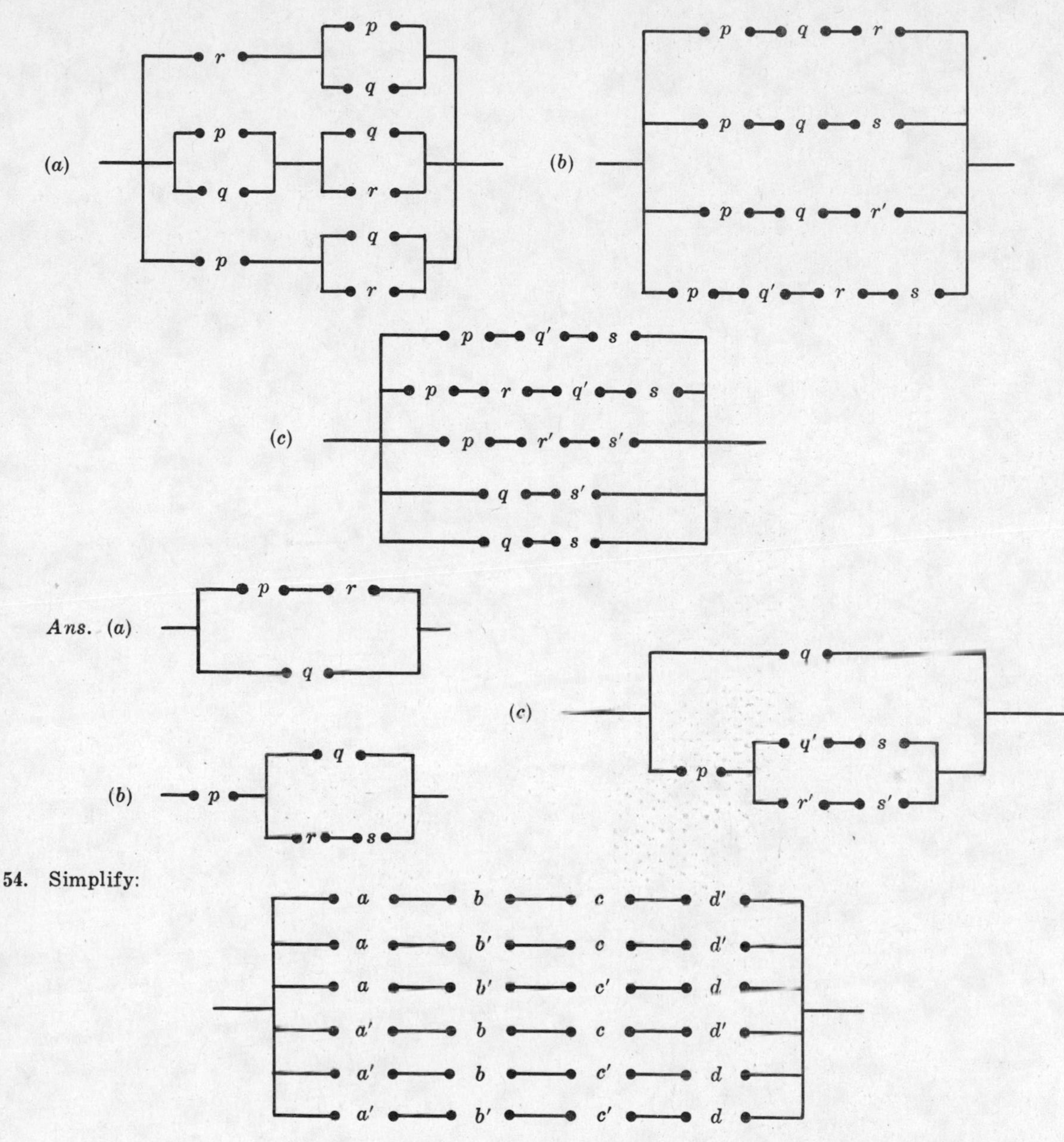

55. Show that the network of Problem 50 will permit a light over a stairway to be turned on or off either at the bottom or top of the stairway.

56. From his garage M may enter either of two rooms. Obtain the network which will permit M to turn on or off the light in the garage either from the garage or one of the rooms irrespective of the position of the other two switches.

Ans.

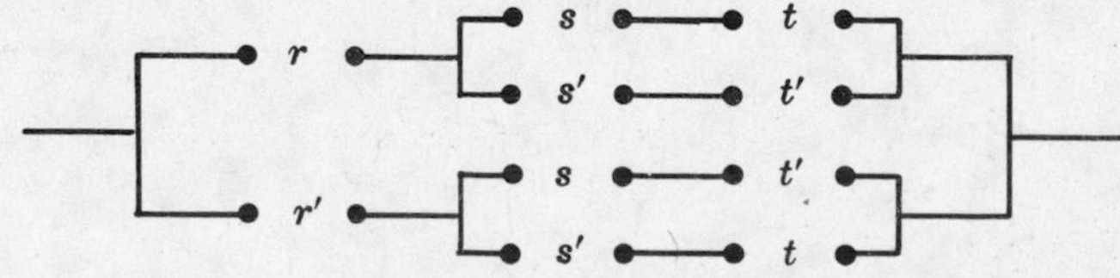

57. Design a network which will permit independent control of a light from any one of four switches.

INDEX

Index of Symbols